D1664334

LASERS FOR ULTRASHORT LIGHT PULSES

Lasers for Ultrashort Light Pulses

by

JOACHIM HERRMANN

BERND WILHELMI

Friedrich-Schiller-Universität Jena

1987

NORTH-HOLLAND

Amsterdam · Oxford · New York · Tokyo

The book is a revised English translation of
"Laser für ultrakurze Lichtimpulse", published in 1984
by Akademie-Verlag Berlin. It has been translated
by Jeffrey Grossman and Wolfgang Rudolph
in co-operation with the authors.

Published in coedition with Akademie-Verlag Berlin, GDR

This book is exclusively distributed in all non socialist countries by

North-Holland Physics Publishing (a division of)
Elsevier Science Publishers B. V.
P. O. Box 103, 1000 AC Amsterdam, The Netherlands

Distributors for the U.S.A. and Canada

Elsevier Science Publishing Company, Inc.
52 Vanderbilt Avenue
New York, NY 10017

Library of Congress Cataloging-in-Publication Data

Herrmann, Joachim, 1931 –
Lasers for ultrashort light pulses.
Revised translation of: Laser für ultrakurze
Lichtimpulse.
Bibliography: p.
Includes index.
1. Laser pulses, Ultrashort. 2. Picosecond pulses.
I. Wilhelmi, Bernd. II. Title.
QC689.5.L37H4713 1987 621.36'6 86-4393

ISBN 0-444-87055-5

With 127 Figures and 10 Tables

Printed in the German Democratic Republic

Preface to the English edition

Since the German edition of this book was published in 1984 the development of lasers for generating ultrashort light pulses and their application in measurement technology and spectroscopy have experienced a large and continuing growth. We therefore thought it desirable to add new, supplementary material to the English version in order to incorporate the new knowledge gained in this field. In particular, the additions concern closer investigation of the production of chirp accompanying the generation of light pulses and the resulting method of chirp compensation for further shortening of light pulses. Furthermore, new material was incorporated into the chapters dealing with the applications of ultrashort light pulses to optoelectronic and electrooptic switching, in the study of transient processes in nonlinear optics and spectroscopy, and for generating short pulses in distributed feedback lasers.

Finally, we would like to thank Mr. Jeffrey Grossman and Dr. Wolfgang Rudolph for the translation of our book. We also gratefully acknowledge the good cooperation with Akademie-Verlag, Berlin.

J. Herrmann, B. Wilhelmi

Introduction

The rapid development of picosecond lasers and picosecond light pulse diagnostics has led to enormous progress in the entire field of ultrafast measuring techniques during the last 15 years (see Fig. 1). Ultrashort light pulses permit novel investigations of extremely rapid physical, chemical and biological phenomena on the picosecond and even femtosecond time scale. The high time resolution attained using picosecond lasers allows new and deeper insights into the nature and especially into the temporal evolution of some extremely fundamental processes in materials, most of which occur on a picosecond time scale. Among the fundamental processes that have been measured are the free decay of molecular vibrations and of orientational fluctuations in liquids, rela-

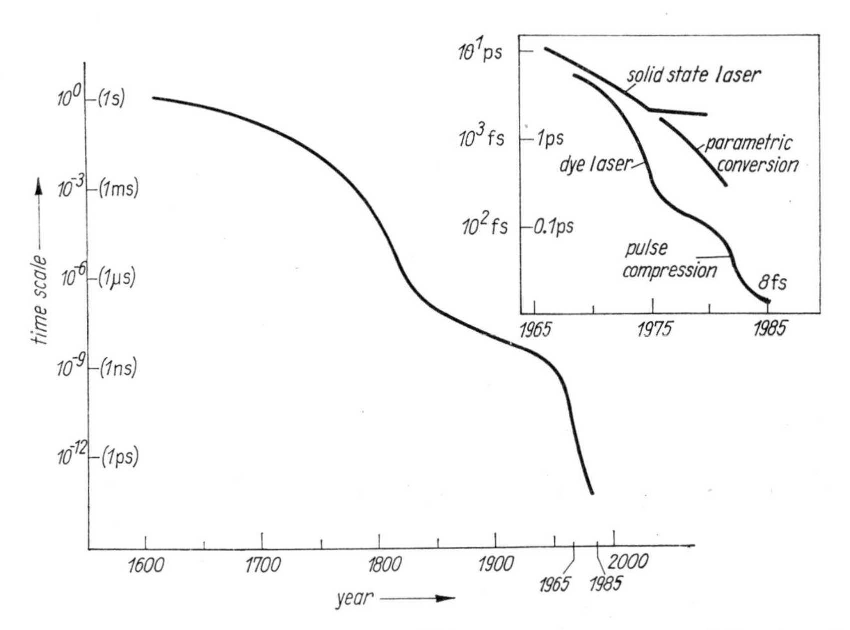

Fig. 1. Decrease of the time scales on which processes were measured directly and of the shortest experimentally generated pulses, respectively, over the last four centuries, where the development of ultrashort pulse generation over the last two decades is shown in greater detail (from Shapiro [16] and Shank [28])

xation processes in small and large molecules, phonon decay, exciton decay and energy migration in solids, charge transfer processes and other nonradiative transfer processes, temperature fluctuation processes, elementary processes of photosynthesis and of other biological phenomena. Moreover picosecond technology has provided new capabilities both for the manipulation of photophysical and photochemical processes and for making extremely fast electrooptical components such as switches, modulators and receivers.

The present rapid expansion of research work on picosecond lasers and their application makes it difficult to survey and comprehend the large number of publications in this field, and an introduction to the subject is made difficult by the absence of any textbook. For this reason, our aim is that the present book should fill this gap. This book differs from the well-known monograph "Ultrashort light pulses", edited by S. L. Shapiro, in that its aim is to give an introduction to the fundamentals of the new field, to its specific experimental and theoretical methods, important applications and present trends.

In our book we do not attempt to achieve monographic comprehensiveness but aim to give a guide to physicists, chemists, biologists, and engineers as well as to students in these fields, who want to become familiar with the generation and measurement of ultrashort light pulses and their application.

The book assumes an elementary knowledge of optics and atomic physics. More detailed theoretical knowledge is only necessary for sections 1.3.2, 4.2, 5.2, 6.2, and 7.2, dealing with the theory of interaction between light and matter and the theory of various types of picosecond lasers. Readers who are less interested in theory may initially omit these sections.

The book begins with three chapters introducing the field of lasers and picosecond phenomena. The first chapter describes some fundamentals of the interaction between light pulses and matter including relaxation processes. The second chapter gives some general fundamentals of laser physics and especially the principle of operation for the generation of ultrashort light pulses. The third chapter deals with important methods for the measurement of ultrashort light pulses and with representative devices used in their application. These three chapters can be omitted by readers who are already familiar with the interaction between light and matter, the physics of continuous wave lasers and ultrafast measuring techniques.

The main part of the book is presented in chapters 4.—7. which describe various methods of the modelocking of lasers. Each of these chapters begins with a simple introduction devoted to the respective principle of operation, followed by a systematic treatment of the laser operation and a computation of optimum conditions and light pulse parameters. Each chapter ends with a discussion of representative experimental methods, devices and results, which are compared with theory. The eighth chapter describes some possible methods of frequency transformation of ultrashort light pulses by nonlinear optical processes and some methods of pulse shortening. Finally, the ninth chapter deals with representative methods and applications of ultrafast spectroscopy.

Contents

0. Notations and Symbols

General Notations

$\underset{\sim}{X}(\omega)$	Fourier transform of $X(t)$
	$\underset{\sim}{X}(\omega) = \int_{-\infty}^{\infty} \mathrm{d}t X(t)\, \mathrm{e}^{-\mathrm{i}\omega t}$
	$X(t) = \frac{1}{2\pi} \int_{-\infty}^{\infty} \mathrm{d}\omega \underset{\sim}{X}(\omega)\, \mathrm{e}^{\mathrm{i}\omega t}$
C.C.	complex conjugate component of a quantity; e.g. $X + \mathrm{C.C.} = X + X^*$
$\langle X \rangle$	ensemble average of a quantity
$\overline{X} = \frac{1}{u} \int_{-u/2}^{u/2} X(t)\, \mathrm{d}t$	time average of a quantity averaged over the time interval u
$\approx$	approximately equal
$\simeq$	of the same order of magnitude
$\lessapprox$	smaller than or approximately equal to
$\gtrapprox$	greater than or approximately equal to
$\ll$	much less than
$\gg$	much greater than
$\propto$	proportional to

Electric Field and Polarization

$\vec{E}(\vec{r}, t)$	electric field
$\vec{e}$	unit vector in the direction of the electric field
$\hat{E}(\vec{r}, t)$	oscillation amplitude of the electric field
	$\vec{E}(\vec{r}, t) = \frac{1}{2} \hat{E}(\vec{r}, t)\, \vec{e}\, \mathrm{e}^{\mathrm{i}\omega t} + \mathrm{C.C.}$
$A(\vec{r}, t)$	wave amplitude of the electric field
	$\vec{E}(\vec{r}, t) = \frac{1}{2} A(\vec{r}, t)\, \vec{e}\, \mathrm{e}^{\mathrm{i}(\omega t - \vec{k}\vec{r})} + \mathrm{C.C.}$
$\vec{P}(\vec{r}, t)$	polarization
$\vec{e}\,'$	unit vector in the direction of the polarization
$\overline{P}(\vec{r}, t)$	wave amplitude of the polarization
	$\vec{P}(\vec{r}, t) = \frac{1}{2} \overline{P}(\vec{r}, t)\, \vec{e}\,'\, \mathrm{e}^{\mathrm{i}(\omega t - \vec{k}\vec{r})} + \mathrm{C.C.}$

Frequently used Symbols

A_{ij}	Einstein coefficient of spontaneous emission for the transition $i \to j$
B_{ij}	Einstein coefficient of absorption ($i < j$) and stimulated (induced) emission ($i > j$), respectively
c	vacuum velocity of light
$\mathscr{E}$	energy
$\mathscr{E}_S$	saturation photon number per area (saturation energy per area/photon energy $\hbar\omega$)
e	elementary electric charge ($e > 0$)
g	line shape function
G	amplification, gain ($I_{out} = I_{in}G$)
g_V	amplification (gain) coefficient
$g = g_V \cdot L = \ln G$	(L — amplifier length)
h, $\hbar$	Planck's constant
i	imaginary unit
I	photon flux density (energy flow density/photon energy $\hbar\omega$)
I_S	saturation photon flux density (saturation intensity/photon energy $\hbar\omega$)
$\vec{k}$	wave number vector
$k = 2\pi/\lambda$	wave number
k	rate constant
k_A	absorption coefficient
k_B	Boltzmann constant
N	Number density
N_i	occupation number density of level i
$\mathscr{N}$	number
n	refractive index
$\vec{r}$	space coordinate
$\vec{r} = (x, y, z)$	
R	reflectivity
$\mathscr{T}$	temperature
T_{ij}	energy relaxation time (longitudinal relaxation time) of the transition $i \to j$
t	time
U	radiation energy per unit volume
$U_\omega(\omega_{21})$	radiation energy per unit volume and frequency at the transition frequency ω_{21}
u	resonator round trip time
v	group velocity
W_{ij}	transition probability ($i \to j$)
$\mathrm{d}W_{ij}/\mathrm{d}t$	transition probability ($i \to j$) per unit time (transition rate)
$W_P = \sigma_{ij} I_P$	pump parameter
ε	permittivity
ε_0	permittivity of free space
η	local time, $\eta = t - z/v$ (retarded time)
λ	wavelength
μ_0	permeability
μ_{ij}	transition matrix element of the dipole operator
$\mu_{ij} = \vec{\mu}_{ij}\vec{e}$	
$\nu = \omega/(2\pi)$	frequency
$\boldsymbol{\rho}$	density operator

ρ_{ij}	elements of the density matrix
σ_{ij}	interaction cross section (transition $i \to j$) (absorption or emission cross section)
τ_L	pulse duration (pulse length, pulse width)
τ_{ij}	phase relaxation time (dephasing time, transverse relaxation time) of the transition $i \to j$
χ	susceptibility
ω	(angular) frequency

1. Fundamentals of the Interaction between Light Pulses and Matter

The basis for the generation and application of ultrashort light pulses is the absorption and emission of photons by matter (atoms or molecules). The temporal evolution for these processes is determined by the characteristics of the directly contributing particles, and by their interaction with one another as well as with other atomic systems — which are often described as thermal baths or dissipative systems.

As an introduction to these problems, we would like to begin with a description of the elementary radiation processes. An overview of fast processes — especially relaxation processes — in gases, liquids and solids follows, as well as a formulation of the basic equations describing the interaction between light pulses and matter that is under the influence of relaxation processes.

1.1 Elementary Radiation Processes

As Einstein recognized in 1916, relatively simple statistical and quantum physical considerations allow us to draw fundamental conclusions about the elementary processes of interaction between radiation and matter. According to the quantum theory atoms possess discrete energy levels. Because of energy conservation in the individual process, only those photons whose energy approximately equals the energy difference between two atomic levels, say levels 1 and 2, whose (angular) frequency is in the vicinity of the resonant frequency

$$\omega_{21} = \frac{1}{\hbar} (\mathcal{E}_2 - \mathcal{E}_1) \tag{1.1}$$

can be absorbed and emitted. For the present, we can restrict ourselves to the treatment of these two energy levels. Via various fundamental processes, the atomic system can convert from the initial to the excited state by absorbing energy, or change from the excited to the initial state by releasing energy (see Fig. 1.1).

1.1.1 Spontaneous Emission

The atom decays from the excited state 2 with the greater energy $\mathcal{E}_2$ to the lower state 1 with emission of a photon with a frequency in the vicinity of ω_{21}. An external radiation field is not necessary for this process to occur. The transition probability per unit time

dW_{21}^{sp}/dt for the spontaneous transition from 2 to 1 — hence, the transition rate — is a constant that is independent of the radiation field and determined only by the characteristics of the atom. This transition rate is called the Einstein coefficient for spontaneous emission A_{21}. Thus, we can write:

$$\frac{d}{dt} W_{21}^{sp} = A_{21}. \tag{1.2}$$

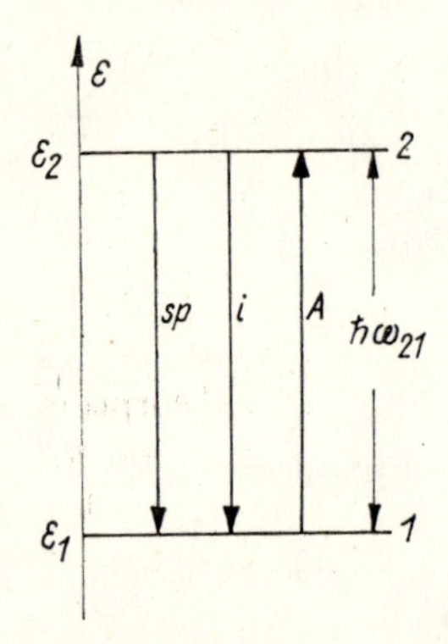

Fig. 1.1. Radiation processes between the atomic levels 1 and 2 where sp denotes the spontaneous emission, i denotes the induced emission and A denotes the absorption

The transition via spontaneous emission with release of photons can only take place within a small frequency interval. The transition probability per unit time and unit frequency $(dW_{21}^{sp}/dt)_\omega$ for the transition with emission of photons in a frequency interval $\omega \cdots \omega + d\omega$ is given by

$$\left(\frac{d}{dt} W_{21}^{sp}\right)_\omega d\omega = A_{21} g(\omega - \omega_{21})\, d\omega \tag{1.3}$$

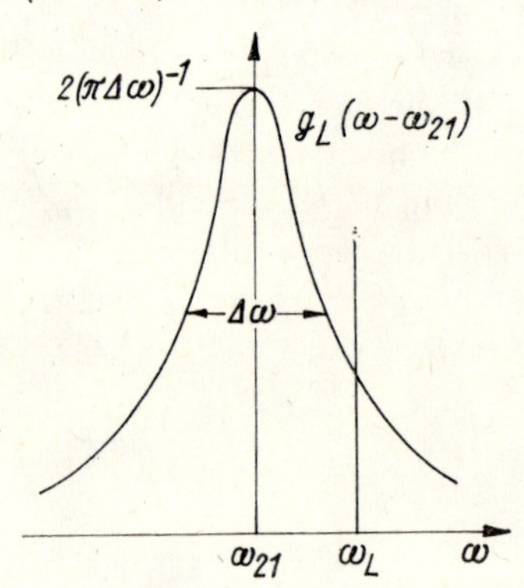

Fig. 1.2. Line shape factor with Lorentzian profile

in which $g(\omega - \omega_{21})$ is a line shape factor. (The integral of $g(\omega)$ from $-\infty$ to $+\infty$ is normalized to 1.) If the atom is not subjected to any other influences, $g(\omega - \omega_{21})$ is determined by the Lorentz function (as shown in Fig. 1.2)

$$g_L(\omega - \omega_{21}) = \frac{\left(\frac{1}{\pi}\right)\frac{\Delta\omega}{2}}{(\omega_{21} - \omega)^2 + (\Delta\omega/2)^2} \tag{1.4}$$

with $\Delta\omega = A_{21}$, see e.g. [10, 11]. Thus, averaging the contributions of many elementary processes one obtains a Lorentzian shaped spectral line of halfwidth A_{21}. If several energy levels lie below the excited level 2, the overall temporal change in the occupation probability of energy level 2 caused by spontaneous transitions into all of the lower energy levels results from the sum of the individual transition probabilities A_{2i}. Hence, we have

$$\frac{\mathrm{d}}{\mathrm{d}t} W_2^{\mathrm{sp}} = \sum_{i(\mathscr{E}_i < \mathscr{E}_2)} A_{2i}. \tag{1.5}$$

The overall line width for the transition $2 \to 1$ as a result of spontaneous emission processes is given by

$$\Delta\omega = \sum_{i(\mathscr{E}_i < \mathscr{E}_2)} A_{2i} + \sum_{j(\mathscr{E}_j < \mathscr{E}_1)} A_{1j}$$

i.e., by the lifetimes of the levels involved in the transition. The line width determined by the radiation lifetime is designated as the natural line width.

1.1.2 Stimulated Emission

In stimulated emission the release of a photon is caused by an external radiation field, where the atom decays from level 2 to level 1 as in spontaneous emission. The temporal change in the occupation probability of the excited state is here proportional to the spectral density (energy per unit volume and frequency) $U_\omega(\omega_{21})$ of the already existing radiation field at the transition frequency ω_{21}. Thus, we have

$$\frac{\mathrm{d}}{\mathrm{d}t} W_{21}^{\mathrm{i}} = B_{21} U_\omega(\omega_{21}). \tag{1.6}$$

The proportionality factor B_{21} is designated as the Einstein coefficient for stimulated emission; it depends solely on the properties of the atomic system.

Note that (1.6) is valid only under the condition that the energy density of the external radiation field is constant within the frequency range of the line width of the atomic transition. Otherwise, the transition probability of the atom is given by

$$\frac{\mathrm{d}}{\mathrm{d}t} W_{21}^{\mathrm{i}} = B_{21} \int_{-\infty}^{\infty} \mathrm{d}\omega\, g(\omega - \omega_{21})\, U_\omega(\omega) \tag{1.7}$$

whereby the Lorentzian function $g_{\mathrm{L}}(\omega - \omega_{21})$ can be inserted in the case of a resting and isolated atom, i.e., one not interacting with other atomic systems. If the external field is monochromatic (more specifically, if its spectral width is small compared to that of $g(\omega - \omega_{21})$), we may write $U_\omega(\omega) = U(\omega_{\mathrm{L}})\, \delta(\omega - \omega_{\mathrm{L}})$, where $\delta(x)$ designates the Dirac δ-function and thus we obtain from (1.7)

$$\frac{\mathrm{d}}{\mathrm{d}t} W_{21}^{\mathrm{i}}(\omega_{\mathrm{L}}) = B_{21} g(\omega_{\mathrm{L}} - \omega_{21})\, U(\omega_{\mathrm{L}}) \tag{1.8}$$

(see Fig. 1.2).

1.1.3 *Absorption*

The atom passes from the lower energy level to the excited state and absorbs a photon in this process. The transition probability per unit time is again proportional to the energy density per unit frequency of the interacting radiation field, and thus we have

$$\frac{\mathrm{d}}{\mathrm{d}t} W_{12}^{\mathrm{A}} = B_{12} U_{\omega}(\omega_{21}), \tag{1.9}$$

where B_{12} is the Einstein coefficient for absorption. Considerations similar to those in the case of stimulated emission can be made regarding the line width.

1.1.4 *Transition Probabilities in Thermal Equilibrium*

Let us now consider the three processes introduced above for the special case of a temperature radiator (black body radiator), in which the atomic systems are in a state of thermal equilibrium with the radiation field. Under these conditions, the number of atoms that decay per unit time from the excited level 2 to the ground level 1 must be equal to the number of atoms involved in the transition in the opposite direction. This requirement leads to

$$N_2[A_{21} + B_{21} U_{\omega}(\omega_{21})] = N_1 B_{12} U_{\omega}(\omega_{21}) \tag{1.10}$$

where N_1 and N_2 are the number of atoms per unit volume in levels 1 and 2, respectively. Since in thermal equilibrium (with non-degenerate levels!) the ratio N_2/N_1 is given by the Boltzmann distribution

$$\frac{N_2}{N_1} = \exp\left\{\frac{-\mathscr{E}_2 + \mathscr{E}_1}{k_{\mathrm{B}}\mathscr{T}}\right\} \tag{1.11}$$

we obtain from (1.10) and (1.11)

$$U_{\omega}(\omega_{21}) = \frac{A_{21}/B_{21}}{\dfrac{B_{12}}{B_{21}} \mathrm{e}^{(\hbar\omega_{21}/k_B\mathscr{T})} - 1} \tag{1.12}$$

for the spectral radiation energy density of the temperature radiator. (Here k_{B} designates Boltzmann's constant and $\mathscr{T}$ the absolute temperature.) On the other hand, from Planck's radiation formula we know the radiation density of the black body radiator, which is

$$U_{\omega}(\omega_{21}) = \frac{\hbar\omega_{21}^3}{\pi^2 c^3} \frac{1}{\mathrm{e}^{(\hbar\omega_{21}/k_B\mathscr{T})} - 1}. \tag{1.13}$$

When we compare these two results, we obtain

$$A_{21} = \frac{\hbar\omega_{21}^3}{\pi^2 c^3} B_{21} \tag{1.14}$$

and

$$B_{12} = B_{21}. \tag{1.15}$$

Hence the Einstein coefficients for stimulated emission and absorption are equal. (With degenerated levels having the degeneracy factors g_1 and g_2 we obtain the general relation $B_{12}g_1 = B_{21}g_2$.) We wish once again to emphasize that the introduction of two distinct emission processes, namely spontaneous and stimulated emission, is necessary in order to obtain the correct radiation formula (1.13). The ratio between stimulated and spontaneous emission decreases with increasing frequency under constant spectral energy density of the given radiation field.

Note that similar considerations can also be carried out for the probabilities of atomic transitions with absorption and emission of photons in the frequency interval $\omega \cdots (\omega + d\omega)$. In particular, we can therefore conclude that the same line shape factor appears in all three processes, which we had already assumed.

Although the relations (1.14) and (1.15) were obtained for the special case of a system in thermal equilibrium, its validity is not bound to this specific assumption. Using quantum theory, it can be shown that these equations also hold for non-equilibrium states.

1.1.5 Rate Equations

At this point, we want to establish a relationship between the Einstein coefficients of absorption and stimulated emission and macroscopic measurable quantities. With this aim we consider the scheme of an experimental set-up outlined in Fig. 1.3 in which the absorption (or gain) of monochromatic light of frequency ω_L and of the photon flux

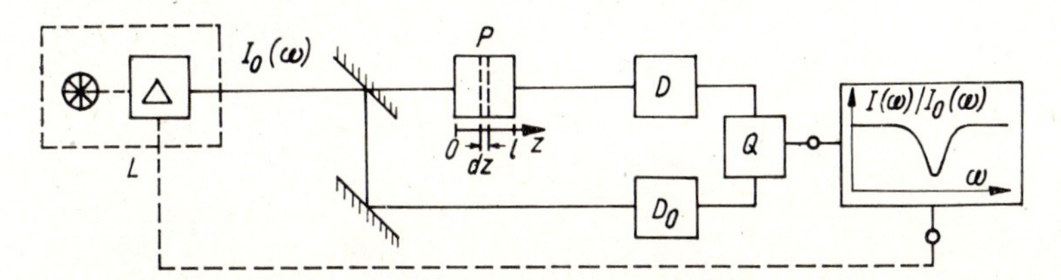

Fig. 1.3. Measurement of the transmission and the absorption coefficient
Monochromatic light of variable frequency ω and intensity $I_0(\omega)$ strikes a sample P of length l. (The approximately monochromatic light can be generated by narrow-band, tunable lasers or filtered out by means of a monochromator from the radiation of a broad-band light source.) Using the radiation detectors D_0 and D, the intensities $I_0(\omega)$ and $I(\omega)$ are measured in front of the sample and behind the sample, respectively. The transmission is defined as the ratio $I(\omega)/I_0(\omega)$ which is calculated electronically. The transmission is represented as the y-coordinate of an x,y-recorder, whose x-coordinate is adjusted proportionally to the frequency ω. The absorption coeffcient is given by $k_A(\omega) = l^{-1} \ln [I_0(\omega)/I(\omega)]$.

density I_L is measured. The photon flux density of a travelling wave gives the number of photons per unit time and area that pass through a reference plane. In the case of a plane wave, the photon flux density is related to the energy density by

$$I_L = \frac{c}{n} \frac{1}{\hbar\omega_L} U(\omega_L), \tag{1.16}$$

where c/n is the phase velocity and $\hbar\omega_L$ is the photon energy. The incident wave is attenuated by absorption processes and amplified by emission processes. Let us assume here that, due to the magnitude of the radiation field, the spontaneous processes can be neglected compared with the stimulated processes. In a thin layer of thickness dz and area A, which contains atoms or molecules which do not interact with one another, the number of photons absorbed per unit time is $[\mathrm{d}W_{12}^{A}(\omega_L)/\mathrm{d}t] \cdot [N_1 A\, \mathrm{d}z]$ and the number of photons created by stimulated emission is $[\mathrm{d}W_{21}^{i}(\omega_L)/\mathrm{d}t] \cdot [N_2 A\, \mathrm{d}z]$.

This amplification does not change the polarization direction and the coherence properties of the wave (cf. [11]). Thus the photon flux density changes according to

$$-\mathrm{d}I_L = \left[\frac{\mathrm{d}}{\mathrm{d}t} W_{12}^{A}(\omega_L)\, N_1 - \frac{\mathrm{d}}{\mathrm{d}t} W_{21}^{i}(\omega_L)\, N_2\right] \mathrm{d}z, \tag{1.17}$$

which, together with (1.8) and a corresponding relation for the absorption probability, leads to

$$-\mathrm{d}I_L = k_A(\omega_L)\, I_L\, \mathrm{d}z \tag{1.18}$$

where

$$k_A(\omega_L) = (N_1 - N_2)\, \sigma_{12}(\omega_L)$$

and

$$\sigma_{12}(\omega_L) = B_{12} \frac{n}{c}\, \hbar\omega_L g(\omega_L - \omega_{21}) \cdot$$

The parameter $k_A(\omega_L)$ is called the absorption coefficient and $\sigma_{12}(\omega_L)$ the absorption cross section of the atomic system. (For $k_A(\omega_L) < 0$, the gain $g_V(\omega) = -k_A(\omega)$ is frequently used.) In these relations, it is assumed that the contributions from individual molecules can be simply summed. If dense gases, liquids and solids are concerned, the validity of this condition must be checked for each case. It is evident that for $N_1 > N_2$ which, for example, is always satisfied in thermal equilibrium, the absorption processes predominate, and accordingly, the incoming radiation is attenuated, whereas if $N_2 > N_1$, the stimulated emission causes amplification. Using the transition probabilities per unit time, we can also calculate the changes in the occupation numbers of the individual atomic levels caused by elementary radiation processes. Due to absorption processes the number of excited systems increases in the time interval dt by $[N_1\, \mathrm{d}t]\, [\mathrm{d}W_{12}^{A}/\mathrm{d}t]$. At the same time it decreases due to stimulated emission processes by $[N_2\, \mathrm{d}t]\, [\mathrm{d}W_{12}^{i}/\mathrm{d}t]$. In addition spontaneous emission causes a decrease in the occupation of the upper level by $[N_2\, \mathrm{d}t]\, [\mathrm{d}W_{21}^{sp}/\mathrm{d}t]$, so that taking into consideration the relations (1.2), (1.6) and (1.9) we obtain

$$\mathrm{d}N_2 = [B_{21} g(\omega_L - \omega_{21})\, (N_1 - N_2)\, U(\omega_L) - A_{21} N_2]\, \mathrm{d}t \tag{1.19}$$

for the total change of the occupation number density of the upper level. Besides the depopulation of the upper level caused by spontaneous emission, there are other processes that are not caused by external radiation fields, which change the upper level occupation — so-called radiationless relaxation processes. In gases these processes are, for example, caused by collisions between molecules; in solids, for instance, by interaction of the atom under consideration with the crystal lattice. In each of these processes energy is exchanged. (Transitions to other levels not directly involved in the radiation processes can also be produced by radiationless relaxation.) To calculate the mean occupation numbers only the total transition probability, which is equal to the sum of the individual transition probabilities, is of interest. In relation (1.19) we must accordingly replace A_{21} by

$$k_{21} = k_{R21} + k_{NR21} \quad \text{with} \quad k_{R21} \equiv A_{21}; \tag{1.20}$$

k_{NR} denotes here the transition rate of the radiationless transitions. The reciprocal value of the total transition probability k_{21} gives the mean lifetime T_{21} of the molecules in the excited level 2. (Notice that this relation is only valid for $\mathscr{E}_2 - \mathscr{E}_1 \gg k_B\mathscr{T}$ which is satisfied for optical transitions.)

The calculation of the occupation of the lower level results in a relation similar to (1.19) with opposite signs, since in the two-level system considered every process that produces a decrease in the occupation of level 2 produces a corresponding increase in the occupation of level 1 and vice versa. Thus, (1.18) to (1.20) result in

$$\frac{d}{dz} I_L = \sigma_{12}(\omega_L)\, I_L[N_2 - N_1], \tag{1.21}$$

$$\frac{d}{dt} N_2 = -\sigma_{12}(\omega_L)\, I_L[N_2 - N_1] - \frac{1}{T_{21}} N_2, \tag{1.22}$$

$$\frac{d}{dt} N_1 = \sigma_{12}(\omega_L)\, I_L[N_2 - N_1] + \frac{1}{T_{21}} N_2. \tag{1.23}$$

If we add the equations (1.22) and (1.23), we obtain $d(N_1 + N_2)/dt = 0$ which by integration leads to

$$N_1 + N_2 = N \tag{1.24}$$

where N is the total number of molecules per unit volume involved in the process.

If we consider the interaction long after the monochromatic radiation field with constant photon flux density I_L has been switched on, we will see that the occupation number densities will have reached a stationary value and, accordingly, their time derivatives will disappear. For the stationary solution, we obtain from (1.22) to (1.24) the density of the occupation number inversion $(N_2 - N_1)$ as a function of the photon flux density I_L:

$$(N_2 - N_1) = \frac{-N}{1 + 2\sigma_{12}(\omega_L)\, I_L T_{21}}. \tag{1.25}$$

For $I_L \to 0$ this difference takes on the value $-N$ and for $I_L \to \infty$ the value 0. For $I_L = I_S = 1/(2\sigma_{12}T_{21})$, where I_S is the so-called saturation photon flux density, the inversion density decreases to half the small signal value. If we insert the inversion density $(N_2 - N_1)$ from (1.25) in (1.21), we obtain a differential equation for I_L, which in general is nonlinear.

Until now, we have only considered a photon flux density as time independent but, in deriving the set of equations (1.21) to (1.24) in the way described, this condition is unnecessary. We wish therefore to repeat the previous treatment in which I_L can now depend not only on the space coordinate z but also on time t. We obtain again the equations (1.21) to (1.24) in which the differential quotients on the left side represent the total derivations with respect to t and z. Providing that the atomic systems rest or move only very slowly and that light energy is transported with the group velocity v_L, we obtain for the relation between the total and partial derivatives

$$\frac{\mathrm{d}}{\mathrm{d}z} I_L(t, z) = \left(\frac{\partial}{\partial z} + \frac{1}{v_L}\frac{\partial}{\partial t}\right) I_L(t, z) \tag{1.26a}$$

and

$$\frac{\mathrm{d}}{\mathrm{d}t} N_i(t, z) = \frac{\partial}{\partial t} N_i(t, z). \tag{1.26b}$$

Together with these two relations, (1.21) to (1.23) represent a set of partial differential equations used to determine the dependence of the photon flux density and occupation number densities on coordinate z and time t. For the special case in which the photon flux density changes only very slowly with time it may be justified to neglect $\left(\frac{1}{v_L}\partial/\partial t\right) I_L$ in favour of the term $(\partial/\partial z)\, I_L$.

The equations obtained in this way are called rate equations. They are relatively simple to derive: using the absorption cross sections and relaxation rates (which are determined experimentally or can be calculated by means of quantum theory — cf. 1.3.3), we write down the change in the occupation numbers of the atomic levels caused by the various processes, such as stimulated and spontaneous emission, absorption and radiationless relaxation. In this manner, we obtain a set of nonlinear partial differential equations that determine the dependence of all quantities on z and t. Often the two-level system considered above is not sufficient. In such cases, at least three or even more levels must be considered in the calculation. To illustrate this, we will write down the rate equations for the three-level system shown in Fig. 1.4 with two participating waves whose frequencies are resonant with two different atomic transitions $1 \to 2$ and $1 \to 3$:

$$\left(\frac{\partial}{\partial z} + \frac{1}{v_1}\frac{\partial}{\partial t}\right) I_1 = -\sigma_{13} I_1 [N_1 - N_3], \tag{1.27a}$$

$$\left(\frac{\partial}{\partial z} + \frac{1}{v_2}\frac{\partial}{\partial t}\right) I_2 = \sigma_{12} I_2 [N_2 - N_1], \tag{1.27b}$$

$$\frac{\partial N_3}{\partial t} = \sigma_{13} I_1 [N_1 - N_3] - \frac{1}{T_{32}} N_3 - \frac{1}{T_{31}} N_3, \tag{1.28a}$$

$$\frac{\partial N_2}{\partial t} = \sigma_{12} I_2 [N_1 - N_2] + \frac{1}{T_{32}} N_3 - \frac{1}{T_{21}} N_2. \tag{1.28b}$$

We can omit one further equation for N_1, because again the total balance $N_1 + N_2 + N_3 = N$ from which N_1 can be determined holds.

It should be pointed out here that in certain cases, rate equations do not completely describe the experimental findings. Indeed, the rate equations follow from a more comprehensive theory under defined conditions (cf. 1.3.2). The most important condition for the validity of rate equations is that the characteristic time duration of the process, which may, for example, be given by the duration τ_L of the light pulses, is large compared to the so-called phase decay time or transverse relaxation time τ_{21} (which we will introduce in 1.2 and 1.3). Since, however, this time is less than 10^{-13} s in most substances in a condensed phase, the rate equation approximation provides a reliable and practically sufficient method of calculation even for subpicosecond pulses in very many processes. In section 1.3 we will return to these problems in greater detail. Rate equations are successfully applied in particular to describe lasers. Therefore we will make frequent use of them in later chapters.

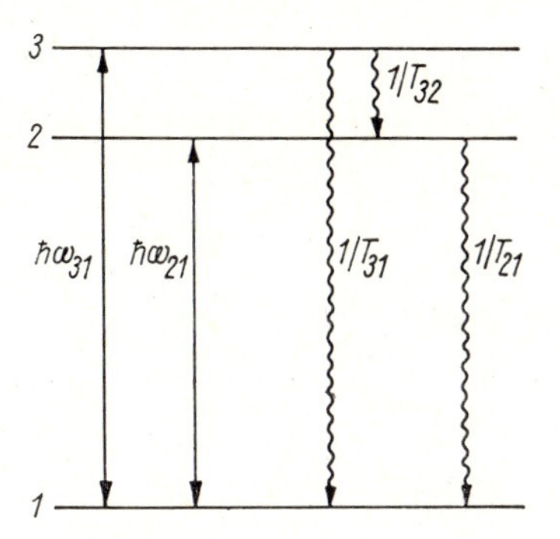

Fig. 1.4. Transitions in a three-level system

1.1.6 *Inhomogeneously Broadened Transitions*

Until now we have assumed that all atomic systems of an ensemble have the same parameters; in particular, all systems were attributed to the same transition frequency ω_{21} and the same line shape function $g(\omega - \omega_{21})$ with the halfwidth $\Delta\omega$. Accordingly, all atoms contribute with equal probability to the absorption or emission of radiation with an arbitrary frequency. Processes that limit the lifetime of the upper level — such as spontaneous emission and deactivating collisions as well as (in comparison to the lifetime) fast stochastic fluctuations in the level spacing — caused e.g. by frequency-modulating collisions or lattice vibrations — result in a line broadening around a fixed center frequency, which is uniform for each individual atomic system. Line broadenings of this kind are denoted as homogeneous.

On the other hand there are processes that lead to a spread in the transition frequencies of the individual systems — i.e. to an inhomogeneous broadening. These processes produce either a temporally constant distribution of the transition frequencies — one example of which is the spatially varying Stark effect in solids which is a result of inhomogeneities in the atomic densities — or a (in comparison to the lifetime) slow fluctuation of the resonance frequencies. One example of the latter case is the Doppler effect in diluted gases, which due to the Maxwell velocity distribution of gas particles, leads to varying effective resonance frequencies of the single particles, where the resonance frequencies are only slowly altered by velocity changing collisions (see e.g. [1.1, 1.2]).

Let us assume that the single particles differ only in their resonance frequencies, while the (homogeneous) halfwidth and the (homogeneous) line profile are equal. In this case, we can calculate an ensemble average — e.g. for the spontaneous emission of photons of frequency ω by summing up the contributions of all particles that are characterized by their individual resonance frequencies. Thus, we have to multiply the expression of equation (1.3) for the rate of spontaneous emission by the statistical distribution function $g_{\mathrm{inh}}(\omega_{21} - \omega_{21}^{(0)})$ of the resonance frequencies ω_{21}, which have the mean resonance frequency $\omega_{21}^{(0)}$ and the halfwidth $\Delta\omega_{\mathrm{inh}}$ called inhomogeneous linewidth. Then, we have to integrate this expression over the resonance frequencies and obtain

$$\left(\frac{\mathrm{d}}{\mathrm{d}t} W_{21}^{\mathrm{sp}}\right)_{\omega} \mathrm{d}\omega = A_{21}\mathcal{G}(\omega - \omega_{21}^{(0)})\,\mathrm{d}\omega \tag{1.29}$$

where

$$\mathcal{G}(\omega - \omega_{21}^{(0)}) = \int \mathrm{d}\omega_{21} g_{\mathrm{inh}}(\omega_{21} - \omega_{21}^{(0)})\, g(\omega_{21} - \omega)\,.$$

$\mathcal{G}(\omega - \omega_{21}^{(0)})$ is a line shape factor that takes into consideration the homogeneous and inhomogeneous line broadening. If the homogeneous line width is small compared to the inhomogeneous line width ($\Delta\omega \ll \Delta\omega_{\mathrm{inh}}$), $\mathcal{G}(\omega - \omega_{21}^{(0)}) \approx g_{\mathrm{inh}}(\omega - \omega_{21}^{(0)})$ (see Fig. 1.5) follows from (1.29).

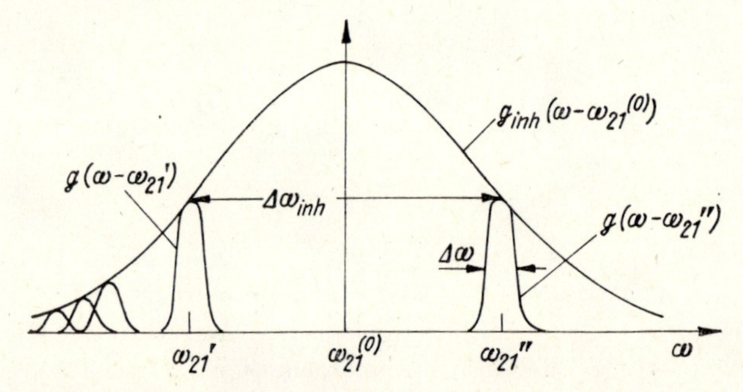

Fig. 1.5. Homogeneous and inhomogeneous contributions to the line profile
If the inhomogeneous line broadening is produced by the Doppler effect, the effective resonant frequency $\omega_{21} = \omega_{21}^{(0)}(1 + v/c)$ depends for absorption and emission on the velocity component v of the atoms in the direction of the light source and in the direction of the radiation detector, respectively. Using Maxwell's velocity distribution we obtain $\omega_{21}' = \omega_{21}^{(0)}(1 - v_{\mathrm{m}}/c)$ and $\omega_{21}'' = \omega_{21}^{(0)}(1 + v_{\mathrm{m}}/c)$ where $v_{\mathrm{m}} = \sqrt{2 \ln 2 k_{\mathrm{B}}\mathcal{T}/M}$ (M = particle mass) and $\Delta\omega_{\mathrm{inh}} = \omega_{21}^{(0)} \times \sqrt{8 \ln 2 k_{\mathrm{B}}\mathcal{T}/Mc^2}$.

Note that for the distribution of the transition frequencies ω_{21} we have to use the statistical equilibrium distribution, if the atomic systems obey this distribution in their initial state. This requirement is met for example for thermal radiators. Similar conclusions are valid for the processes of stimulated emission and absorption in thermal equilibrium, as long as this is not substantially disturbed by the radiation processes under consideration. Under these conditions, homogeneous and inhomogeneous line broadening processes have equal effects and therefore cannot be directly distinguished from one another with measurements. However, if we observe an atomic ensemble under the influence of strong radiation fields, we must in our calculation consider non-equilibrium distributions caused by the interaction between light and matter. As a simple example we consider a gas, all particles of which are in the ground state of energy $\mathcal{E}_1$ without irradiation by light and which accordingly exhibits no spontaneous emission. Some of the atoms are then excited into state 2 by an intensive, narrow-band light wave of center frequency ω_{L}, whereby the transition $1 \to 2$ is inhomogeneously broadened as a result of the Doppler effect. Particles whose velocity component in the direction of the light source leads to an effective resonance frequency $\omega_{21} = \omega_{21}^{(0)} \cdot (1 + v/c)$ in the vicinity of ω_L (see Fig. 1.6a) are more likely to reach the excited state. Hence the spontaneous emission also does not occur with the spectral width $\Delta\omega_{inh}$ of the equilibrium distribution, but only in a narrower interval around ω_L, whose minimal width is equal to the homogeneous line width $\Delta\omega$. In this manner, the homogeneous line width can also be experimentally determined under suitable conditions, and thus for sufficiently small pressure even the natural line width can be measured.

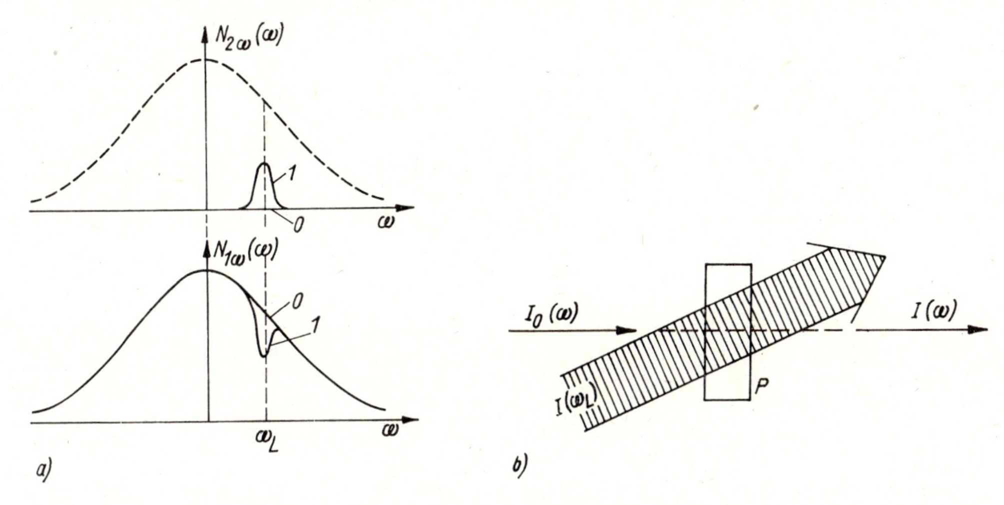

Fig. 1.6. Changes of the occupation numbers in inhomogeneously broadened systems acted on by radiation

a) Dependence on frequency of the occupation numbers of the ground and excited states per unit volume and frequency for the case without irradiation (0) and with irradiation (1); the dotted curve designates the maximum achievable occupation number density in the excited state.
b) Probe beam configuration for the measurement of the absorption saturation ($I(\omega_L)$ is the intensity of the pump laser at frequency ω_L, $I_0(\omega)$ and $I(\omega)$ are the intensities of the probe light at frequency ω at the input and at the output of the sample P, respectively).

Similar results can be obtained from so-called saturation spectroscopy, in which as described before through absorption processes a strong light wave of frequency ω_L reduces the occupation number difference between atoms in the ground and excited states having a velocity component that results in an effective transition frequency near ω_L (compare Fig. 1.6a). By this means, for a probe beam of variable frequency ω that is passed through the sample (see Fig. 1.6b), a reduced absorption is measured in the immediate vicinity of ω_L. Thus the absorption line does not become homogeneously saturated, but saturation only occurs in a narrow range. Therefore by the saturation of the absorption we can also distinguish between homogeneously and inhomogeneously broadened lines and measure the homogeneous line profile of single ensemble members within an inhomogeneously broadened line (see [12, 1.1]). Table 1.1 provides several typical values for the halfwidths of homogeneously and inhomogeneously broadened lines.

Table 1.1 Characteristic values and examples for linewidths

Process/Example	Type of broadening	$\frac{\Delta\omega}{2\pi} \Big/ \mathrm{Hz}$	$\Delta\omega/\omega$
Radiation damping	homogeneous	$0 \cdots 10^9$	$0 \cdots 10^{-6}$
Ne, $\lambda = 0.6328\ \mu m$		2×10^7	4×10^{-8}
Ar^+, $\lambda = 0.4880\ \mu m$		1.1×10^8	1.8×10^{-7}
Rhodamine 6 G, $\lambda \approx 0.600\ \mu m$		2×10^8	4×10^{-7}
CO_2, $\lambda = 10.6\ \mu m$		4×10^2	10^{-12}
Collisional broadening	homogeneous		
Ne, $\lambda = 0.6328\ \mu m$			
He:Ne = 5:1		1×10^6/Pa	
Doppler broadening	inhomogeneous		
Ne, $\lambda = 0.6328\ \mu m$, $\mathcal{T} = 300$ K		1.7×10^9	3×10^{-6}
Influence of phonons and inhomogeneities in solids and liquids	homogeneous, inhomogeneous	$\gtrsim 10^{10}$	$\gtrsim 10^{-5}$
Cr^{3+}: ruby, $\lambda = 0.694\ \mu m$, $\mathcal{T} = 300$ K	homogeneous	3.6×10^{11}	8×10^{-4}
Nd^{3+}: garnet, $\lambda = 1.0648\ \mu m$, $\mathcal{T} = 300$ K	homogeneous	2.0×10^{11}	6×10^{-4}
Nd^{3+}: glass, $\lambda = 1.06\ \mu m$, $\mathcal{T} = 300$ K	inhomogeneous	1.0×10^{13}	3×10^{-2}
Rhodamine 6 G, $\lambda \approx 0.6\ \mu m$	homogeneous	$\simeq 10^{14}$	$\simeq 10^{-1}$

1.2 Fast Processes in Matter

Many processes of motion in microphysics and chemistry are very fast. From table 1.2 it is evident that the times with which these processes can be characterized vary widely in range; in particular the nanosecond (ns) and picosecond (ps) regions ($10^{-9} - 10^{-12}$ s) contain important information about the systems under investigation (compare e.g. [1.3–1.6, 10]).

In the following, we want to discuss the significance of these characteristic times using several examples. The experimental investigation of the processes under discussion with respect to the determination of the characteristic times is carried out, in most cases, by measuring the time dependence of a macroscopic parameter in an ensemble of atomic systems, which was brought before the measurement into a state of nonequilibrium by a short-time excitation (see Fig. 1.7). We then observe the return of the ensemble to the initial state or the transition into a new stable or relatively stable state.

1.2.1 Internal Transitions and Relaxation Processes

Let us begin with transitions in the molecules (or atoms) of diluted gases. If no radiation acts on the system and the collisions between gas particles are very seldom, then the deactivation of an excited particle can only occur by spontaneous emission. The relaxa-

Table 1.2. Representative values for the characteristic times of some physical and chemical processes
($2\pi/\omega_{ij}$ is the inverse transition frequency (oscillation period); T is the lifetime of a system in the excited state; τ is the phase decay time; $1/k$ is the inverse rate parameter of a reaction)

Process/Transition	State of aggregation	Typical ranges of the characteristic time scales/s		
		$2\pi/\omega_{ij}$	T/s	τ/s
Transitions of atomic systems				
Free molecular rotation	gas (10^5 Pa)	10^{-12}—10^{-10}	10^{-10}—10^{-7}	10^{-11}—10^{-9}
Reorientation of molecules in a viscous environment	liquid	—	—	10^{-12}—∞
Molecular vibration	gas (10^5 Pa)	10^{-14}—10^{-12}	10^{-9}—10^{0}	10^{-11}—10^{-9}
	condensed	10^{-14}—10^{-12}	10^{-12}—10^{-3}	10^{-13}—10^{-11}
Electronic transitions		10^{-16}—10^{-14}	10^{-13}—10^{-2}	10^{-14}—10^{-10}
Reactions		$1/k$		
Elementary step of a monomolecular reaction		10^{-12}—∞		
Diffusion controlled reaction	liquid (concentration: 1 mol/liter)	10^{-10}—∞		
Important laser transitions	Lifetime of the upper laser level/s			
He-Ne ($\lambda = 0.6328$ μm)	gas	22×10^{-9}		
Ar^+ ($\lambda = 0.4880$ μm)	gas	9×10^{-9}		
CO_2 ($\lambda = 10.6$ μm)	gas	2.4×10^{-3}		
Ruby ($\lambda = 0.6943$ μm)	crystal	3.0×10^{-3}		
Nd:YAG ($\lambda = 1.0648$ μm)	crystal	0.2×10^{-3}		
Nd:glass ($\lambda = 1.06$ μm)	glass	$(0.06—0.8) \times 10^{-3}$		
F_A-centers in KCl ($\lambda = 2.7$ μm)	crystal	8×10^{-8}		
Rhodamine 6 G ($\lambda \approx 0.6$ μm)	solution	5×10^{-9}		

tion rate is, in accordance with the explanation in section 1.1, given by the Einstein coefficient $k_{21} = A_{21}$. In table 1.3, typical values for A_{21} for electronic as well as vibrational and rotational transitions are given. The table also contains the transition frequencies ω_{21} and the transition moments μ_{21} of the molecules, which are given in section 1.3 in relation to the Einstein coefficients. The Einstein coefficients are proportional to $|\mu_{21}|^2$ as well as to ω_{21}^3. The transition moments describe the magnitude of the oscillating dipole moments which are responsible for the emission or absorption. For strong electron transitions, the transition moments lie in the order of the product of elementary charge and molecular diameter, that is in the order of some 10^{-29} As m. The angular frequency of typical electronic transitions lies in the ultraviolet range in the order of 10^{16} s^{-1}, the Einstein coefficient of strong emission transitions at 10^9 s^{-1}.

With molecular vibrations, the largest transition moments are about 1 to 2 orders of magnitude smaller, because the effective vibrating charges are usually smaller than the elementary charge and the vibrational amplitudes are very much smaller than the molecular dimensions. With transition frequencies being more than one order of magnitude smaller than that for electronic transitions we obtain Einstein coefficients of strong vibrational transitions in the order of $10^3\,s^{-1}$. In rotational transitions, the transition moment is identical with the permanent dipole moment and can, accordingly, take on values of up to about 10^{-28} As m; with angular frequencies of $10^{13}\,s^{-1}$, values of around $10^2\,s^{-1}$ are obtained for A_{21}.

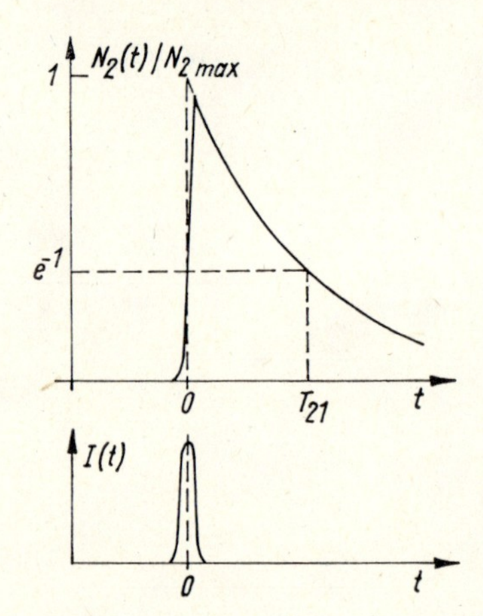

Fig. 1.7. Change in the occupation number of the excited level through absorption of an intense, short light pulse and subsequent spontaneous emission ($I(t)$ is the intensity of the excitation pulse, $N_2(t)$ is the occupation number of the upper level per unit volume as a function of time).

Table 1.3. Typical parameters of strong electronic, vibrational, and rotational transitions of molecules

(Δ_{21} is the (angular) transition frequency; λ_{21} is the transitioon wavelength; μ_{21} is the transition moment; A_{21} is the Einstein coefficient for spontaneous emission and σ_{12} denotes the absorption cross section in the line maximum for transitions in liquids and solids)

Transition	Δ_{21}/s^{-1}	$\lambda_{21}/\mu m$	μ_{21}/Asm	A_{21}/s^{-1}	σ_{12}/m^2
Electronic	10^{16}	0.2	10^{-29}	10^9	10^{-20}
Vibrational	3×10^{14}	6	10^{-30}	10^3	10^{-22}
Rotational	10^{13}	200	10^{-28}	10^2	—

The transition moments and the Einstein coefficients given above were estimated for particularly strong transitions. Depending on specific conditions in actual molecules the values can deviate and be considerably lower, for example, when the transition is for certain reasons of symmetry very unprobable. Thus, for instance, a transition between the triplet and singlet electron levels of an atom or molecule is much less probable (often by a factor of 10^{-4} or less) than a transition within the singlet or triplet systems. We speak here of an intercombination prohibition. A level that can be reached from the ground level only by such transitions is designated as metastable (see e.g. [1.6]).

Besides radiation processes collisions can also deactivate excited molecules of a gas. The corresponding relaxation rate k_{NR} is proportional to the number of collisions per unit time, and hence, to the gas pressure. The probability for deactivation by a collision decreases with increasing energy of the transition. Whereas this probability for rotational transitions lies in the order of 1 to 10^{-1}, it decreases with vibrational transitions to values from 10^{-3} to 10^{-10}. In this connection let us note that a complete conversion of the entire vibrational energy into translational energy of the collision partners (vib $\rightarrow$ trans) is much less probable than a deactivation of the primarily excited vibration through other vibrations (vib $\rightarrow$ vib) or through rotations (vib $\rightarrow$ rot), whereby only a small energy difference is converted into translational energy. In general, the number of possible relaxation paths greatly increases with the size of the molecule, and the corresponding relaxation times decrease (see e.g. [1.4]).

In the condensed phase, the interactions of atoms and molecules among themselves are considerably greater than in the gas phase. These interactions inhibit free rotation and lead to very fast deactivations of molecular vibrations and electron excitations. The deactivation rate of the transitions between the first excited electronic level and the ground level can achieve values of $10^{12}\ s^{-1}$, and values of $10^{14}\ s^{-1}$ for transitions between excited electronic levels. This electronic deactivation is caused by the vibrational motions in the surroundings of the excited molecule. Therefore, with electronic deactivation, vibrations of the molecule or of the lattice, so-called phonons, are usually excited. The relaxation rate k_{vib} for vibrational levels can assume values of up to $10^{12}\ s^{-1}$ in the electron ground state and up to $10^{14}\ s^{-1}$ in the excited states.

In Fig. 1.8, as an example, the lowest electronic levels of an organic molecule and the corresponding transitions are depicted. From this figure and the given values of the relaxation rate the following becomes apparent. If one such molecule is excited into a higher vibrational level of the S_1-level, it will relax very rapidly, i.e. in several 10^{-13} s, to the vibrationless state through vibrational relaxation processes. (If the energy of the vibrational quanta are not large compared to $k_B\mathcal{T}$, but comparable with $k_B\mathcal{T}$, then instead of a vibrationless state an equilibrium distribution in the occupation of the vibrational levels of the S_1-state will be reached.) After a typical lifetime of about 10^{-8} s, the molecule will pass from the vibrationless state of the S_1-level into the excited vibrational levels of the electronic ground state. This vibrational excitation decays in several picoseconds. Thus, the step in this entire process that determines the overall velocity is, in general, the electronic relaxation $(S_1, 0) \rightarrow (S_0, v)$. With this temporal behavior of the deactivation we will generally observe fluorescence emitted only from the vibrationless S_1-state, in which the molecule lives for a relatively long time. The fluorescent quantum efficiency, which gives the number of emitted photons in relation to the number of exciting photons, can be written as

$$\eta_F = \frac{(k_R)_{(S_1,0)\rightarrow(S_0,v)}}{(k_R)_{(S_1,0)\rightarrow(S_0,v)} + (k_{NR})_{S_1\rightarrow S_0}}$$

and with $(k_R)_{(S_1,0)\rightarrow(S_0,v)} \ll (k_{NR})_{S_1\rightarrow S_0}$ by

$$\eta_F = \frac{(k_R)_{(S_1,0)\rightarrow(S_0,v)}}{(k_{NR})_{S_1\rightarrow S_0}}.$$

The fluorescence from an excited vibrational level that is denoted as hot fluorescence

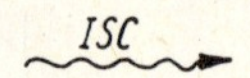

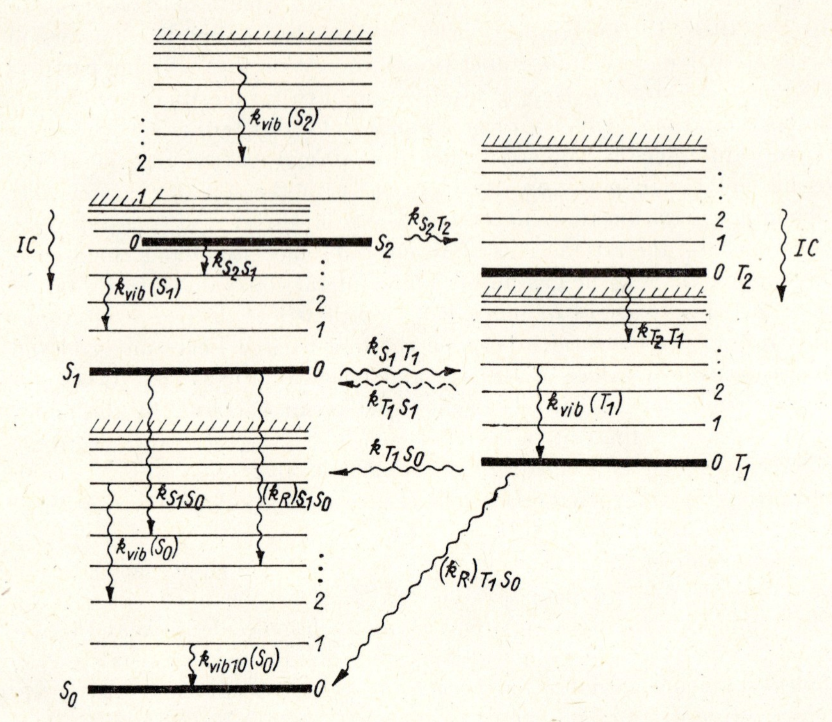

Fig. 1.8. Scheme of the energy levels of an organic molecule and the relaxation transitions between these levels. On the left side, the three lowest singlet levels are depicted, on the right side, the two lowest triplet levels. The vibrational levels are superimposed over the electronic levels. For the purpose of simplification vibrational levels of one normal vibration were represented for every electronic level. In actuality, large molecules have many normal vibrations (N atoms have $3N - 6$) that give a large number of vibrational levels and the vibrational relaxation is determined by the transitions between the different normal vibrations. The transitions within the singlet system and within the triplet system are designated as internal conversion (IC), those between the two systems are designated as intersystem crossing (ISC). The following are typical values for the relaxation rates of the plotted transitions:

$k_{S_1S_0}$: 10^8 s^{-1}, $k_{S_2S_1}$: 10^{12} s^{-1},

$k_{T_2T_1}$: 10^{10} s^{-1}, $k_{S_2T_2}$ and $k_{S_1T_1}$: 10^7 s^{-1}, $k_{T_1S_0}$: 10^4 s^{-1}, $k_{vib}(S_0)$: 10^{12} s^{-1}, $k_{vib}(S_1)$: 10^{13} s^{-1}.

has however (for $\mathcal{E}_{vib} \gg k_B \mathcal{T}$) only a fluorescent quantum efficiency of

$$\eta_F = \frac{(k_R)_{(S_1,v')\to(S_0,v)}}{(k_R)_{(S_1,v')\to(S_0,v)} + k_{vib}(S_1)} \approx \frac{(k_R)_{(S_1,v')\to(S_0,v)}}{k_{vib}(S_1)},$$

because the upper vibrational level is very rapidly depopulated by vibrational relaxation within the electronic state S_1 with the rate $k_{vib}(S_1)$ (compare e.g. [1.6]). If we excite a molecule into the S_2-level or an even higher level, it will normally relax very rapidly and nonradiatively into the vibrational ground state of the S_1-level and from there, in part by radiation and in part nonradiatively into the electronic ground level. Therefore, we usually only observe a fluorescence coming from the level $S_1(v = 0)$; this behavior is called the Vavilov rule.

Not only the rates of radiation transitions, but also those of nonradiative transitions can, due to certain prohibitions, deviate considerably from the typical values of strong "allowed" transitions. This holds, for example, for the transitions between the singlet and triplet systems of organic molecules shown in Fig. 1.8. Thus, in the lowest triplet level a molecule can exist for a relatively long time (up to a few seconds) not only in a collisionless environment, but also under the influence of strong interaction processes in the condensed phase. Such relatively long-living, metastable levels can be found also with ions, which are incorporated in crystal lattices. For example, the upper laser level of the ruby-laser (i.e. the ^2E-level of the $^2E \rightarrow {}^4A$-transition of the Cr^{3+}-ion, which corresponds to a wavelength of $\lambda = 0.694\ \mu m$) has a lifetime of 3×10^{-3} s at room temperature; the Nd^{3+}-ions in the YAG crystal have in the upper laser level a lifetime of 0.2×10^{-3} s [4].

1.2.2 Photochemical Reactions

Through radiative excitation, molecules may change via monomolecular reactions or participate together with other partners in reactions that do not take place or proceed very slowly in the ground state. The simplest reaction equations for these two types of processes are

$$A + (\hbar\omega) \rightarrow A^* \underset{k_{A^*B^*}}{\rightleftarrows} B^* \underset{k_{B^*B}}{\rightleftarrows} B \tag{1.30}$$

and

$$A + B + (\hbar\omega) \rightarrow A^* + B \underset{k_{A^*B,(AB)^*}}{\rightleftarrows} (AB)^* \underset{k_{(AB)^*,AB}}{\rightleftarrows} AB. \tag{1.31}$$

Besides the "forward reactions" for which the velocity constants or transition rates are given, the "reverse reactions" are represented by dashed lines. If the reverse reactions are negligible, the rate equations for N_{A*}, N_{B*} and N_B are (compare (1.21))

$$\begin{aligned} \frac{d}{dt} N_{A*} &= -k_{A*B*} N_{A*}, \\ \frac{d}{dt} N_{B*} &= k_{A*B*} N_{A*} - k_{B*B} N_{B*}, \end{aligned} \tag{1.32}$$

which lead to

$$\begin{aligned} N_{A*}(t) &= N_{A*}(0)\, e^{-k_{A*B*}t}, \\ N_{B*}(t) &= \frac{k_{A*B*}}{k_{A*B*} - k_{B*B}} N_{A*}(0)\, [e^{-k_{B*B}t} - e^{-k_{A*B*}t}], \end{aligned} \tag{1.33}$$

and

$$N_B(t) = N_{A*}(0) - N_{B*}(t) - N_{A*}(t).$$

Also, dealing in general with more complex problems, we can usually establish sets of rate equations for the particle numbers of various molecules in different states, having the same structure as (1.32).

A special case for the reaction of excited molecules is energy transfer. In this case, the excitation energy is transferred according to

$$A^* + B \underset{k_{DA}}{\rightleftarrows} A + B^*. \tag{1.34}$$

In the forward reaction, A is designated as the donor and B as the acceptor. First the transfer can take place during a collision, i.e. during interim formation of a complex (AB)*, and hence by a local interaction between the two partners. The formation of such complexes is determined in liquids by diffusion processes in connection with the initial distribution of the donors and acceptors in space (see e.g. [1.7]). Secondly, the energy transfer can result from nonlocal interaction processes, called the Förster-transfer [1.8]. In the simplest case the Förster-transfer is caused by the interaction between induced dipole moments in the donors and acceptors. Accordingly, the transfer rate between the donor and an acceptor is proportional to R_{DA}^{-6}, where R_{DA} is the distance between donor and acceptor. In order to obtain the deactivation law for the donor, we must sum up the transfer probabilities from the donor to all the acceptors statistically distributed around the donor. For a homogeneous probability distribution of the acceptors in space and neglecting the diffusion processes, we obtain the Förster law for the decrease of the number of excited donor molecules N_D

$$N_D(t) = N_D(0) \exp\left\{-\frac{t}{T_D} - 6.276 R_0{}^3 N_A \sqrt{\frac{t}{T_D}}\right\}. \tag{1.35}$$

In this case, T_D is the relaxation time of the donor without consideration of the acceptor influence; R_0 is the so-called Förster-transfer radius, which can be determined experimentally or calculated from the parameters of the donor and acceptor transition and N_A is the number of acceptors per unit volume. In favorable combinations of donors and acceptors, R_0 can assume values up to about 5 nm. This means that the energy can be transferred with high probability within the time $t = T_D$ to acceptors inside a sphere of this radius. Thus, energy transfer is possible over a distance of several molecular diameters. It is worth noting that from the superposition of individual contributions of the form $e^{-k_{DA}(R_{DA})t}$ follows a nonexponential deactivation law in the form of (1.35). Immediately after the excitation the energy transfer leads to a very fast deactivation due to the transfer to acceptors that are located near the donor. For greater times, the exponential decrease described by the term $\exp\{-t/T_D\}$ prevails. From the example of the Förster-transfer, it is evident that the relaxation processes do not inevitably lead to an exponential decay of the excitation.

1.2.3 Phase Relaxation Processes

In our examples up to now we have only very roughly described the interaction between light pulses and matter. In regard to the atomic system, we have only given the occupation probabilities for various levels and the temporal evolution of these probabilities; for the light pulses, the center frequency, the bandwidth, the pulse duration, and the photon flux density were sufficient. In addition to occupation probabilities, other physical parameters in single atomic systems or in ensembles can change under irradiation by light. Using a simple model, let us discuss these changes for parameters such as the induced dipole moment and the polarization which depend on the electric field strength of the light waves and can therefore contain phase information.

We consider an ensemble with M atomic systems which are located in a volume element whose spatial extension is small compared to the light wavelength. Regarding the light waves, we make use of the fact that we can describe very well the radiation field of a laser (as opposed to that of thermal light sources) by a classical electromagnetic wave with a non-fluctuating amplitude. All atoms in this ensemble are acted on by the electrical field

$$E_L(t) = \frac{1}{2}\,\hat{E}_L(t)\exp\{i\omega_L t\} + \text{C.C.} \tag{1.36}$$

or

$$E_L(t) = |\hat{E}_L(t)|\cos\{\omega_L t + \varphi_L(t)\},$$

where ω_L is the center frequency of the light pulse and $\hat{E}_L(t)$ is a (generally complex) pulse envelope or slowly varying amplitude function of modulus $|\hat{E}_L(t)|$ and phase $\varphi_L(t)$, which determines the temporal shape of the pulse. In every atomic system the field will generate a co-vibrating dipole moment

$$\mu(t) = \frac{1}{2}\,\hat{\mu}(t)\,e^{i\omega_L t} + \text{C.C.}, \tag{1.37}$$

in which the amplitude function $\hat{\mu}(t)$ depends on the influencing electric field and the parameters of the individual atoms as well as on their interaction with one another and with thermal "baths". Since all atoms are excited with equal phase, all the dipole moments oscillate in phase, whereby the M dipole moments per unit volume (if initially all the atoms are in the same state) sum up to a polarization

$$P(t) = \frac{1}{2}\,\hat{P}(t)\,e^{i\omega_L t} + \text{C.C.} \tag{1.38}$$

where

$$\hat{P}(t) = \frac{1}{\Delta V}\sum_{i=1}^{M}\mu_i(t) = \frac{1}{\Delta V}\,M\hat{\mu}(t).$$

This time dependent polarization of the ensemble can be measured by using appropriate experimental methods (compare 3. and 9.). If the frequency of the exciting field coincides with (or lies in the immediate vicinity of) the resonant frequency ω_{21} of the atomic systems, the dipole moment will continue to oscillate with this resonant frequency after a fast turn-off of the laser field at time t_0. In the course of this process, however, the amplitude of these oscillations is damped. Experimentally and also from quantum theory we obtain the time law

$$\hat{\mu}(t) = \hat{\mu}(t_0)\,e^{-(t-t_0)/\tau_{21}}, \tag{1.39}$$

where, since we are describing the relaxation of a phase-sensitive oscillation process, τ_{21} is designated as the phase decay time (or, in accordance with the terminology of high frequency spectroscopy, the transverse relaxation time — compare e.g. [1.9, 1.10]). If all particles of the ensemble possess the same resonant frequency and thus the ensemble is homogeneously broadened, the same decay law applies for the polarization, which is likewise a phase-sensitive quantity. The phase decay time is connected with the homogeneous line width $\Delta\omega$, e.g. with a Lorentz profile we have the relation

$$\tau_{21} = 2/\Delta\omega. \tag{1.40}$$

Only in the case of pure radiation damping in two level systems is a simple relation between the lifetime T_{21} (which, according to the terminology of high frequency spectroscopy, is also denoted the longitudinal relaxation time) and the phase decay time satisfied, namely $\tau_{21} = 2T_{21}$, (compare e.g. [11]). In systems subjected to strong interaction with other atomic particles, τ_{21} is usually considerably shorter than T_{21} (compare Table 1.2). In a gas, we can easily explain these facts about the interaction processes. Whereas only strong collisional influences contribute to deactivation, for changing the phase of an oscillation even a small frequency modulation resulting from the weak influence of a passing particle suffices. In liquids and solids, the circumstances are very similar; e.g. for electrons that have been excited into the conduction band the direction of their momentum changes very rapidly through interaction with the lattice, whereas deactivation to the valence band takes considerably longer.

If the resonant frequencies of atoms in the ensemble differ, this means that the transition is inhomogeneously broadened, the polarization of the ensemble decays by a process other than the damping of the single dipole moments. Whereas dipole moments superimpose in-phase until time t_0 because they are driven by the same electric field, the fast oscillating dipoles run ahead of the slow ones in their phase after the field has been switched off. Accordingly, the polarization will already decay after a time τ_{inh} of the order of the reciprocal inhomogeneous line width, which is

$$\tau_{\mathrm{inh}} \simeq 1/\Delta\omega_{\mathrm{inh}}. \tag{1.41}$$

A state is thus produced in which the polarization of a macroscopic ensemble disappears, although the oscillation of single dipole moments has not yet been destroyed (compare 8.4.2).

1.3 Basic Equations for the Description of the Interaction between Light Pulses and Matter

We have already pointed out in section 1.1 that rate equations do not in every case provide an adequate description of interaction processes. The full quantum-theoretical description of the atomic systems and of the electromagnetic field outlines the general foundation on the basis of which any interaction processes of radiation with matter can be described. For many phenomena, however, a semi-classical description leads to results that coincide well enough with the empirical findings. The atomic systems are described here by means of quantum mechanics, while the radiation field is described on the basis of the classical Maxwell theory. Such methods of calculation are well-suited to the description of radiation fields of high power, as long as one is not interested in the buildup of the wave out of the spontaneous process, i.e. out of the noise (see also e.g. [11]).

1.3.1 *Wave Equation*

In order to describe the electromagnetic field we begin with Maxwell's theory. The electric field in an optical medium (with permeability $\mu = 1$ and conductivity $\sigma_L = 0$) can be calculated using the wave equation (compare e.g. [1.11, 1.12])

$$\text{rot rot } \vec{E} + \frac{1}{c^2}\frac{\partial^2}{\partial t^2}\vec{E} = -\mu_0 \frac{\partial^2}{\partial t^2}\vec{P}. \tag{1.42}$$

For the description of linear and nonlinear optical processes knowing the polarization $\vec{P}$ is important. Through the influence of external fields (as previously explained in section 1.2.3) dipole moments in the molecules are induced, which give rise to additional electric fields. Consequently, the polarization of the medium becomes, in general, a function of the electric field $\vec{P} = \vec{P}(\vec{E})$, whose form is determined by the properties of the molecules with which the electric field interacts. In accordance with the common procedures, we separate from the entire polarization the contribution $\vec{P}^{L}$ that results only from the non-resonant interaction of the medium with the electric field and which is linearly dependent on the field strength. Thus we obtain

$$\vec{P}(t) = \vec{P}^{L}(t) + \vec{P}'(t). \tag{1.43}$$

where $\vec{P}'(t)$ is composed of the linear resonant term $\vec{P}^{LR}$ as well as of the nonlinear contribution $\vec{P}^{NL}$. The Fourier transform $\tilde{\vec{P}}^{L}(\omega)$ of $\vec{P}^{L}(t)$ is related to the Fourier-transformed field strength $\tilde{\vec{E}}(\omega)$ by

$$\tilde{\vec{P}}^{L}(\omega) = \varepsilon_0[\varepsilon(\omega) - 1]\,\tilde{\vec{E}}(\omega), \tag{1.44}$$

The quantity $\varepsilon(\omega)$ thus represents the linear optical permittivity of the material at frequency ω neglecting the contribution from a possible resonant transition with $\omega_{ij} \approx \omega$. Whereas $\varepsilon(\omega)$ changes only slowly with the frequency, the resonant contribution depends strongly on ω in the vicinity of ω_{ij}.

The electromagnetic field can usually be represented as a superposition of various quasi-monochromatic and quasi-plane waves:

$$\vec{E} = \sum_l \vec{E}_l = \frac{1}{2}\sum_l A_l(\vec{r}, t)\,\vec{e}_l\,e^{i(\omega_l t - \vec{k}_l \vec{r})} + \text{C.C.}, \tag{1.45}$$

In comparison to the exponential functions that oscillate fast in time and space, the amplitude functions $A_l(\vec{r}, t)$, which for light pulses may also be denoted as envelopes, depend only weakly on time and space coordinates. The vector $\vec{e}_l$ signifies the unit vector of the polarization direction of the l-th partial wave. The representation described here is a generalization of the concept of the plane, monochromatic waves for which the complex amplitudes A_l, are constant. The weak time-space dependency of the amplitude functions can describe a finite time duration of the radiation field and its finite spectral line width. An example of such a radiation field is the wave packet depicted in Fig. 1.9.

The polarization of the medium oscillates with nearly the same frequency as the field by which it is driven. Like the field we can therefore describe $\vec{P}(\vec{r}, t)$ as:

$$\vec{P} = \frac{1}{2}\sum_l \vec{e}_l' P_l(\vec{r}, t)\,e^{i(\omega_l t - \vec{k}_l \vec{r})} + \text{C.C.}. \tag{1.46}$$

In the following investigation, we choose the coordinate system such that the medium occupies the half space with $z > 0$ and the waves propagate in the z direction $(\vec{k}_l = (0, 0, k_l))$. Under this condition, it is often possible to neglect the dependence of the amplitude functions $A_l(\vec{r}, t)$ and $\overline{P}_l(\vec{r}, t)$ on x and y and merely to consider their change in the propagation direction z and in time t. This means we assume that the cross sections of the light beams do not change and that the various components of the field propagate nearly colinearly. We insert the relations (1.45) and (1.46) into the wave equation (1.42). Due to the fast oscillation of the exponential functions in the polarization and field strength mentioned above, each individual term of the sum over l in the

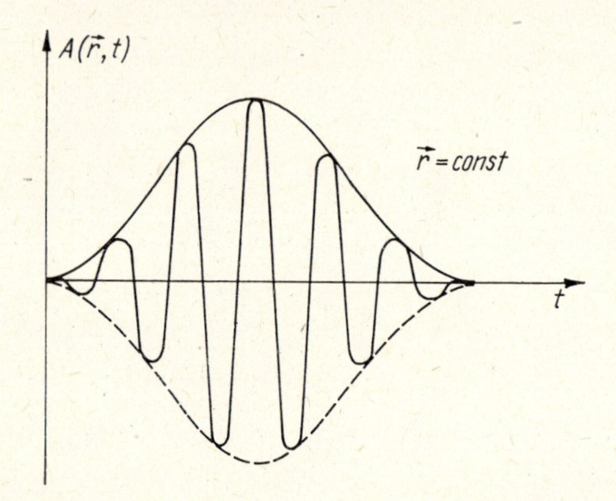

Fig. 1.9. Amplitude function (or pulse envelope) $A(\vec{r}, t)$ and carrier wave of a wave packet

wave equation must approximately vanish, in order that the wave equation is satisfied for all values of z and t. For the amplitude function A_l we thus obtain (for $\vec{e} = \vec{e}\,'$)

$$2\mathrm{i}k_l \frac{\partial A_l}{\partial z} + \left(k_l^2 - \frac{\omega_l^2}{c^2}\right) A_l + \frac{2\mathrm{i}\omega_l}{c^2} \frac{\partial A_l}{\partial t} + \frac{1}{c^2} \frac{\partial^2 A_l}{\partial t^2} - \frac{\partial^2 A_l}{\partial z^2}$$

$$= \mu_0 \left(\omega_l^2 \overline{P}_l - 2\mathrm{i}\omega_l \frac{\partial \overline{P}_l}{\partial t} - \frac{\partial^2 \overline{P}_l}{\partial t^2}\right). \tag{1.47}$$

According to (1.43) we now separate the linear, non-resonant part from the polarization, whereby we take into account the dispersion of the medium up to the second order in $(\omega - \omega_l)$. We expand $\varepsilon(\omega)$ at the center frequency ω_l, insert this expansion into (1.44) and transform this relation back into the time domain. Thus we obtain

$$\overline{P}_l^{\mathrm{L}}(z, t) = \varepsilon_0 \chi(\omega_l)\, A_l(z, t) - \mathrm{i}\varepsilon_0 \left(\frac{\mathrm{d}}{\mathrm{d}\omega} \chi\right)_{\omega_l} \frac{\partial}{\partial t} A_l(z, t) - \frac{1}{2} \varepsilon_0 \left(\frac{\mathrm{d}^2}{\mathrm{d}\omega^2} \chi\right)_{\omega_l} \frac{\partial^2}{\partial t^2} A_l(z, t), \tag{1.48}$$

where we have introduced the linear susceptibility $\chi(\omega) = (\varepsilon(\omega) - 1)$. The quantities $\mathrm{d}\chi/\mathrm{d}\omega$ and $\mathrm{d}^2\chi/\mathrm{d}\omega^2$ can be rewritten as

$$\left(\frac{\mathrm{d}\chi}{\mathrm{d}\omega}\right)_{\omega_l} = \left(\frac{\mathrm{d}\varepsilon}{\mathrm{d}\omega}\right)_{\omega_l} = \left[c^2 \frac{\mathrm{d}}{\mathrm{d}\omega} \left(\frac{k(\omega)}{\omega}\right)^2\right]_{\omega_l} = 2c^2 \frac{k(\omega_l)}{\omega_l^2} \left[\frac{1}{v_l} - \frac{n_l}{c}\right], \tag{1.49a}$$

and

$$\left.\frac{\mathrm{d}^2\chi}{\mathrm{d}\omega^2}\right|_{\omega_l} = 2c^2 \left[-\frac{4k_l}{\omega_l^3 v_l} + \frac{1}{\omega_l^2 v_l^2} + \frac{k_l k_l''}{\omega_l^2} + \frac{3k_l^2}{\omega_l^4}\right], \tag{1.49b}$$

whereby we have inserted the phase velocity $c/n_l = \omega_l/k_l$, the group velocity $v_l = (\mathrm{d}\omega/\mathrm{d}k)$ and the reciprocal dispersion of the group velocity

$$k_l'' = \left(\frac{\mathrm{d}^2k}{\mathrm{d}\omega^2}\right)_{\omega_l} = \frac{\lambda_l^3}{2\pi c^2}\left(\frac{\mathrm{d}^2n}{\mathrm{d}\lambda^2}\right)_{\lambda_l}.$$

Using (1.43) and (1.48) we then obtain from (1.47)

$$2\mathrm{i}k_l\left(\frac{\partial}{\partial z}A_l(z,t) + \frac{1}{v_l}\frac{\partial}{\partial t}A_l(z,t)\right) + \left(\frac{1}{v_l^2} + k_lk_l''\right)\frac{\partial^2}{\partial t^2}A_l(z,t) - \frac{\partial^2}{\partial z^2}A_l(z,t)$$

$$= \mu_0\omega_l^2\overline{P}_l'(z,t) \tag{1.50}$$

where $\overline{P}_l' = \overline{P}_l^{\mathrm{LR}} + \overline{P}_l^{\mathrm{NR}}$

and where we have made use of the relations $\omega|\overline{P}_l^{\mathrm{NL}}| \gg \left|\frac{\partial\overline{P}_l^{\mathrm{NL}}}{\partial t}\right|$, $\omega\,|P_l^{\mathrm{LR}}| \gg \left|\frac{\partial P_l^{\mathrm{LR}}}{\partial t}\right|$.

We consider now the pulse development in a coordinate system $\eta_l = t - \frac{z}{v_l}$, $z = z$ that travels with the group velocity v_l. Due to the weak change of the amplitude function compared to the fast oscillating exponential functions in (1.45) we can assume furthermore the relations

$$\left|\frac{\partial^2A_l}{\partial z^2}\right| \ll k_l\left|\frac{\partial A_l}{\partial z}\right|, \qquad \left|\frac{\partial^2A_l}{\partial z\,\partial\eta_l}\right| \ll k_lv_l\left|\frac{\partial A_l}{\partial z}\right|.$$

Therefore the terms $\frac{\partial^2A_l}{\partial z^2}$ and $\frac{\partial^2A_l}{\partial z\,\partial\eta_l}$ occuring in equation (1.50) can be neglected. Then, in place of (1.50) we can write

$$\frac{\partial A_l}{\partial z}(z,\eta_l) - \frac{\mathrm{i}}{2}k_l''\frac{\partial^2A_l}{\partial\eta^2}(z,\eta_l) = -\frac{\mathrm{i}\mu_0\omega_l^2}{2k_l}\overline{P}_l'(z,\eta_l). \tag{1.50'}$$

The term proportional to k_l'' causes with very short pulses and very long samples a pulse distortion even without the action of the nonlinear polarization. Such a significant pulse distortion results above a critical sample length $L_\mathrm{D} \sim \frac{\tau_\mathrm{L}^2}{|k_l''|}$. Under the most common conditions the sample length L is small compared to L_D ($L \ll L_\mathrm{D}$); accordingly, in this case the term with the second derivative with respect to η_l is negligible. Summarizing, the approximations undertaken in equation (1.50′) are called the slowly varying envelope approximation (SVEA). These conditions, however, can be violated (i.e. $L \gtrsim L_\mathrm{D}$ holds with the propagation of very short pulses along long optical paths. In this case, the second term on the left side of (1.50′) must be taken into consideration (compare section 8.3).

1.3.2 *Quantum-Statistical Description of Atomic Systems*

The various processes of light-matter interaction are considerably influenced by the relaxation of the participating atoms or molecules. This means that the influence of the environment on the atomic system, the interaction of which with the radiation field is considered, is of great importance and must be taken into account. Let us consider in a gas, for example, one specific molecule whose properties are mainly given by the electronic and nuclear structure of an isolated molecule; the ambient particles, however, exert an influence on the molecule considered — e.g. through their stochastic translational motion — which provides important modifications in the interaction between light and matter. The influence of thermal lattice vibrations on atomic systems in solids and the spontaneous emission of light by atoms are other examples of relaxation mechanisms. All these processes are irreversible, and they are characterized by the fact that the dynamic system, whose interaction with the radiation field is our main interest later on and which has only comparatively few degrees of freedom, is coupled to a dissipative system — also called thermal bath or bath — which represents the environment and which as a consequence has a very large number of degrees of freedom. Thus, the Hamiltonian

$$\mathbf{H} = \mathbf{H}^0 + \mathbf{H}^\mathrm{W} + \mathbf{H}^\mathrm{R} \tag{1.51}$$

of the system consists of three parts:
$\mathbf{H}^0$ represents here the Hamiltonian of the isolated atom or molecule with the eigenstates $|\psi_n{}^0\rangle$ and the energy eigenvalues $\mathcal{E}_n{}^0$, which satisfy the equation

$$\mathbf{H}^0 |\psi_n{}^0\rangle = \mathcal{E}_n{}^0 |\psi_n{}^0\rangle . \tag{1.52}$$

$\mathbf{H}^\mathrm{W}$ describes the interaction between the atomic system and the radiation field, whereas $\mathbf{H}^\mathrm{R}$ represents the coupling of the dynamic to the dissipative system. Both of these perturbations are assumed to be weak.

Since, in general, not all individual systems of an ensemble are found to be in the same state $|\psi_n\rangle$, in order to characterize the whole system, we introduce the probability w_n with which an individual system is in state $|\psi_n\rangle$ (mixed state). For the description of an ensemble of this type it is advisable to convert from the eigenstates $|\psi_n\rangle$ to the density operator $\boldsymbol{\rho}$ which is defined as

$$\boldsymbol{\rho} = \sum_n w_n |\psi_n\rangle \langle\psi_n| . \tag{1.53}$$

By

$$\langle \mathbf{A} \rangle = \mathrm{Tr}\,(\boldsymbol{\rho}\mathbf{A}) = \sum_n w_n \langle\psi_n |\mathbf{A}| \psi_n\rangle \tag{1.54}$$

the expectation value $\langle \mathbf{A} \rangle$ of an arbitrary observable $\mathbf{A}$ averaged over the ensemble can be calculated by means of the density operator $\boldsymbol{\rho}$.

From the Schrödinger equation we obtain

$$i\hbar \frac{\mathrm{d}\boldsymbol{\rho}(t)}{\mathrm{d}t} = [\mathbf{H}, \boldsymbol{\rho}] \tag{1.55}$$

as equation of motion for the density operator $\boldsymbol{\rho}$.

The density operator has the properties

$$\boldsymbol{\rho}^+ = \boldsymbol{\rho},$$

$$\mathrm{Tr}\,\boldsymbol{\rho} = 1,$$

$$\mathrm{Tr}\,\boldsymbol{\rho}^2 \leqq 1.$$

We pass over now to the matrix representation of the operator ρ where we make use of the set of eigenstates $|\psi_n{}^0\rangle$ of the Hamiltonian $\mathbf{H}^0$: $\varrho_{mn} = \langle\psi_m{}^0|\, \rho\, |\psi_n{}^0\rangle$. After a somewhat tedious calculation procedure, for whose detailed representation [11] should be referred to, we obtain from (1.51), (1.52) and (1.55) the following set of density matrix equations:

$$\frac{\mathrm{d}}{\mathrm{d}t}\,\varrho_{kl} + \mathrm{i}\omega_{kl}\varrho_{kl} + \frac{1}{\tau_{kl}}\,\varrho_{kl} = \frac{1}{\mathrm{i}\hbar}\sum_m (H^{\mathrm{W}}_{km}\varrho_{ml} - \varrho_{km}H^{\mathrm{W}}_{ml}) \quad \text{für} \quad k \neq l \tag{1.56a}$$

and

$$\frac{\mathrm{d}}{\mathrm{d}t}\,\varrho_{nn} + \sum_m (k_{nm}\varrho_{nn} + k_{mn}\varrho_{mm}) = \frac{1}{\mathrm{i}\hbar}\sum_m (H^{\mathrm{W}}_{nm}\varrho_{mn} - \varrho_{nm}H^{\mathrm{W}}_{mn})\,. \tag{1.56b}$$

These equations take into account the influence of the dissipative system on the dynamic one, which causes the relaxation. Here $\omega_{kl} = \frac{1}{\hbar}\,(\mathscr{E}_k^{(0)} - \mathscr{E}_l^{(0)})$ are the transition frequencies of the dynamic system; the corresponding term in (1.56) containing these frequencies results from $\mathbf{H}^0$ in (1.51); τ_{kl}^{-1} and k_{kl} characterize transition probabilities resulting from dissipative processes, which in (1.51) are represented by the term $\mathbf{H}^{\mathrm{R}}$. The calculation of these transition probabilities, which can be carried out using the matrix elements of $\mathbf{H}^{\mathrm{R}}$ shall not be further considered here. In the practical use of equation (1.56) the parameters τ_{kl}^{-1} and k_{kl} are mostly considered as phenomenological parameters whose exact numerical values can be experimentally determined. The processes that determine k_{mn} are combined with an energy loss of the dynamic systems concerned, whereas the transverse relaxation times τ_{mn} are caused not only by energy loss processes, but also by phase changing processes, and delineate the destruction of given phase relations in the dynamic systems (compare 1.2.).

Thus the left-hand sides of (1.56) describe the evolution of the dynamic system under the influence of dissipative systems. In addition, the terms on the right-hand side of (1.56) take into account the influence of the radiation field on the atomic system, where H^{W}_{km} represents the matrix elements of the interaction operator $\mathbf{H}^{\mathrm{W}}$. In dipole approximation, we have

$$H^{\mathrm{W}}_{km} = -\vec{\mu}_{km}\vec{E}\,, \tag{1.57}$$

where $\vec{\mu}$ is the operator of the electric dipole moment. (For one-electron systems $\vec{\mu} = -e\vec{r}$ holds.) The density matrix given by (1.56) depends only on the variables of the dynamic system; in regard to the variables of the dissipative system ensemble averages have already been taken before, in which fluctuations of the dynamic system caused by its coupling to thermal baths, are not taken into account. In reality, however, the dissipative system affects stochastically the dynamic system, and causes not only damping (relaxation) but also fluctuation. We can subsequently consider this fluctuation phenomenologically as in the Langevin method of statistical mechanics (compare e.g. [1.13]), by adding a stochastic term $\mathscr{N}_{ik}$ to the right-hand side of (1.56a) for the nondiagonal elements of the density matrix. This phenomenological approach can be established in a strictly deductive sense. Here the stochastic term $\mathscr{N}_{ik}$ can be treated as "white noise" and the correlator is given by

$$\langle \mathcal{N}_{ik}(t)\,\mathcal{N}^*_{ik}(t')\rangle = \frac{2}{\tau_{ik}}\,\delta(t-t')\,, \tag{1.58}$$

where the strength of the fluctuation follows from the fluctuation-dissipation theorem [1.14] and is indicated by the prefactor of Dirac's δ-function in (1.58).

From the density matrix the polarization of the medium can now be calculated according to the general rule (1.54) as the quantum statistical expectation value of the dipole operator $\vec{\boldsymbol{\mu}}$:

$$\vec{P} = N\,\mathrm{Tr}\,(\boldsymbol{\rho}\vec{\boldsymbol{\mu}}) = N\sum_{i,k}\varrho_{ik}\vec{\mu}_{ki}\,, \tag{1.59}$$

in which N is the number density of molecules actively participating in the process, which are supposed to be identical. Accordingly, the equation for the field amplitudes (1.50) and those for the density matrix elements (1.56) are coupled by relations (1.57) and (1.59). In general, a nonlinear relationship exists here between the polarization $\vec{P}$ and the field strength $\vec{E}$, which represents the basis for the description of a variety of nonlinear processes in optics.

1.3.3 *Treatment of Two-Level Systems*

To conclude this section, we would like to deal with the simplest model of an atomic system, namely the two-level system, which is assumed to be nearly resonant with an electromagnetic field ($\omega_\mathrm{L} \approx \omega_{21}$). Here we will also show how the rate equations introduced in section 1.1 are related to the more comprehensive equations of motion for the density matrix. Simultaneously, with this consideration, we shall determine the Einstein coefficients B_{12}, B_{21} of absorption and stimulated emission empirically introduced in section 1.1 and thus by using (1.14) also the Einstein coefficient A_{21} of spontaneous emission as well as the line shape factor $g_\mathrm{L}(\omega-\omega_{21})$. With the knowledge of the eigenfunctions of the atomic system these quantities are now at least in principle calculable.

The density matrix equations for a two-level system, according to (1.56) and (1.57), are

$$\frac{\partial\varrho_{12}}{\partial t} + \left(\mathrm{i}\omega_{12} + \frac{1}{\tau_{12}}\right)\varrho_{12} = \frac{\mathrm{i}}{\hbar}\,\vec{\mu}_{12}(\varrho_{22}-\varrho_{11})\,\vec{E}\,, \tag{1.60}$$

$$\frac{\partial\varrho_{22}}{\partial t} + k_{21}\varrho_{22} = -\frac{\mathrm{i}}{\hbar}\,(\vec{\mu}_{12}\varrho_{21}\vec{E} - \vec{\mu}_{21}\varrho_{12}\vec{E}\,)\,, \tag{1.61}$$

$$\varrho_{11} + \varrho_{22} = 1\,, \quad \varrho_{21} = \varrho^*_{12}\,. \tag{1.62}$$

The polarization is, according to (1.59), determined by the relation

$$\vec{P}^\mathrm{NL} + \vec{P}^\mathrm{LR} = N(\varrho_{12}\vec{\mu}_{21} + \varrho_{21}\vec{\mu}_{12})\,. \tag{1.63}$$

After splitting off fast varying factors according to

$$\vec{E} = \frac{1}{2}\,A_\mathrm{L}\vec{e}_\mathrm{L}\,\mathrm{e}^{\mathrm{i}(\omega_\mathrm{L}t-k_\mathrm{L}z)} + \mathrm{C.C.}, \qquad \varrho_{12} = \tilde{\varrho}_{12}\,\mathrm{e}^{\mathrm{i}(\omega_\mathrm{L}t-k_\mathrm{L}z)} \tag{1.64}$$

and after neglecting fast oscillating terms which do not make any essential contribution in the equation of motion (rotating wave approximation), we obtain for $\tilde{\varrho}_{12}$

$$\frac{\partial \tilde{\varrho}_{12}}{\partial t} + \left[\frac{1}{\tau_{21}} + \mathrm{i}(\omega_L - \omega_{21})\right] \tilde{\varrho}_{12} = \frac{\mathrm{i}}{2\hbar} (\vec{\mu}_{12}\vec{e}_L) (\varrho_{22} - \varrho_{11}) A_L . \quad (1.65)$$

The transition to rate equations is possible in cases in which the term with the time derivative in (1.65) is negligible compared with the remaining terms. If the characteristic time scale on which the essential processes occur is determined by the laser pulse duration τ_L, we can make the order-of-magnitude estimation $\partial \tilde{\varrho}_{12}/\partial t \simeq \tilde{\varrho}_{12}/\tau_L$, from which we obtain for the validity of rate equations the condition

$$\tau_L \gg \tau_{21} . \quad (1.66\text{a})$$

Moreover, the fields must not be too strong, so as to avoid fast oscillations (so-called Rabi oscillations) of the molecular parameters (compare 8.4). This condition leads to the inequality

$$\frac{A_L(\vec{\mu}_{12}\vec{e}_L)}{2\hbar} \ll \frac{1}{\tau_{21}} . \quad (1.66\text{b})$$

Under these conditions, we can insert the steady state solution for $\tilde{\varrho}_{12}$ from (1.65) into (1.61). Multiplying the resulting equation by the number density N of the molecules participating in the process and using the relations $N_2 = N\varrho_{22}$ and $N_1 = N\varrho_{11}$, we obtain

$$\frac{\partial N_2}{\partial t} + k_{21} N_2 = -\sigma_{12}(\omega_L) (N_2 - N_1) I_L \quad (1.67)$$

where

$$\sigma_{12}(\omega_L) = \frac{\pi \omega_L n}{c \varepsilon_0 \hbar} \langle |\vec{\mu}_{21}\vec{e}_L|^2 \rangle_0 \, g(\omega_L - \omega_{21}) \quad (1.68)$$

and

$$g(\omega_L - \omega_{21}) = \frac{1}{\pi \tau_{21} \left[(\omega_L - \omega_{21})^2 + \dfrac{1}{\tau_{21}^2}\right]} . \quad (1.69)$$

In (1.67) we introduced the photon flux density $I_L = c\varepsilon_0 |A_L|^2/2\hbar\omega_L n$. Furthermore, in (1.68) we have taken into consideration that the orientation of the dipole moments $\vec{\mu}_{21}$ of various molecules can differ, and consequently an average $\langle \ldots \rangle_0$ with respect to all orientations of the molecular dipole moments must be taken. If we additionally use the relationships between the absorption cross section $\sigma_{12}(\omega_L)$ and the Einstein coefficient B_{12} according to (1.18), we obtain

$$B_{12} = \frac{\pi}{\varepsilon_0 \hbar^2} \langle |\vec{\mu}_{21}\vec{e}_L|^2 \rangle_0 . \quad (1.70)$$

The dependence of the other Einstein coefficients A_{21} and B_{21} on the atomic parameters results from (1.70) in connection with (1.14) and (1.15). Inserting the steady state solution of (1.65) into the expression for the polarization of (1.63), we obtain from (1.50) for the field amplitude the equation

$$\frac{\partial A_L}{\partial z} + \frac{1}{v_L} \frac{\partial A_L}{\partial t} = \frac{1}{2} \sigma_{12}(\omega_L) A_L(N_2 - N_1) [1 - \tau_{21} i(\omega_L - \omega_{21})]^{-1}. \tag{1.71}$$

Provided that $\tau_L \gg L/v_L$ (L — sample length), the time derivative can be neglected in (1.71) in view of the remaining terms. If we multiply (1.71) by A_L^* and add the complex conjugate equation, then we obtain for the photon number flux density the equation (1.21), whereas (1.67) is identical with (1.22).

2. Fundamentals of Lasers for Ultrashort Light Pulses

Before we deal individually with the various methods of generating ultrashort light pulses in the following chapters, we wish to discuss here as a necessary basis the general principle of lasers, the method of optical pumping of lasers and the properties of laser resonators, of important active materials and of some representative lasers. At the end of the chapter the principle of modelocking (mode coupling, mode synchronization) and the most important methods of realizing this principle will be briefly introduced.

2.1 The Laser Principle

As described in chapter 1, emission and absorption processes occur in the resonant interaction between electromagnetic fields and matter. The change of the photon flux density (I_L), which is caused by the induced emission and absorption, is according to (1.21)

$$dI_L = \sigma_{21} I_L (N_2 - N_1)\, dz\,. \tag{2.1}$$

As we can see from (2.1), the wave is attenuated, if $N_2 < N_1$, and amplified if $N_2 > N_1$. As we know, in thermodynamic equilibrium, the occupation of the two levels 1 and 2 in an atomic system is determined by the Boltzmann distribution

$$\frac{N_2^e}{N_1^e} = \exp\left[-\frac{\mathscr{E}_2 - \mathscr{E}_1}{k_B \mathscr{T}}\right]. \tag{2.2}$$

Therefore, in thermodynamic equilibrium $N_2 < N_1$ is always found, i.e., the sample composed of such atomic systems attenuates the radiation at frequency $\nu = (\mathscr{E}_2 - \mathscr{E}_1)/h$, since the absorption outweights the emission. If, however, the occupation numbers can be successfully changed so that there are more atoms present in the higher state than in the lower, i.e. a population inversion occurs, then the sample functions as an amplifier. The medium in which an occupation inversion can take place is called an active medium. There is a set of known processes, such as optical pumping, collisional excitation, excitation by electronic current and chemical pumping, by means of which we can generate an occupation inversion. In the following, we will give particular attention to the optical pumping, since the most important lasers for ultrashort light pulses are excited by means of this process.

By constructing a feedback loop — analogous to the amplifiers with feedback in broadcast engineering — the amplifier, which consists of the active medium in an inverted state, can be forced to work as a light oscillator having stable field strength amplitude. We may achieve such a feedback in the optical wavelength range, for instance, by mounting parallel to each other two mirrors M_1 and M_2 of high reflectivity, which accordingly form a Fabry-Perot resonator, and between which the laser medium M is positioned alongside the resonator axis (Fig. 2.1). Since, according to (2.1), the change in the photon flux density is proportional to the already present photon flux density, self-starting oscillations may in this case occur due to a kind of avalanche process: due to multiple reflections by the mirrors the noise intensity generated in the resonator by spontaneous emission is repeatedly amplified in the active medium, whereby during each roundtrip only a part of the radiation generated, in this way leaves the resonator through the partially reflecting mirror M_2.

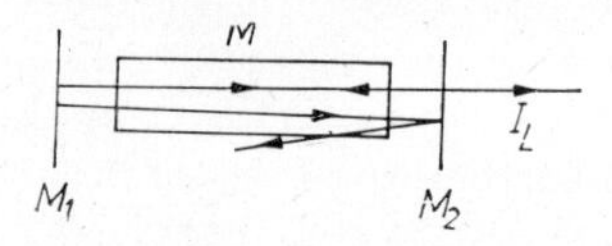

Fig. 2.1. Scheme of a laser resonator consisting of the active medium M and the mirrors M_1, M_2

The typical properties of a laser, such as high monochromacity, low divergence, and high spectral energy density of the radiation, are determined to a great extent by the combined action of resonator and active medium. The selection of the propagation direction of laser radiation comes about in such a manner that only those waves whose direction coincides with or deviates only little from the resonator axis experience a large amplification. All other waves will leave the laser resonator after a certain number of reflections on the mirrors and therefore no longer participate in the amplification process. Let us now discuss the conditions under which a self-starting generation of radiation is possible. Such oscillation can only start in a laser if the amplification by the active medium compensates for the losses which arise, for instance, by the escape of radiation at the mirrors. At the beginning of the generation processes, the intensity of the radiation $I_{\mathrm{L}}(z)$ is still small so that the dependence of the occupation number inversion on the intensity $I_{\mathrm{L}}(z)$ can be neglected. The amplification of radiation on one passage through the active medium of length L^{a} is then in accordance with (2.1) given by the relation

$$I_{\mathrm{L}}(L^{\mathrm{a}}) = \exp\{\sigma_{21}(N_2 - N_1)\, L^{\mathrm{a}}\}\, I_{\mathrm{L}}(0). \tag{2.3}$$

The amplification compensates for the losses from the mirrors precisely if the relation

$$I_{\mathrm{L}}(0) = R_1 R_2 I_{\mathrm{L}}(2L^{\mathrm{a}}) \tag{2.4}$$

between $I_{\mathrm{L}}(0)$ and the intensity after a double passage through the medium $I_{\mathrm{L}}(2L^{\mathrm{a}})$ holds, where R_1 and R_2 denote the reflectivity of the two mirrors. Thus the gain coefficient $g_{\mathrm{v}} = \sigma_{21}(N_2 - N_1)$ must exceed a certain threshold value, which, using (2.3) and (2.4), is given by the relation

$$(g_{\mathrm{v}})_{\mathrm{Thr}} = \frac{1}{2L^{\mathrm{a}}}\left[\ln\left(\frac{1}{R_1 R_2}\right) + \varkappa\right]. \tag{2.5}$$

On this occasion, we took into account that not only losses from the mirrors but also other linear losses occur, which for instance can be caused by scattering at inhomogeneities of the optical media, by diffraction or non-resonant absorption. The influence of these losses is taken into consideration in (2.5) with an additional term $\tilde{\varkappa}$. Occasionally, we will summarize the total losses arising from partial reflection, absorption, diffraction and scattering as "linear optical absorption" and will characterize it by the linear loss coefficient $\varkappa = \tilde{\varkappa} - \ln(R_1 R_2)$.

2.2 Generation of Occupation Number Inversion by Optical Pumping

Let us consider more closely the generation of occupation inversion by optical pumping. Through intense irradiation of light from a pump source, molecules are raised into the excited state. Depending on the type of laser and its application, gas discharges, e.g. flash lamps, light-emitting diodes, other lasers or even sun light can be used. The essential characteristics of the pump and laser processes, depending on the type of laser, can be understood from a three or four-level scheme (Fig. 2.2). Let us consider first a three-level laser, as it is realized using the ruby laser (Fig. 2.2a). By irradiating the laser material with optical pump radiation of photon flux density I_P, the molecules are excited from the ground state 1 into the upper state 3. Most molecules fall from this level into the relatively long-living level 2 by fast, radiationless processes. In order that laser action can start, the life-time of this level must be long compared to the relaxation time of the transition $3 \rightarrow 2$, i.e.

$$T_{32} \ll T_{21}. \tag{2.6}$$

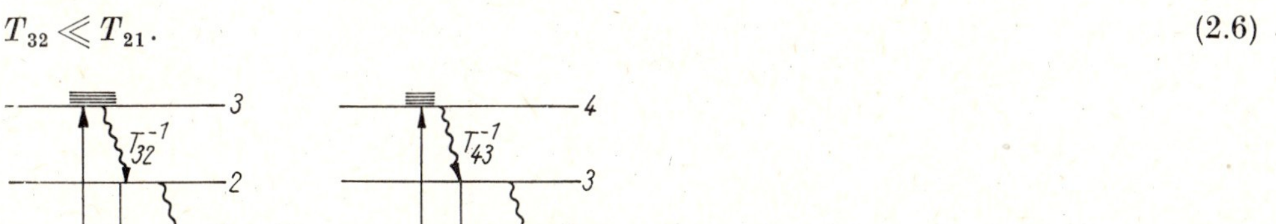

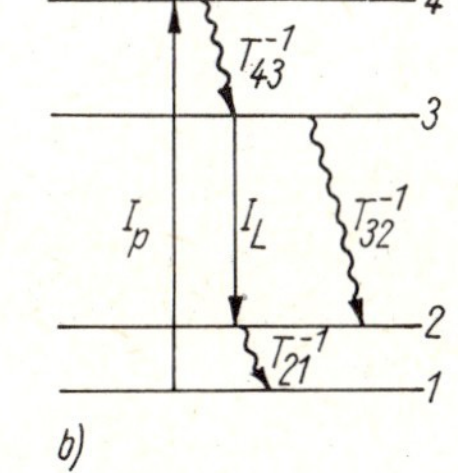

Fig. 2.2. Energy level scheme for a three-level laser (a) and a four-level laser (b)

Under these conditions, it is possible that upon exceeding a certain threshold value of pump radiation level 2 obtains a higher occupation, which is a necessary requirement for the laser process. The dynamics of this process can be described using rate equations, as they were written down for three-level systems with homogeneously broadened transitions in equations (1.27) to (1.28). I_1 is to be identified here with the photon flux density of the pump wave I_P, and I_2 with that of the laser wave I_L. If we use relation (2.6) in (1.28a), we can estimate

$$N_3 \approx W_P T_{32} N_1 \ll N_1, N_2, \tag{2.7}$$

where the pump rate $W_P = \sigma_{13} I_P$ was introduced. The occupation of the pump level 3 is thus negligibly small. For the occupation inversion, i.e. the occupation number difference (per unit volume) $\Delta N = N_2 - N_1$, we obtain from (1.28b)

$$\frac{\partial}{\partial t} \Delta N = -2\sigma_{13} I_{\mathrm{L}} \Delta N - \left(W_{\mathrm{P}} + \frac{1}{T_{21}}\right) \Delta N + \left(W_{\mathrm{P}} - \frac{1}{T_{21}}\right) N, \tag{2.8}$$

where $N = N_1 + N_2$.

We consider now the dependence of the occupation inversion on the pump rate and the relaxation time with steady-state excitation, which is also called continuous wave (cw) excitation. W_{P} is a constant parameter, and in (2.8) we can set $\partial \Delta N / \partial t = 0$. For the occupation inversion under steady-state conditions

$$\Delta N = \frac{\left(W_{\mathrm{P}} - \frac{1}{T_{21}}\right) N}{W_{\mathrm{P}} + \frac{1}{T_{21}} + 2\sigma_{12} I_{\mathrm{L}}} \tag{2.9}$$

follows from (2.8). In order for there to be a positive occupation inversion, i.e. $\Delta N > 0$, the pump rate W_{P} must exceed the critical value

$$(W_{\mathrm{P}})_{\mathrm{cr}} = \frac{1}{T_{21}}. \tag{2.10}$$

Thus for three-level lasers only materials having a relatively long relaxation time T_{21} are of interest. If the critical value for the pumping power is exceeded, then stimulated emission begins to surpass absorption: In this case the number of photons increases rapidly in the resonator, if additionally the amplification coefficient is greater than the value given by (2.5). In the resonator, high photon number flux densities I_{L} arise, which produce a decrease in the occupation inversion. A diminished occupation inversion yields however smaller amplification of radiation, so that after many round trips in the resonator a value of the photon flux is attained that compensates precisely for the out-coupling losses and the inner losses $\varkappa$. Let us calculate this intensity value under steady-state conditions. For the photon flux density of the laser, we can use the relation (1.27b) with the occupation number difference (2.9) and obtain

$$\frac{\mathrm{d}I_{\mathrm{L}}}{\mathrm{d}z} = \frac{g_0 I_{\mathrm{L}}}{1 + \frac{I_{\mathrm{L}}}{I_{\mathrm{S}}}}, \tag{2.11}$$

in which we have introduced the small signal gain coefficient

$$g_0 = \frac{T_{21} W_{\mathrm{P}} - 1}{1 + W_{\mathrm{P}} T_{21}} \sigma_{12} N \tag{2.12}$$

and the saturation intensity

$$I_{\mathrm{S}} = \frac{T_{21} W_{\mathrm{P}} + 1}{2\sigma_{12} T_{21}}. \tag{2.13}$$

For the purpose of simplification, we have neglected the inner losses $\varkappa$ in comparison with the outcoupling losses. The integration of (2.11) supplies the relation

$$\ln \frac{I_L(L^a)}{I_L(0)} + \frac{1}{I_S}\left(I_L(L^a) - I_L(0)\right) = g_0 L^a, \tag{2.14}$$

where L^a is the length of the active laser rod. If the amplification compensates precisely for the outcoupling losses we can use (2.4) and obtain

$$I_L(0) = \frac{[2g_0 L^a + \ln R_1] I_S R_1}{1 - R_1} \tag{2.15}$$

where $R_2 = 1$ has been assumed. Since $I_L(0)$ must be positive, (2.15) yields the condition

$$2g_0 L^a \geqq \ln \frac{1}{R_1}, \tag{2.16}$$

which corresponds to the already derived relation (2.5) for $\varkappa = 0$ and $R_2 = 1$.

At this point, the relations for four-level lasers shall be considered (Fig. 2.2b). In the four-level system, certain disadvantages of the three-level laser, which occur because the lower level of the laser transition is the ground level of the molecule, are eliminated. In the same way as in three-level systems, the first step consists in the excitation of the active material into a strongly broadened level, i.e. level 4, from which the molecules pass into level 3 by fast, radiationless relaxation processes. The laser transition now takes place, however, not from this level into the ground level, but into another excited level, i.e. level 2. Here again fast radiationless transitions into the ground level occur. Thus the relaxation times must satisfy the conditions

$$T_{43} \ll T_{31}, T_{32}, \qquad T_{21} \ll T_{31}, T_{32}. \tag{2.17}$$

The occupation numbers of levels 4 and 2 prove again to be negligibly small. From a set of rate equations analogous to (1.27), (1.28),

$$N_4 \approx T_{43} W_P N_1, \qquad N_2 \approx \left(T_{21}\sigma_{32} I_L + \frac{T_{21}}{T_{32}}\right) N_3 \tag{2.18}$$

can be evaluated, with which we can obtain the equation

$$\frac{\partial \Delta N}{\partial t} = -\sigma_{32} I_L \Delta N + W_P(N - \Delta N) - \left(\frac{1}{T_{32}} + \frac{1}{T_{31}}\right) \Delta N \tag{2.19}$$

for the population inversion, i.e. the occupation number difference (per unit volume) $\Delta N = N_3 - N_2 \approx N_3$. Under steady state conditions, the population inversion for a four-level system is thus given by

$$\Delta N = \frac{W_P N}{W_P + \sigma_{32} I_L + \dfrac{1}{T_{32}} + \dfrac{1}{T_{31}}}. \tag{2.20}$$

According to (2.20) we always have $\Delta N > 0$. Thus in the four-level system, positive values of the population inversion can be produced with arbitrarily small pump rates under the conditions mentioned above. This follows from the fact that according to condition (2.17), the occupation of the lower laser-level is always near to zero. For this reason, the four-level laser offers considerable advantages in comparison with the three-level laser, and the majority of lasers used at present work according to the four-level scheme. For the photon flux density of laser radiation, we can derive the rate equation

$$\frac{dI_L}{dz} = \sigma_{13} I_L \Delta N \tag{2.21}$$

which, using (2.20), leads to relation (2.11) with changed small signal amplification coefficients g_0 and changed saturation intensity I_S, for which

$$g_0 = \frac{W_P N \sigma_{32}}{W_P + \frac{1}{T_{32}} + \frac{1}{T_{31}}} \tag{2.22}$$

and

$$I_S = \frac{W_P + \frac{1}{T_{32}} + \frac{1}{T_{31}}}{\sigma_{32}} \tag{2.23}$$

hold. For the steady-state photon flux density in the resonator, we obtain again relation (2.15), in which g_0 and I_S are now given by (2.22) and (2.23).

2.3 Optical Resonators

2.3.1 Open and Closed Resonators

In this section, the feedback of radiation in passive optical resonators, also referred to as optical cavities, will be considered. A resonator of this kind is usually open, that is to say it is only partially closed, i.e., as shown in Fig. 2.1, it does not possess side walls, but only two opposing mirrors. However, for the purpose of an approximate description of resonators consisting of two plane mirrors, we can begin with a closed, box-shaped resonator of length L with ideally reflecting walls. Length L is parallel to the z axis of our coordinate system, and the edges of the two-square cover surfaces of length $2a$, where $2a \ll L$, are parallel to the x and y axes, respectively. In such resonators the wave fields that propagate approximately along the z axis, so called paraxial waves, differ only slightly from the corresponding waves of the open laser resonator. As we know, the wave equation together with the appropriate boundary conditions provides standing waves for closed cavity resonators, the electric field strength $\vec{E}(t, \vec{r})$ of which is given by

$$\begin{aligned} E_x &= E_x^0 \cos k_x x \sin k_y y \sin k_z z \sin \omega t, \\ E_y &= E_y^0 \sin k_x x \cos k_y y \sin k_z z \sin \omega t, \\ E_z &= E_z^0 \sin k_x x \sin k_y y \cos k_z z \sin \omega t. \end{aligned} \tag{2.24}$$

The components k_x, k_y, k_z of the wave number vector $\vec{k}$ are given by

$$k_x = \frac{m_x \pi}{(2a)}, \quad k_y = \frac{m_y \pi}{(2a)}, \quad k_z = \frac{m_z \pi}{L} \tag{2.25}$$

as a result of the boundary condition, where m_x, m_y, m_z are positive integers, and the wave equation provides

$$\omega = \pi c \sqrt{\left(\frac{m_x}{2a}\right)^2 + \left(\frac{m_y}{2a}\right)^2 + \left(\frac{m_z}{L}\right)^2} \tag{2.26}$$

for the discrete spectrum of the eigenfrequencies $\omega = c \cdot k$ of the resonator. Such standing waves of the resonator are also called resonator modes or cavity modes. For paraxial rays, evidently $|k_z| \gg |k_x|$, $|k_y|$ holds true, from which we can conclude $m_z \gg m_x$, m_y. Under this condition we obtain in approximation

$$\omega = \pi c \frac{m_z}{L} \left[1 + \frac{1}{2} \left(\frac{L}{2a}\right)^2 \frac{m_x^2 + m_y^2}{m_z^2}\right] \tag{2.27}$$

from (2.26). From (2.27) we learn immediately that the frequency difference of two modes, which coincide in m_x and m_y and differ in m_z by 1, is approximately equal to

$$\delta\omega = \frac{c\pi}{L}. \tag{2.28}$$

These two modes differ from one another only in the z-dependence of their field distribution. We therefore also call this frequency difference the mode spacing of two adjacent axial modes or longitudinal modes. By way of contrast, we call the frequency spacing of modes with equal m_z, but with m_x or m_y differing by 1, the mode distance of neighboring transverse modes, since their field distribution differs from one another only in reference to their x or y-dependence or both. From (2.27) we obtain, for example,

$$\delta\omega_t = \delta\omega \frac{m_x + \frac{1}{2}}{8\mathcal{N}} \tag{2.29}$$

for two modes which differ in the value of m_x by 1, where $\mathcal{N} = a^2 m_z / 2L^2 = a^2 / L\lambda$ signifies a dimensionless number called the Fresnel number. Usually $\mathcal{N} \gg 1$ holds for typical laser systems, so that the frequency spacing between transverse modes is very much smaller than that between longitudinal modes.

Until now we have considered all the modes that may principally be found in the laser resonator. Of course, not all the frequencies given by relation (2.26) are excited in every case. For example, the only modes that appear in a laser are those which can simultaneously be amplified in its active medium. Since, according to (2.28), the spacing between two axial modes at a resonator length of $L = 1$ m lies in the order of $\Delta\omega/2\pi \simeq 10^8$ Hz and, on the other hand, the line widths of laser transitions in various active media take on values from $\Delta\omega/2\pi \simeq 10^9$ Hz (in gases at low pressure) to $\Delta\omega/2\pi \simeq 10^{13}$ Hz—10^{14} Hz (in dyes and solids), it is possible that depending on the type of laser only a few or a huge number up to some ten thousand axial modes can be amplified in the laser resonator.

For many purposes, we are interested in working with a defined, smallest possible number of modes or even with a single mode. As for the transverse modes, this aim can be attained relatively simply owing to the differences in their diffraction losses. For instance, we affix an additional diaphragm in the resonator, which causes the diffractional losses of the higher transverse modes to increase greatly. The selection of individual axial modes can be achieved by using frequency selective elements, e.g. an additional Fabry-Perot-etalon, in the resonator cavity.

For the generation of ultrashort light pulses, however, we must make an effort to work with the highest possible number of axial resonator modes. We must therefore use laser materials that possess the broadest possible gain profile and avoid any suppression of axial modes.

The field strength given in (2.24) as a function of space and time shows that, once a resonator has been excited, its field does not decay again even after an indefinite period of time. In open resonators and even in actual closed resonators, there are losses that cause a decrease in intensity as time progresses. We wish now to describe these losses phenomenologically, using as an analogy the description of actual mechanical oscillators and actual electrical resonator circuits. Accordingly, the decay of the radiation can be described as follows: we represent the contribution of each mode μ to the field strength as product of a space-dependent mode function $u_\mu(\vec{r})$ and a time-dependent coefficient $\tilde{E}_\mu(t)$ as given by

$$E(\vec{r}, t) = \sum_\mu \tilde{E}_\mu(t)\, u_\mu(\vec{r}) \tag{2.30}$$

and add a damping term $\frac{2}{T_A} \dot{\tilde{E}}_\mu(t)$ to the differential equation for $\tilde{E}_\mu(t)$ which follows from the wave equation. We obtain

$$\ddot{\tilde{E}}_\mu(t) + \frac{2}{T_A} \dot{\tilde{E}}_\mu(t) + \omega_\mu{}^2 \tilde{E}_\mu(t) = 0. \tag{2.31}$$

T_A is designated here as the relaxation time of the field amplitude. Let us assume, that the attenuation of the field during one vibration period is small and that we therefore have $1/T_A \ll \omega_\mu$. Then the solution of (2.31)

$$\tilde{E}_\mu(t) = E_0\, e^{i\omega_\mu t}\, e^{-t/T_A} \tag{2.32}$$

for excitation before $t = 0$ displays an exponential decay for $t > 0$. This description of the one mode field corresponds to the consideration of an equivalent electrical resonator circuit with resistive losses (see Fig. 2.3). Here, too, the amplitude decreases exponentially with a time constant $T_A = 2T_{\mathscr{E}}$ and

$$\mathscr{E}(t) = \mathscr{E}(0)\, e^{-t/T_{\mathscr{E}}} \tag{2.33}$$

holds true for the decrease of the stored energy. The losses in resonator circuits and resonator cavities are frequently characterized by the resonator quality-factor or Q-factor

$$Q_\mu = \omega_\mu T_{\mathscr{E}}. \tag{2.34}$$

(In general, the Q-factor is defined as the number of vibrations by which the stored energy has dropped to the ($e^{2\pi}$)-th portion.) In the frequency domain now, when taking into account losses, the mode is no longer characterized by an infinitely precise frequency value, but by a resonance line of finite half-width $\Delta\omega_R$. With exponential temporal decay of the field, as explained below Fig. 2.3, a Lorentzian line results where

$$\Delta\omega_R = \frac{1}{T_{\mathscr{E}}} = \frac{2}{T_A}. \tag{2.35}$$

If there are several causes of field attenuation involved, each of which individually leads to an exponential decay for the resonator energy, then the effective total attenuation can also be characterized by a simple exponential relation. Then,

$$Q_{\mu\text{eff}}^{-1} = \sum_i Q_{\mu i}^{-1} \tag{2.36}$$

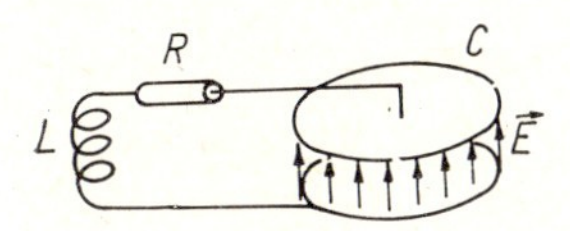

Fig. 2.3. Electrical resonant circuit consisting of the inductivity L, the capacity C and the ohmic resistance R

After exciting an electric d.c. field of amplitude $\bar{E}$ until time $t = 0$, the field at time t is given by $E(t) = \frac{1}{2}\bar{E}\, e^{-t/2T_{\mathscr{E}}}\, e^{-i\omega_0 t} + \text{C.C.}$ where $T_{\mathscr{E}} = L/R$ and $\omega_0 = \sqrt{1/CL - (R/2L)^2}$. The Fourier transform of $E(t)$ is given by

$$\underset{\sim}{E}(\omega) = \frac{1}{2}\bar{E}\,\frac{1}{i(\omega_0 - \omega) + 1/(2T_{\mathscr{E}})}$$

and the spectral density by

$$|\underset{\sim}{E}(\omega)|^2 = \frac{1}{4}\bar{E}^2\,\frac{1}{(\omega - \omega_0)^2 + (\Delta\omega_R/2)^2}$$

where $\Delta\omega_R = 1/T_{\mathscr{E}}$.

holds true for the total Q-factor. Accordingly, the sum of the various contributions gives us the total line width. In the metal cavity resonator, such losses can occur as a result of the finite conductivity of the walls as well as the energy release through holes in the walls. In open two-mirror resonators, as already described, the release of energy through the mirrors (having a reflectivity of $R < 1$) and at the side of the mirrors, as well as inner losses, cause the decrease in the field energy stored within these resonators. In the following the dependence of such losses on the resonator parameters will be more precisely characterized.

2.3.2 *Plane Fabry-Perot Resonator*

For open two-mirror resonators, the analogy with closed resonators as well as with electrical resonator circuits built up of lumped elements could appear to be somewhat artificial. Therefore, we wish to verify the main results of this treatment considering a plane Fabry-Perot resonator with infinitely extended mirrors. A plane wave of frequency ω and amplitude $\bar{E}_0$ travels from the left into the Fabry-Perotresonator of optical length L. The wave vector forms an angle θ with the normal on the mirror surface (see Fig. 2.4.). The amplitude $\bar{E}$ of the field behind the resonator is the result of the superposition of all the partial waves, which arise under repeated reflection at the two mirrors (see e.g. [2.1]). Thus we obtain

$$\frac{\bar{E}}{E_0} = \mathsf{T}\, \mathrm{e}^{\mathrm{i}\delta'}(1 + R\, \mathrm{e}^{\mathrm{i}\delta} + R^2\, \mathrm{e}^{2\mathrm{i}\delta} + \dots + R^n\, \mathrm{e}^{\mathrm{i}n\delta} + \dots) = \mathsf{T}\, \mathrm{e}^{\mathrm{i}\delta'} \frac{1}{1 - R\, \mathrm{e}^{\mathrm{i}\delta}}$$

$$\text{where} \quad \delta = \frac{2\pi}{\lambda} \cdot 2L\cos\theta, \quad \delta' = \frac{2\pi}{\lambda} L \frac{1}{\cos\theta}, \quad \mathsf{T} = \sqrt{\mathsf{T}_1\mathsf{T}_2}, \quad R = \sqrt{R_1R_2}. \tag{2.37}$$

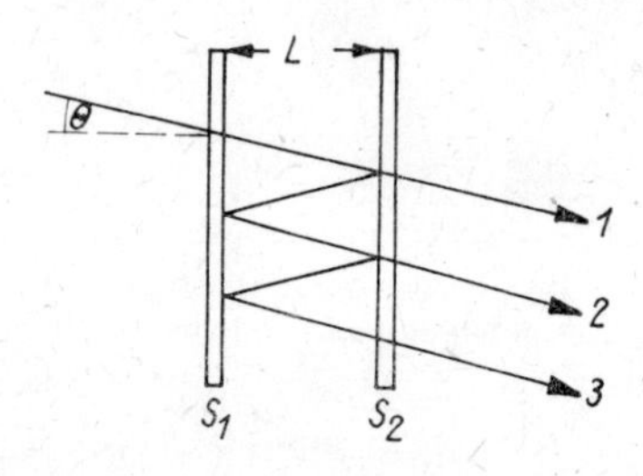

Fig. 2.4. Plane Fabry-Perot resonator
(S_1, S_2 are the mirrors, L is the optical resonator length and θ is the angle of incidence)

T_1, T_2, R_1, R_2 are here the transmission and reflectivity, respectively, of the two mirrors: L is the optical length of the resonator. For the total transmission T_{FP} of the Fabry-Perot resonator, we obtain

$$\mathsf{T}_{\mathrm{FP}} = \left|\frac{\bar{E}}{E_0}\right|^2 = \left(\frac{\mathsf{T}}{1-R}\right)^2 \frac{1}{1 + \dfrac{4R}{(1-R)^2}\sin^2(\delta/2)} \tag{2.38}$$

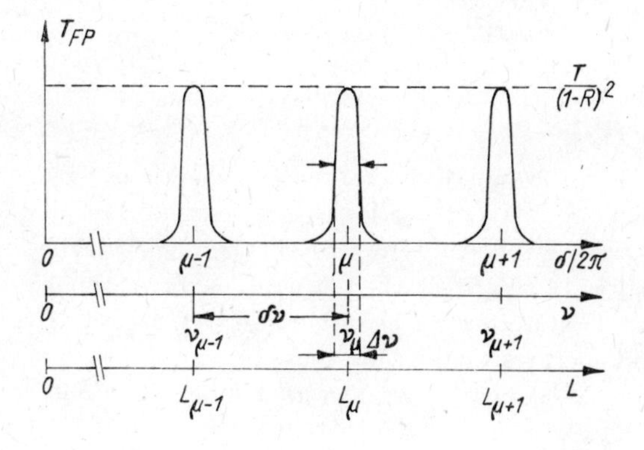

Fig. 2.5. Transmission of the Fabry-Perot resonator

from (2.37). Figure 2.5 shows the transmission as a function of the phase difference δ, which in particular is proportional to the frequency if the resonator length L is constant and proportional to the resonator length if the frequency ν is constant. The transmission peaks appear at the frequencies

$$\nu_\mu = \frac{\mu c}{2L\cos\theta} \quad (\mu \text{ is an integer}). \tag{2.39}$$

For plane waves propagating in the direction of the interferometer axis, we thus obtain the same frequency spacing ($\nu_\mu - \nu_{\mu-1} = \delta\nu$) as given in (2.38) for the longitudinal modes. The halfwidth of the transmission peaks is given by

$$(\Delta\nu)_\mathrm{R} = \delta\nu \frac{1-R}{\pi\sqrt{R}}. \qquad (2.40)$$

The ratio between the mode spacing $\delta\nu$ and this halfwidth $(\Delta\nu)_\mathrm{R}$, i.e. $(\delta\nu)/(\Delta\nu)_\mathrm{R} = \mathcal{F}$, is designated as the finesse of the interferometer. In our consideration of ideal plane and infinitely extended mirrors, the reflectivity R alone determines the finesse. Roughness of the surface as well as diffractional losses, which occur with mirrors of finite diameter, reduce the total finesse. We can sum up the reciprocal finesses, which arise from various origins, to obtain an approximation of the total finesse (similar to the addition of the reciprocal Q-factors). With high-quality mirrors, the finesse which originates from roughness can amount to several hundred. The diffractional losses can be kept small by suitable choice of the resonator geometry. Thus, it is possible to reach values of the total finesse above 100, if the reflectivity of the mirrors is higher than 99 percent.

Let us now consider the decrease of the electromagnetic field after a sudden turn-off of the continuous, monochromatic irradiation along the interferometer axis at time $t = -L/c$. For $t > 0$ the amplitude at the resonator output is no longer given by (2.37), because at least the first partial wave in this series no longer contributes to the result. At positive values $t_j = j(2L)/c$ $(j = 0, 1, 2, \ldots)$ of time the output field strength is given by

$$\frac{\bar{E}}{\bar{E}_0} = \mathsf{T}\,\mathrm{e}^{\mathrm{i}\delta'} \sum_{p=j}^{\infty} (R\,\mathrm{e}^{\mathrm{i}\delta})^p = \mathsf{T}\,\mathrm{e}^{\mathrm{i}\delta'}(R\,\mathrm{e}^{\mathrm{i}\delta})^j \frac{1}{1 - R\,\mathrm{e}^{\mathrm{i}\delta}} \qquad (2.41)$$

from which

$$\frac{|\bar{E}|^2}{|\bar{E}_0|^2} = R^{2j}\left(\frac{\mathsf{T}}{1-R}\right)^2 \frac{1}{1 + \dfrac{4R}{(1-R)^2}\sin^2(\delta/2)} \qquad (2.42)$$

follows for the square of the field strength. The time dependence of $|\bar{E}|^2$ is solely contained in the factor $R^{2t_jc/2L} = \exp\left\{\frac{c}{2L}(\ln R)\, t_j\right\}$. This means that the decrease in the light intensity $I(t)$ at the output of the interferometer and hence in the radiation energy, which is stored in the resonator, can be approximately represented by a simple exponential law,

$$I(t) = I(0)\,\mathrm{e}^{-t/T_\mathscr{E}} \quad \text{or} \quad \mathscr{E}(t) = \mathscr{E}(0)\,\mathrm{e}^{-t/T_\mathscr{E}} \qquad (2.43)$$

where $T_\mathscr{E} = -(2L/2c \ln R)$, in agreement with the simple phenomenological description from (2.32). For long resonators with high-reflectivity mirrors, the release times for the stored energy are rather long; e.g. for $2L = 3$ m and $R = 0.99$ we obtain $T_\mathscr{E} \approx 500$ ns: such high values are found, for instance with resonators of high power He—Ne lasers. Short resonators with low-reflectivity mirrors, such as those common with semiconductor lasers, have small $T_\mathscr{E}$ values; for example, we obtain $T_\mathscr{E} = 7$ ps for the resonator life time with $L = 3$ mm and $R = 0.5$.

2.3.3 *Diffraction Theory of Open Resonators*

Until now in our treatment of open resonators we have assumed conditions that are by no means always satisfied in experiments. A more precise description of the losses in open resonators requires, for example, taking diffraction effects into account. Since the exact treatment of this problem by solving the complete set of the Maxwell's equations is extremely complicated, we will restrict ourselves to a description of the equivalent diffraction problem of electromagnetic waves using Kirchhoff's formulation of Huygens' principle. For this description, the so-called quasi-optical nature of the problem has to be assumed; thus it is required that

— the dimensions of the resonator are large compared to the wave length,
— the fields in the resonator are transverse in close approximation,
— the distance between the two interferometer mirrors is large compared to the mirror diameter and that
— the mirrors are only slightly curved.

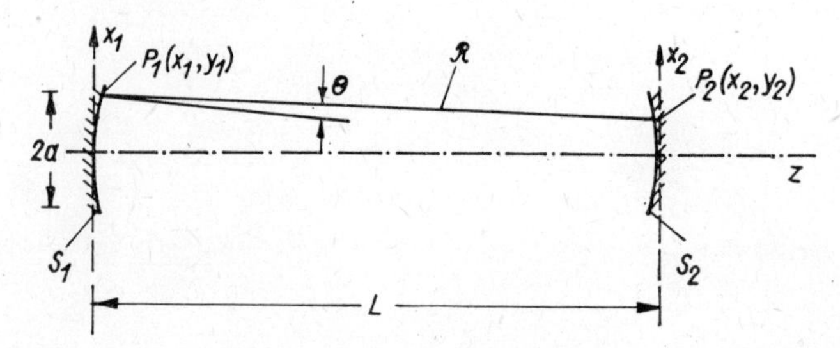

Fig. 2.6. Open resonator with spherically curved mirrors S_1 and S_2. The mirror diameters are $2a$; $P_1(x_1, y_1)$ $P_2(x_2, y_2)$ are points on the mirrors 1 and 2, respectively; L is the optical resonator length, θ is the angle between the connecting line P_1P_2 and the surface normal at point P_1, $\mathcal{R}$ is the separation between P_1 and P_2.

Then, the equations for the transverse field components E_x, E_y, H_x, H_y are decoupled, and we can proceed on the basis of separate, scalar equations, for example, for one component

$$E(x, y, z, t) = \frac{1}{2} \hat{E}(x, y, z) \, \mathrm{e}^{\mathrm{i}\omega t} + \text{C.C.}$$

of the electric field strength whose polarization direction is not explicitly designated. (An extensive description of the diffraction theory for open resonators can be found e.g. in [2.2].)

Let us now consider the field distribution $\hat{E}_2(P_2)$ on mirror 2 (see Fig. 2.6) as a result of the diffraction of the electromagnetic field distribution $\hat{E}_1(P_1)$ at mirror 1. According to Kirchhoff's law we obtain

$$\hat{E}_2(P_2) = \frac{\mathrm{i}k}{4\pi} \int_1 \mathrm{d}S_1 \, \frac{\mathrm{e}^{-\mathrm{i}k\mathcal{R}}}{\mathcal{R}} \big((1 + \cos\theta) \, \hat{E}_1(P_1)\big). \tag{2.44a}$$

Conversely, the electromagnetic field on the second mirror produces by diffraction a distribution on the first mirror according to

$$\hat{E}_1(P_1) = \frac{\mathrm{i}k}{4\pi} \int\limits_2 \mathrm{d}S_2 \, \frac{\mathrm{e}^{-\mathrm{i}k\mathcal{R}}}{\mathcal{R}} \big((1 + \cos\theta')\, \hat{E}_2(P_2)\big). \tag{2.44b}$$

Resonator modes are defined now as self-consistent field distributions. After one full round trip in the resonator, i.e. after two diffraction processes, a self-consistent field distribution should reproduce itself until apart from a position-independent factor. Let us simplify our treatment and assume a symmetrical resonator consisting of two equal mirrors — in which the self-consistency of the field distribution has to be assumed after only half a round trip, i.e. on the path from one mirror to the other. Thus the field $\hat{E}(P)$ must satisfy the integral equation

$$\gamma \hat{E}(P) = \frac{\mathrm{i}k}{4\pi} \int \mathrm{d}S' \, \frac{\mathrm{e}^{-\mathrm{i}k\mathcal{R}}}{\mathcal{R}} \, (1 + \cos\theta)\, \hat{E}(P'). \tag{2.45}$$

Here the integral is taken over the field at one of the two mirrors; γ is the position-independent, complex factor. The integral equation (2.45) describes the field strength distribution of the resonator modes. The various solutions of this equation, which in general are only to be numerically found by suitable iteration procedures, can be characterized by three mode indices m, n, q, as we have already mentioned in the previous section. Because of their transverse electromagnetic structure, their eigen solutions are denoted as TEM_{mnq} modes. (Note that electromagnetic fields in resonators are, in general, not purely transverse. In our study, the transversality results from the particular assumptions given at the beginning of this section.) Figure 2.7 shows the transverse mode structure, i.e. the field distribution in the (x,y)-plane, for some low indexed modes. Each eigen solution (i.e. mode) possesses a complex eigen value

$$\gamma_{mnq} = \beta_{mnq} \exp{(\mathrm{i}\psi_{mnq})}. \tag{2.46}$$

The real quantity β_{mnq} describes the decrease in the field strength amplitude through diffractional loss and is denoted accordingly as the amplitude damping factor. ψ_{mnq} denotes the phase shift of the mode compared with a plane wave of equal frequency. In general, the modulus and phase of γ depend, in the small frequency range of interest (e.g. the halfwidth of a laser transition), only weakly on the mode index q, which mainly determines the frequency of the mode, whereas γ depends more on m and n, which delineate the field distribution in the x,y-plane. In passing from one mirror to the other, the modulus $|\overline{E}|$ of the amplitude is to be replaced by $|\overline{E}| \times \beta$ and the phase $\varphi = (-\mathrm{i}kz + \Phi)$ by $\big(-\mathrm{i}k(z + L) + \Phi + \psi\big)$, and in a full round trip in the resonator by $|\overline{E}|\,\beta^2$ and $\big(-\mathrm{i}k(z + 2L) + \Phi + 2\psi\big)$, respectively. In one half cycle the radiation power $\mathcal{P}$ falls to $\mathcal{P}\beta^2$, for which reason the quantity $\mathcal{P}(1 - \beta^2)$ is designated as the power loss and $(1 - \beta^2)$ as the power loss factor. Figure 2.8 shows as an example the power loss factor and the phase shift of the lowest transverse modes for resonators built up of plane-parallel round mirrors and for confocal resonators consisting of round mirrors. It is recognized that the fundamental modes, for which $m, n = 0$ holds, distinguish themselves by having the lowest diffractional losses. From this fact, we obtain possibilities for the selection of transverse modes, in that we increase the diffractional losses of the higher modes with the use of an appropriate diaphragm in the resonator. Thus, by introducing an active material in the resonator, we can exceed the laser threshold for the transverse fundamental mode, though not for the higher ones.

Let us now consider the resonator eigenfrequencies, which are related to the phase shift ψ_{mn}. In particular, for the confocal resonator the phase shift ψ_{mn} can be provided in analytic form [2.2] as

$$\psi_{mn} = \frac{\pi}{2}(1 + m + 2n). \tag{2.47}$$

The precise values of the eigenfrequencies of the cavity modes can be found from the following consideration. After one full round trip in the resonator, the phase of the electric field strength may only change by an integral multiple of 2π, i.e.

$$\Delta\varphi = -k_{mnq} \cdot 2L + 2\psi_{mn} = -2\pi q. \tag{2.48}$$

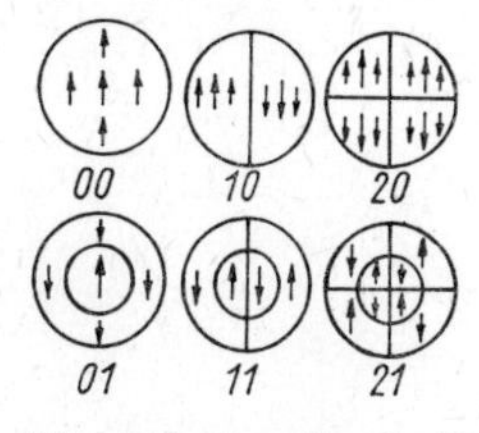

Fig. 2.7. Amplitude distribution of the electric field of various transverse modes on a laser mirror at a fixed instant in time ("instantaneous photograph")
Lengths and directions of the arrows describe the field at this moment. Furthermore, the nodal lines are plotted, i.e. the lines on which a sign reversal takes place and the field vanishes. The first and the second mode index (m and n) give the number of nodal lines with respect to the angle and radial coordinate.

Using the relation $k_{mnq} = 2\pi\nu_{mnq}/c$, the resonance frequency of the TEM_{mnq}-mode is found to be

$$\nu_{mnq} = \frac{c}{2L}\left(q + \frac{1}{\pi}\psi_{mn}\right). \tag{2.49}$$

If for ψ_{mn} we insert values for the plane-parallel Fabry-Perotinterferometer from Figure 2.8b), we see that the eigenfrequencies for large Fresnel numbers are given to a close approximation by (2.27).

If the mirrors have a reflectivity of $R < 1$, the total power loss in one cycle is found to be the sum of power losses by diffraction and reflection. Thus, for small losses during one round trip, we obtain in approximation the relation

$$I^{[K+1]} = R^2\beta_{mn}^4 I^{[K]} \tag{2.50}$$

between the radiation intensites $I^{[K]}$ and $I^{[K+1]}$ in the Kth and the $(K+1)^{\text{th}}$ cycle, respectively. If we divide the intensity change $\Delta I = I^{[K+1]} - I^{[K]}$ by the round-trip time $\Delta t = 2L/c$, we obtain the difference equation

$$\frac{\Delta I}{\Delta t} = -(1 - R^2\beta_{mn}^4)\, I \frac{c}{2L}, \tag{2.51}$$

from which, for small losses per cycle, i.e. $(1 - R^2\beta_{mn}^4) \ll 1$, we can pass over to the differential equation

$$\frac{dI}{dt} = -\frac{1}{T_{\mathscr{E}}} I \tag{2.52}$$

where

$T_{\mathscr{E}} = 2L/c(1 - R^2\beta_{mn}^4)$.

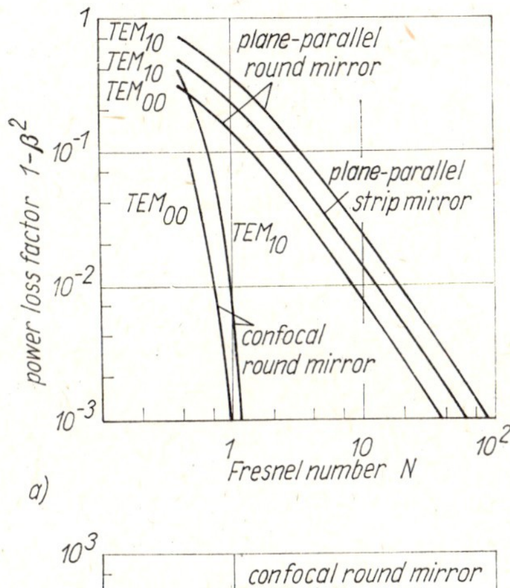

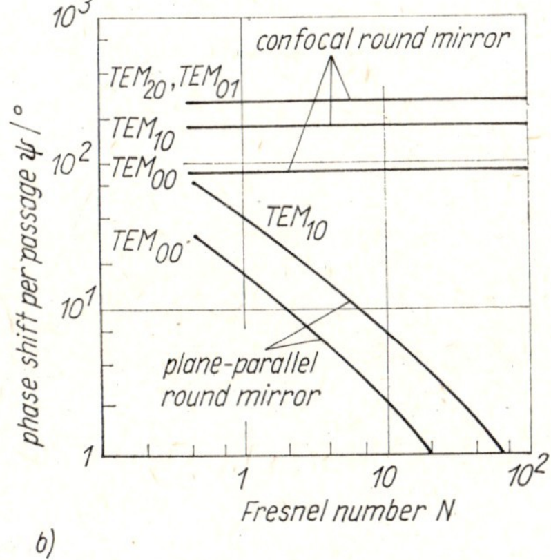

Fig. 2.8. Power loss factor $(1 - \beta^2)$ per passage (a) and phase shift ψ per passage (b) as functions of the Fresnel number $\mathscr{N} = a^2/(L \cdot \lambda)$ for several transverse modes TEM_{mn} in resonators consisting of plane-parallel round mirrors and in confocal resonators ($R_1 = R_2 = L$) consisting of round mirrors

2.3.4 *Description of the Field Distribution Inside and Outside Laser Resonators by Means of Gaussian Beams*

The description of laser radiation inside and outside the resonator using Kirchhoff's diffraction formula is connected by tedious numerical calculations, particularly when dealing with complex optical arrangements. Therefore, we want to outline another method, in which we neglect the losses originating from diffraction of radiation at finite apertures but take into account the transverse mode structure of the field. Here, we follow [2.2] and start from the time-free wave equation

$$\Delta\hat{E} + k^2\hat{E} = 0. \tag{2.53}$$

In the investigation of a light beam that propagates in the z-direction, we separate the fast z-dependence by writing

$$\hat{E}(x, y, z) = A(x, y, z)\,\mathrm{e}^{-\mathrm{i}kz} \tag{2.54}$$

and we obtain

$$\left(\frac{\partial^2}{\partial x^2} + \frac{\partial^2}{\partial y^2}\right) A(x, y, z) - 2\mathrm{i}k \frac{\partial A(x, y, z)}{\partial z} = 0 \tag{2.55a}$$

neglecting the second derivative of the slowly varying amplitude $A(x, y, z)$ with respect to z. For the present, let us investigate the field distribution of the transverse fundamental mode TEM_{00}, in which A depends only on the distance r from the optical axis and on z. Thus

$$\left(\frac{\partial^2}{\partial r^2} + \frac{1}{r}\frac{\partial}{\partial r}\right) A(r, z) - 2\mathrm{i}k \frac{\partial A(r, z)}{\partial z} = 0 \tag{2.55b}$$

holds. For this symmetrical field distribution we make the ansatz

$$A(r, z) = \exp\left(-\mathrm{i}P(z) - \frac{\mathrm{i}k}{2q(z)}\, r^2\right), \tag{2.56}$$

i.e., for the radial dependence of the field strength a Gaussian distribution is assumed, where the beam diameter and the phase may alter under propagation. We insert now (2.56) in the time-free wave equation and separately require its satisfaction for the terms proportional to r^0 and r^2. From here

$$P'(z) = -\frac{\mathrm{i}}{q(z)} \tag{2.57a}$$

and

$$q'(z) = 1 \tag{2.57b}$$

follow, where derivatives with respect to the propagation coordinate z are designated by primes. The solution of (2.57 b) is $q(z) = q_0 + z$. We represent the complex beam parameter $q(z)$ for practical reasons in the form of

$$\frac{1}{q(z)} = \frac{1}{R(z)} - \frac{\mathrm{i}\lambda}{\pi w^2(z)}, \tag{2.58}$$

which is equivalent to

$$\hat{E}(r, z) = A(0, 0)\,\mathrm{e}^{-\mathrm{i}kz}\,\mathrm{e}^{-\mathrm{i}\left(P(z) + \frac{k}{2R(z)} r^2\right)}\,\mathrm{e}^{-\frac{r^2}{w^2(z)}}. \tag{2.59}$$

The value $2w$ denotes the beam diameter (see Fig. 2.9a). For the surface of constant phase of $\hat{E}(r, z)$ (see Fig. 2.9b), which crosses the z-axis at z_0,

$$k(z - z_0) = \frac{k}{2}\left(\frac{\partial^2 z}{\partial r^2}\right)_{r=0} r^2 = -k\,\frac{1}{2R(z_0)}\, r^2 \tag{2.60}$$

follows from (2.59). This means that $R(z)$ gives the radius of curvature of the phase surface at position z.

One distinguishable site is where the phase front is plane $\left(R(z) \to \infty\right)$ and the beam contracts to the minimum diameter $2w_0$ in the so-called beam waist. Frequently, one sets the origin of the z-coordinate at the beam waist (see Fig. 2.9b) and obtains with $R \to \infty$ from (2.58) a value for q_0, which is inserted in the solution of (2.57b). This procedure yields

$$q(z) = \frac{\mathrm{i}\pi w_0^2}{\lambda} + z. \tag{2.61}$$

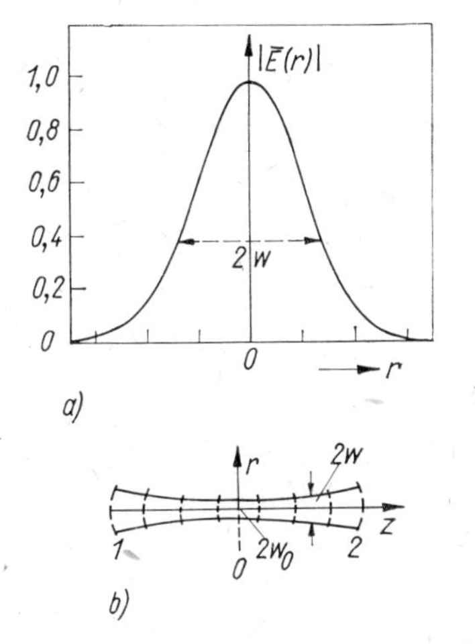

Fig. 2.9. Field strength amplitude of a Gaussian beam
(a) Modulus of the field strength amplitude as a function of the radial coordinate r (in arbitrary units)
(b) Schematic representation of the Gaussian beam
The surface of constant phase can be represented near the z-axis approximately by spherical segments; accordingly their section with the x, z plane can be represented by circular segments. The radius of curvature of the phase surfaces changes with z. The point of the beam with plane phase front was defined as the origin of the z-axis. The modulus of the field strength amplitude decreases from the maximum value on the axis with increasing distance from the axis as outlined in (a). The curve, at which the amplitude has fallen to the value $|\overline{E}(0, z)|/\mathrm{e}$, is represented by a solid line in b). (At points 1 and 2 the amount of the radius of curvature is equal to $2\,|z|$. If at these points mirrors are mounted, which have the same radius of curvature as the phase surfaces, we have a confocal resonator, where the common focal point of the mirrors is at $z = 0$.)

Together with (2.58) we now have the relations needed to determine $w(z)$ and $R(z)$:

$$R(z) = z\left(1 + \left(\frac{\pi w_0^2}{\lambda z}\right)^2\right), \tag{2.62a}$$

$$w^2(z) = w_0^2\left(1 + \left(\frac{\lambda z}{\pi w_0^2}\right)^2\right). \tag{2.62b}$$

The beam contour, represented by the function $w(z)$, is thus a hyperbola whose asymptote includes an angle $\theta = \arctan\,(\lambda/\pi w_0)$ with the z-axis.

By using the solution for $q(z)$ we get from (2.57a)

$$P(z) = -\varphi(z) - \mathrm{i} \ln \frac{w(z)}{w_0} \tag{2.63}$$

where

$$\varphi(z) = \arctan \frac{z\lambda}{\pi w_0^2}.$$

Thus, we finally obtain

$$\hat{E}(r, z) = \frac{w_0}{w} \, \mathrm{e}^{-\mathrm{i}(kz - \varphi(z))} \, \mathrm{e}^{-\left(\frac{1}{w^2(z)} + \mathrm{i}\frac{k}{2R(z)}\right) r^2} \tag{2.64}$$

for the field strength amplitude. The square of the factor $\left(\frac{w_0}{w}\right)$ describes the decrease in intensity of the beam on the z-axis due to its widening; $\varphi(z)$ is the phase difference between the laser beam and a plane wave of the same frequency on the z-axis. The propagation of the fundamental mode, which is of particular significance in laser physics has thus been sufficiently described. All other transverse propagation modes, which together with the fundamental mode build a complete orthogonal set of solutions of the wave equation (2.55a), can be treated in similar manner. These higher transverse modes can be characterized by the same parameters $w(z)$, $R(z)$ and $P(z)$. Of course, the space dependence of the fundamental mode is here superimposed by a complicated transverse structure.

Let us now treat the transformation of propagation modes by lenses and mirrors. An ideal lens leaves the transverse field distribution of a beam mode unchanged, i.e. the fundamental mode and also higher transverse modes are preserved in this process. The lens changes only the beam parameters $R(z)$ and $w(z)$. An ideal thin lens of focal distance f transforms a spherical wave with radius of curvature R_1 into one whose radius of curvature is R_2 as described by

$$\frac{1}{R_2} = \frac{1}{R_1} - \frac{1}{f}. \tag{2.65}$$

(The radius of curvature is taken to be positive, if the wave front, looking at it from $z = \infty$, is convex.) The phase front of the fundamental mode, which can be approximately described by a spherical surface, is transformed in the same manner. Since the beam diameter is the same on both sides of the thin lens,

$$\frac{1}{\tilde{q}_2} - \frac{1}{\tilde{q}_1} = \frac{1}{R_2} - \frac{1}{R_1} = -\frac{1}{f} \tag{2.66}$$

holds for the q-parameters $\tilde{q}_1$, $\tilde{q}_2$ directly in front of and behind the lens. In the free space in front of and behind the lens $q(z)$ changes in accordance with (2.57b). If we start with a beam of beam parameter q_1 at distance L_1 left of the lens and want to determine the parameter q_2 at distance L_2 right of the lens, then

$$\tilde{q}_1 = q_1 + L_1, \tag{2.67a}$$

$$\tilde{q}_2^{-1} = \tilde{q}_1^{-1} - f^{-1} \tag{2.67b}$$

and

$$q_2 = \tilde{q}_2 + L_2, \tag{2.67c}$$

hold, which lead to

$$q_2 = \frac{Aq_1 + B}{(Cq_1 + D)} \tag{2.68}$$

where

$$A = 1 - \frac{L_2}{f}, \quad B = L_1 + L_2 - \frac{L_1 L_2}{f},$$

$$C = -\frac{1}{f} \quad \text{and} \quad D = 1 - \frac{L_1}{f}.$$

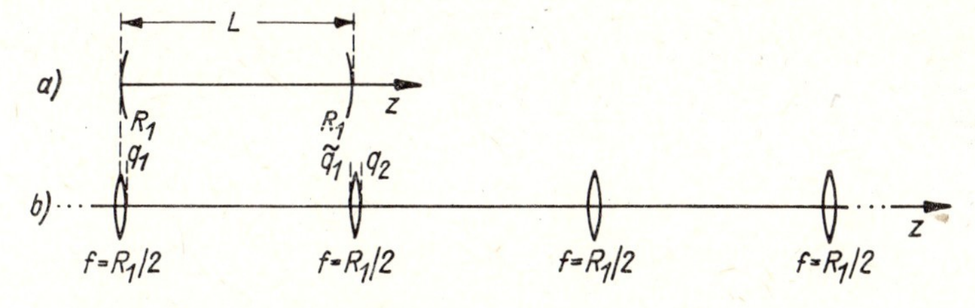

Fig. 2.10. Symmetrical resonator
a) Symmetrical resonator of length L with spherical mirrors having radii of curvature $R_1 = R_2$
b) Schematic representation of the "folded up" equivalent lens arrangement

Taking as a basis the calculated effect of a thin lens on propagation modes, let us now investigate the radiation fields in open resonators, whose mirrors are large compared to the beam diameters. The radiation field in the resonator is represented by wave beams that propagate between the mirrors. For a resonator mode, the beam parameters should by definition be the same after one complete round trip. The self-consistency condition is now used to calculate the mode parameters. As in the previous section, for purposes of simplification we will first deal with a symmetrical resonator configuration, which is schematically represented in Fig. 2.10a; Fig. 2.10b shows an equivalent "folded up" structure built up of an infinite series of thin lenses. In a resonator of this kind the self-consistency condition must already be satisfied after half a cycle. Thus, using the symbols from figure 2.10b,

$$q_1 = q_2 \tag{2.69}$$

must hold. (2.66), (2.67c) and (2.69) lead to

$$\frac{1}{q_1} = -\frac{1}{2f} \pm \sqrt{\frac{1}{4f^2} - \frac{1}{fL}}. \tag{2.70}$$

It follows from the definition equation (2.58) for the beam parameter q that a real value of this parameter signifies an infinite value of the beam diameter $2w$, and thus an instability of the resonator mode. Therefore, according to (2.70), we must require $L < 2f$ for stable configurations with which the root in

$$\frac{1}{q_1} = -\frac{1}{2f} - \mathrm{i}\sqrt{\frac{1}{fL} - \frac{1}{4f^2}} \tag{2.71}$$

becomes real. (The positive sign before the root does not need to be taken into account, since according to (2.58) it would lead to an imaginary value for the beam diameter.) By comparing (2.71) with (2.58) we can immediately recognize that, directly behind the thin lens or the mirror, the radius of curvature of the wave front coincides with the radius of curvature $R_1 = 2f$ of the mirror, and that the beam radius of the fundamental mode is given by

$$w_1^2 = \frac{\lambda R_1}{\pi} \frac{1}{\sqrt{2\frac{R_1}{L} - 1}}. \tag{2.72a}$$

In the center of the resonator, where for reasons of symmetry the beam waist must lie,

$$w_0^2 = \frac{\lambda}{2\pi}\sqrt{L(2R_1 - L)} \tag{2.72b}$$

holds. The relations for the beam diameter take a particularly simple form for the confocal resonator, where $L = R_1 = 2f$. They are

$$w_1 = \sqrt{\frac{\lambda L}{\pi}}, \quad w_0 = \sqrt{\frac{\lambda L}{2\pi}} = \frac{w_1}{\sqrt{2}}. \tag{2.72c}$$

It is quite obvious from the relations dealt with up to now in connection with the depiction of the field distribution of the confocal resonator in Fig. 2.9b that, within the applied approximations, for each resonator an equivalent confocal resonator can be found. The confocal resonator in this figure had been introduced such that in a general representation of Gaussian beams we have placed mirrors at specific points ($z = L/2 = R_1/2$) whose radius of curvature coincides with that of the phase front of the light beam. From the treatment of the self-consistency condition we see that the given field strength distribution of the Gaussian beams is not changed by these mirrors. Instead of the confocal mirrors, however, we can also affix mirrors at other points on the z-axis, which leave the field distribution unchanged at the point where their radius of curvature coincides with that of the field. The configuration need not be symmetric here. Since all of these various resonator arrangements lead to the same field distribution, we say that they are equivalent. Due to the simple, easy to survey characteristics of confocal resonators for a given, in general, asymmetric resonator we often look for the equivalent confocal resonator. This method is not limited to two mirror resonators; we can also determine equivalent confocal resonators for more complex configurations consisting of many lenses or mirrors where, however, the region of space in which the field distributions should coincide must be precisely prescribed.

By calculating the phase shift of individual modes after one round trip in the resonator and taking into account the selfconsistency requirement, we can also determine the eigenfrequencies of the modes. This point, however, will not be dealt with further here (cf. [2.2]). The same results, which have already been given in the previous section for the example of the confocal resonator, can be obtained by use of this procedure.

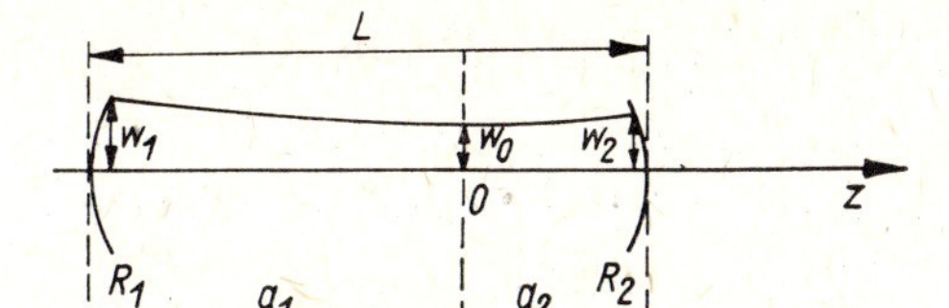

Fig. 2.11. Unsymmetrical two-mirror resonator

Similar to the manner discussed in connection with symmetrical resonators, the modes and mode parameters of asymmetric resonators can be determined by considering the equivalent confocal resonator (see Fig. 2.11). Here we obtain

$$w_1^4 = \left(\frac{\lambda R_1}{\pi}\right)^2 \frac{(R_2 - L)\,L}{(R_1 - L)(R_1 + R_2 - L)}, \tag{2.73a}$$

$$w_2^4 = \left(\frac{\lambda R_2}{\pi}\right)^2 \frac{(R_1 - L)\,L}{(R_2 - L)\,(R_1 + R_2 - L)}, \tag{2.73b}$$

$$w_0^4 = \left(\frac{\lambda}{\pi}\right)^2 \frac{L(R_1 - L)\,(R_2 - L)\,(R_1 + R_2 - L)}{(R_1 + R_2 - 2L)^2} \tag{2.73c}$$

for the beam radii w_1, w_2 and w_0 at the mirrors 1,2 and in the waist and, for the distances of the waist from the mirrors, we have

$$a_1 = \frac{L(R_2 - L)}{R_1 + R_2 - 2L}, \tag{2.73d}$$

$$a_2 = \frac{L(R_1 - L)}{R_1 + R_2 - 2L}. \tag{2.73e}$$

(Note, the waist can now lie inside or outside of the resonator.)

2.3.5 *The Three-Mirror Resonator*

Until now, the radiation fields were investigated in the simplest types of open resonators, the two-mirror resonators. Frequently, it becomes necessary in laser physics and particularly in the generation of ultrashort light pulses to use more complex resonator configurations.

It may be necessary, for instance, that the light beam has waists at various points of the laser resonator, the diameters of which vary independently of each other. This case can occur, for instance, if besides the active material another component, say a saturable absorber or another nonlinear optical sample, has to be positioned within the resonator whose properties depend critically on the intensity of the radiation field (compare 6. and 7.). Of particular significance is the multimirror resonator in connection with the continuously pumped (continuous wave or cw) dye laser. In order to attain the necessary high pumping intensities ($\gtrapprox 1$ MW/cm^2) with the available rather low average pumping power (≈ 1 W) and get optimum use out of it, pump and laser beam must have waists of equal and very small diameter in the same position. In order to fulfil this requirement in laser resonators of optional length, three-mirror resonators are frequently used (see Fig. 2.12). One can imagine the three-mirror resonator of Fig. 2.12 created from a two-mirror resonator in which one end mirror is replaced by the mirror combination M_1, M_2. This mirror combination produces a strong contraction of the laser beam at the position of the active medium. In addition, the mirror M_2 causes the pump beam to be focussed into the active medium. The mirror M_3 is either very slightly curved or flat; the distance d_2 is large compared to d_1.

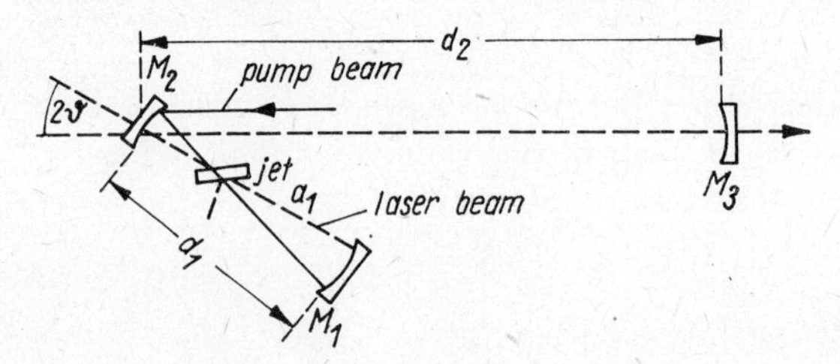

Fig. 2.12. Three-mirror resonator of a continuously pumped dye laser (Only the optical axes of the pump and laser beams have been drawn in. The active material, a free flowing dye solution, the jet, is located at the waist of both beams. In order to work with a defined polarization direction and with the smallest possible loss for this polarization direction, the jet is arranged relative to the laser beam at the Brewster angle. The mirrors M_1 and M_2 have relatively small — in general equal — radii of curvature. The distances between the mirrors M_3 and M_2 as well as between M_2 and M_1 are denoted by d_2 and d_1, respectively, and the distance of the laser waist from the mirror M_1 is denoted by a_1.)

A precise calculation of the waist position and size can be performed by using the described theory of Gaussian beams (cf. [2.3]). First, the stability of the laser resonator has to be investigated by considering the equivalent two-mirror resonator. Stability is found within the following limits of the mirror distance d_1:

$$R_1 + \frac{R_2}{2} + \delta_{\min} \leqq d_1 \leqq R_1 + \frac{R_2}{2} + \delta_{\max} \tag{2.74}$$

where

$$\delta_{\min} = \frac{R_2^2}{2(2d_2 - 2R_3 - R_2)}$$

and

$$\delta_{\max} = \frac{R_2^2}{2(2d_2 - R_2)}.$$

Thus, the stability range is

$$\varDelta = \delta_{\max} - \delta_{\min} = \frac{-R_3R_2^2}{(2d_2 - R_2)(2d_2 - 2R_3 - R_2)}. \tag{2.75}$$

Particularly for $d_1 \ll d_2$ and $R_3 \to \infty$ we obtain from (2.75)

$$\Delta = \frac{R_2^2}{4d_2}. \tag{2.76}$$

For the distance a_1 between the waist of the laser beam and the mirror M_1

$$a_1 = \frac{d(\bar{R}_3 - d)}{R_1 + \bar{R}_3 - 2d} \tag{2.77}$$

holds, where

$$d = d_1 - \frac{d_2 R_2}{2d_2 - R_2}$$

and

$$\bar{R}_3 = \frac{R_3 R_2^2}{(2d_2 - R_2)(2d_2 - 2R_3 - R_2)}.$$

The radius w_0 of this waist can be determined from

$$w_0^4 = \left(\frac{\lambda}{\pi}\right)^2 \frac{d(R_1 - d)(\bar{R}_3 - d)(R_1 + \bar{R}_3 - d)}{(R_1 + \bar{R}_3 - 2d)^2}. \tag{2.78}$$

As stated in the description of figure 2.12, the laser beam passes the dye-jet at the Brewster angle. Because of the oblique path through the jet having the optical thickness $n \cdot b$ the laser beam undergoes an astigmatic deformation. The reflection by mirror M_2 of radius of curvature R_2, which "folds" the laser beam, produces, however, an opposite astigmatic distortion whereby compensation of astigmatism is made possible. This compensation requires the fulfilment of the relation

$$R_2 \sin \vartheta \tan \vartheta = 2 \frac{n^2 - 1}{n^4} \sqrt{n^2 + 1}\, b \tag{2.79}$$

between the "folding" angle 2ϑ and the thickness b of the jet. In resonators that are corrected for astigmatism, Gaussian beams of good quality with waists of about 10 μm in diameter can be generated, where the overlap with the pump beam is almost complete.

2.4 Description of the Active Materials of some Important Lasers

In the following, several laser-active materials will be described which are frequently employed for the generation and amplification of ultrashort light pulses and to which we will refer in the later chapters dealing with the various methods of modelocking.

2.4.1 Nd: YAG Lasers

The Nd:YAG laser belongs to the group of optically pumped solid state lasers. Yttrium aluminium garnet crystals ($Y_3Al_5O_{12}$), in which the Nd^{3+} ions are doped in concentrations of up to about 1.5 percent by volume, serve as laser active material. (Larger dopings are not possible due to the different volumes of Nd^{3+} and Y^{3+} ions.) YAG crystals have a cubic lattice and are accordingly optically isotropic. Fig. 2.13a shows an energy level scheme of Nd^{3+} ions in the crystal field. It is evident from the left side of the diagram that we are dealing with a four-level laser. The levels $^4F_{3/2}$ and $^4I_{11/2}$ function as the upper and lower laser level, respectively. Above $^4F_{3/2}$, a large series of pump levels or pump bands are located from which the excited ions pass quickly over into the

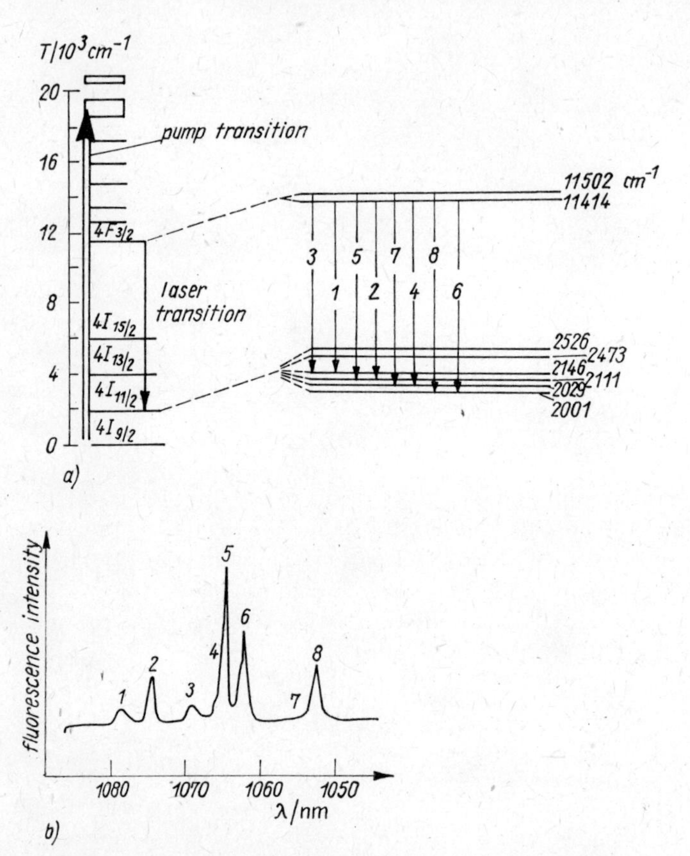

Fig. 2.13. Laser active transitions of a Nd:YAG crystal

a) Energy level scheme

b) Fluorescence intensity (in arbitrary units) versus wavelength

upper laser level by interacting with the lattice. The lower laser level has a distance from the ground level which is large compared to $k_B\mathcal{T}$, and therefore it is nearly unoccupied in thermal equilibrium. The levels $^4F_{3/2}$ and $^4I_{11/2}$ show a splitting in the crystal field, through which several transitions are made possible which are shown on the right side of Fig. 2.13a. (The corresponding splitting of the other levels has not been entered in the diagram.) The strongest laser transition is observed at 1.0641 μm. The cross section of the transition amounts to 8.8×10^{-19} cm², the fluorescence lifetime of the upper level is 230 μs and the fluorescence efficiency is 0.995. Due to the effect of the lattice vibrations, the transitions are homogeneously broadened at room temperature. Because of the highly ordered structure of the crystal, the inhomogeneous broadening is negligibly small whereas, in the Nd:glass systems, it dominates. The main laser active transition has a linewidth of $\Delta\nu \approx 120$ GHz. The Nd:YAG laser can be advantageously pumped with Krypton arc lamps, because their emission bands coincide well with the pump levels. Fig. 2.14 shows the scheme of a pump arrangement. The pump arrangement consists of a double elliptical reflector made of highly reflective material. A cylindrical YAG rod is located in the common focal line. The two Krypton arc lamps are centered in the two other focal lines. By means of this arrangement, a higher pump efficiency can be achieved. To cool the system rod and lamps are doused with water. Because of the good thermal conductivity of the material, its relaxation parameters and efficient cooling, it is possible to operate YAG laser continuously with high power (up to 10^2 W) or at high pulse frequencies (up to about 100 Hz) with pulse energies of 0.1 to 1 J.

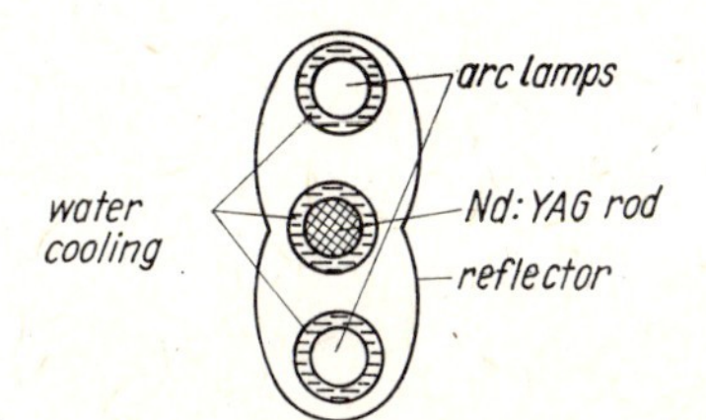

Fig. 2.14. Pump arrangement with a double elliptical reflector

The YAG crystal has a high refractive index ($n(1.064\ \mu m) = 1.818$). Therefore, comparatively high reflection of the laser radiation appears at the end faces. This can be considerably reduced either by dielectric antireflection coatings or by grinding the end faces of the rods at the Brewster angle with respect to the optical axis. Frequently, however one also accepts these losses — which is possibly due to the large gain of the material — but one must then at least grind the end faces toward one another at a small wedge angle ($\approx 1°$) in order to prevent the two end faces from working as a laser resonator or as an additional frequency selective subresonator within the main cavity.

Various methods of generating ultrashort light pulses are successfully employed using Nd:YAG lasers. Methods of active modelocking when using acoustooptical or electrooptical modulators are advantageously applied with continuously pumped lasers (continuous wave or cw operation (compare 4.). The pulse-pumped Nd:YAG laser is mostly passively modelocked in order to generate ultrashort light pulses (compare 7.). However, combinations of active and passive modelocking are also advantageously applied.

Nd:YAG lasers in the continuous wave and pulse-wave regime are used frequently as light sources for generation of higher harmonics, for parametric generators (compare 8) and for pumping of dye lasers (compare 5. und 6.).

2.4.2 Argon and Krypton Ion Gas Lasers

Ion gas lasers are the most efficient continuous wave lasers in the visible part of the spectrum. The output power varies according to the type between 10^{-1} W (small air-cooled discharge tube) and 10^2 W (long water-cooled high performance discharge systems). The efficiency is about 10^{-3}. The most important representatives of ion gas lasers are the argon and krypton ion gas lasers. The most intense laser transitions are given in table 2.1. The laser emission is due to transitions between the levels of (mostly singly) ionized argon or krypton (see Fig. 2.15). The upper laser level is typically a state of the $4p$-group, the lower laser level of the $4s$-group. In continuously operated lasers, the exci-

Table 2.1. Efficient laser transitions of argon ion lasers and krypton ion lasers

λ/nm	Type of laser	λ/nm	Type of laser
799.3	Kr^+	501.7	Ar^+
752.5	Kr^+	496.5	Ar^+
676.4	Kr^+	488.0	Ar^+
647.1	Kr^+	476.5	Ar^+
568.2	Kr^+	457.9	Ar^+
530.9	Kr^+	351.1 – 363.8	Ar^+
514.5	Ar^+	337.4 – 356.4	Kr^+

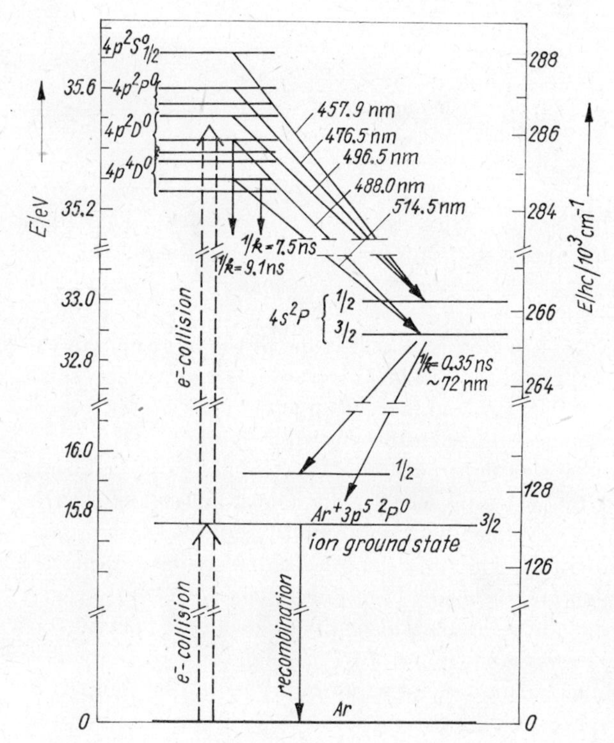

Fig. 2.15. Energy level scheme of the argon ion laser

tation of the upper laser level can occur through two-stage electron collisions in a strong current low-pressure arc discharge (e.g. under current densities of about $10^6 - 10^7$ A/m^2, a gas pressure of 70 Pa and tube diameters of 2.5 mm). The maximum current density is limited by the maximum load of the discharge tube. Such materials as graphite and beryllium oxide, which are distinguished by their good thermal conductivity and resistivity under impinging ions, are suitable. In most setups the energy of these impinging ions is reduced by the application of axial magnetic fields of the order of 10^4 to 10^5 A/m, whereby the efficiency of the laser process simultaneously increases (see Fig. 2.16).

The ionization of the rare gas results from the first collision of the two-stage excitation, where mostly ions in the ground state are generated. Upon the second collision the ion can be excited into the upper laser level. Due to this two-step process the occupation inversion increases with the square of the current density. The emptying of the lower laser level reults from spontaneous emission at the wavelength $\lambda = 72$ nm.

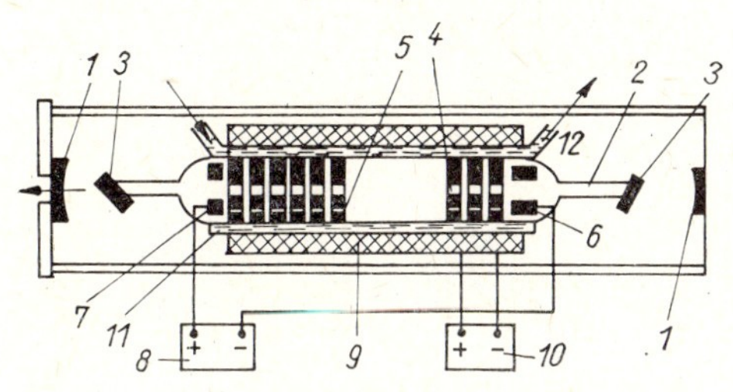

Fig. 2.16. Scheme of an argon ion laser consisting of the resonator mirrors (1) and the gas discharge tube (2) with the Brewster windows (3). The discharge tube consists of the discharge channel (4), the channel for the backward flow (5), the cathode (6) and the anode (7). (8) is the power source of the discharge, (9) is the magnetic coil, (10) is the power supply of the magnetic field, (11) shows a segment of the laser made e.g. of graphite or BeO and (12) is the cooling system (water).

Typical parameters are the following. The diameter of the discharge channel is 1 to 10 mm, the gas pressure is 1 to 100 Pa with pure argon. The discharge current takes on values of 30 to 300 A and the magnetic field of 1 to 8×10^4 A/m. The beam diameter is 0.5 to 2 mm at the output mirrors. The efficiency amounts to about 10^{-3}.

The channel (5) is necessary in addition to the discharge channel for continuously pumped lasers in order to let the gas stream back to the cathode. (Note that the electrons transfer some part of their axial momentum to the neutral atoms, which as a consequence move in the direction of the anode and cause a pressure difference between the anode space and the cathode space.)

Under typical operation conditions, homogeneous and inhomogeneous broadening are of nearly the same order of magnitude in ion gas lasers. In the argon-ion laser (transition at $\lambda = 514$ nm) the Doppler width is at about 3.5 GHz; the homogeneous line width, which is substantially produced by the Stark effect as a result of high electron densities ($\simeq 10^{20}$ m^{-3}) and by spontaneous emission, lies between 0.5 and 0.8 GHz. (The natural line width is 0.46 GHz.) The large homogeneous broadening leads to strong competition between the modes, which without special measures can easily lead to considerable amplitude fluctuations in the multimode operation. A special effect is produced in ion gas lasers by the relatively high drift velocity of the ions ($v_{\text{drift}} \simeq 10^2$ m/s). It causes the gain profile in the laser to split into two Doppler curves separated by a distance of the order of 0.5 GHz.

Ultrashort light pulses are generated in the continuous wave ion gas laser mostly by active modelocking using acoustooptical modulators (compare 4.). The modelocking can however also be passively achieved — e.g. when using dye solutions [2.5].

Modelocked ion gas lasers are of particular importance as pump sources for dye lasers operated in the regime of synchronous pumping (compare 5.).

2.4.3 *Dye Lasers*

Organic dyes in solution distinguish themselves through large absorption and emission cross sections and wide spectral bands. They are well-suited as active materials for lasers with adjustable and tunable wave lengths (see table 2.2). The energy level scheme

Table 2.2. Parameters of several laser dyes

Dye	Center of the fluorescence line/nm	Operating range of the laser/ nm	Pumping region/ nm	Favorable concentrations/ $mol \cdot l^{-1}$
Carbostyril 165	445	419–485	350–365	2.5×10^{-3}
Coumarin 2	450	435–485	340–365	3×10^{-3}
Coumarin 1	470	450–495	350–365	3×10^{-3}
Coumarin 102	495	470–515	400–420	3×10^{-3}
Coumarin 30	515	495–545	400–420	1×10^{-3}
Coumarin 7	535	505–565	400–420	5×10^{-3}
Coumarin 6	538	521–551	458–514	12.5×10^{-3}
Sodium				
Fluorescein	552	538–573	458–514	2.7×10^{-3}
R 110	570	540–600	458–514	12.5×10^{-3}
R 6 G	590	570–650	458–514	2×10^{-3}
RB	630	601–675	458–514	2×10^{-3}
R 101/R 6 G	645	620–690	458–514	1.5×10^{-3} R 101 1.5×10^{-3} R6G
Cresyl-violet/R 6 G	695	675–708	458–514	2.4×10^{-3}
Nile-blue	750	710–790	647	1×10^{-3}
Oxazine 1(4)	750	695–801	647–672	0.6×10^{-3}
DEOTC-P(4)	795	765–875	647–672	0.6×10^{-3}
HITC-P(4)	875	840–940	647–672	0.74×10^{-3}

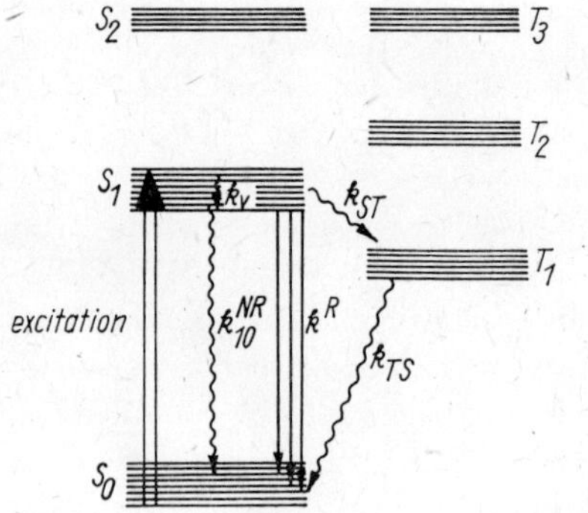

Fig. 2.17. Transitions by radiation and relaxation in dye molecules

that already has been discussed in chapter 1. is depicted in Fig. 2.17 with the respective pump, laser and relaxation transitions. The electronic levels of the singlet and triplet systems are superimposed by vibrational levels. Due to the large number of vibrational modes and the pronounced line broadening in liquids, the individual vibrational transitions are not at all resolved for the most part, but a homogeneous spectral band appears. The dye laser is characterized mostly as an effective four-level laser. Thereby, the excitation of various excited vibrational levels of the S_1 state is produced by pump light according to the Franck-Condon principle. The vibrational deactivation in the S_1 state happens extraordinarily fast ($k_V^{-1} \simeq 10^{-13}$ s), i.e. the molecules accumulate almost immediately at the lower edge of the band. From here they can pass over by fluorescence to various vibrational levels of the S_0 state. If the terminal levels of these transitions are at a distance of more than $k_B\mathcal{T}$ from the ground state, their occupation can be neglected in thermodynamic equilibrium. Since the emptying of these levels through vibrational

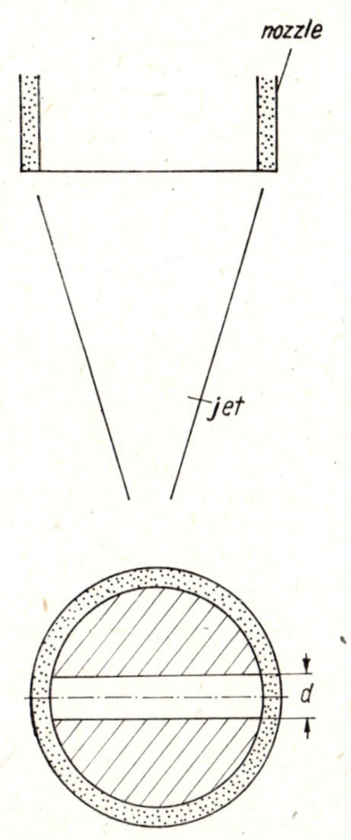

Fig. 2.18. Nozzle used to produce a liquid jet stream

relaxation happens very fast, all preconditions for typical four-level lasers are satisfied. Under any circumstances then, if molecules are in the vibrationless state of the S_1 level, the inversion condition is satisfied and radiation in the spectral range of the fluorescence transitions can be amplified. The fluorescence lifetimes of suitable laser dyes amount to 10^{-8}—10^{-9} s; the fluorescence efficiency is close to one. However, the transitions to the triplet system, particularly under continuous excitation, are very unwholesome for effective laser operation, because due to the long lifetime here the molecules can accumulate in the T_1 level and thus fall out of the laser process. Moreover, fluorescence radiation can be absorbed by the $T_1 - T_x$-transitions. Therefore, one endeavors to use dyes with very small quantum yields for the singlet-triplet transition. Furthermore, one tries to reduce the lifetime in the T_1 level by the addition of triplet quenchers, molecules that receive the excitation energy and give it quickly to the solvent in the form of heat. Because these measures are, in general, not sufficient under continuous laser operation, we must exchange the dye in the excitation volume very fast. This occurs by fast pumping of the dye in sample cells or by fast flowing of the dye solution in a free liquid jet stream, the so-called dye jet. By appropriate shaping of the outlet nozzle, sufficiently high pressure and solvents of sufficiently high viscosity (preferably ethylene glycol), a laminar flow with high optical homogeneity and sufficiently high velocity ($\simeq$ 10 m/s) can be produced. The thickness of the jet stream is mostly chosen between 10 μm and about 0.2 mm, depending on the requirements. Continuously pumped dye laser systems of this kind are produced by the American firms Spectra Physics and Coherent and by the Center of Scientific Instruments of the Academy of Science of the GDR [2.7, 2.8].

For the pulsed pumping of dye lasers, flash lamps, nitrogen lasers, excimer lasers and solid state lasers are suitable, as well as the harmonics of solid state lasers, particularly the second, third and fourth harmonics of the Nd:YAG laser. For continuous pumping, efficient ion gas lasers as well as fundamental waves and harmonics of Nd:YAG lasers are mainly used. The particular suitability of the argon and krypton ion lasers follows from a comparison of tables 2.1 and 2.2.

Ultrashort light pulses are generated in distributed feedback dye lasers (compare 2.8), by synchronous pumping of dye lasers (compare 5.) as well as by passive mode locking (compare 6.). The shortest pulses ($\sim$ 30 fs), which until now were directly generated in lasers, are produced with passive synchronization.

For the near infrared part of the spectrum (0.8 μm—3.8 μm) we can use so-called color center lasers in place of dye lasers. The broad fluorescence bands of certain color centers (e.g. of F_2^+ centers) in alkali halide crystals also allow spectral tuning as well as the generation of very short pulses. Pump mechanism and laser construction are similar to that of dye lasers. Krypton ion lasers and Nd:YAG lasers are particularly suitable as pump light sources (see e.g. [2.14] and references given there).

2.4.4 *Semiconductor Lasers*

As laser-active materials semiconductors are becoming ever more important, because miniaturized and integrable light sources with favorable parameters can be constructed by means of them. By varying the composition of mixed crystals as well as by changing external parameters such as temperature and pressure, the wavelength of the laser transition can be suitably adjusted and tuned.

Until now we have considered light amplification in systems of nearly interaction-free atoms or molecules, whose levels can be described in close approximation by one-electron excitations; in thermal equilibrium the occupation numbers of these levels follow from the Boltzmann statistics. When dealing with semiconductors we have to take into account the energy band structure and occupation number densities according to the Fermi-Dirac statistics. Fig. 2.19 shows schematically the valence and conduction band of a semiconductor as well as optical transitions within and between these bands. The plotted interband transition is a possible laser transition; both intraband absorptions lead to additional losses which, particularly in semiconductors with indirect transitions, greatly disturb and inhibit laser operation. Therefore, semiconductors with direct transitions between the valence and the conduction band such as gallium arsenide (GaAs) are preferred.

The requirement that in laser materials the gain through stimulated emission should be larger then the losses through absorption leads for semiconductors to conditions differing greatly from those for materials that have been handled up to now, as can be easily understood. In the formerly treated interaction-free one-electron systems the transition probabilities were dependent only on the occupation of the initial level, e.g. the upper laser level. In semiconductors, however, due to the Pauli principle, we must bear in mind that the corresponding transition can only take place if the upper level is occupied and the lower level is unoccupied. Accordingly,

$$\frac{\mathrm{d}W_{12}^{\mathrm{a}}}{\mathrm{d}t} \propto N(\mathscr{E}_1)\,P(\mathscr{E}_2), \tag{2.80a}$$

$$\frac{\mathrm{d}W_{21}^{\mathrm{i}}}{\mathrm{d}t} \propto N(\mathscr{E}_2)\,P(\mathscr{E}_1) \tag{2.80b}$$

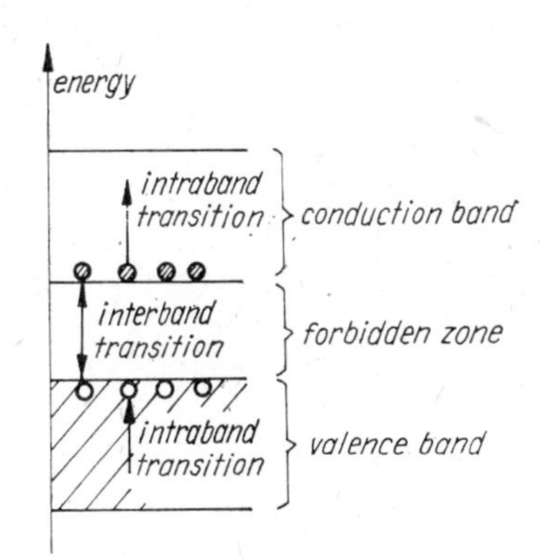

Fig. 2.19. Intra- and interband transitions in semiconductors

holds for the transition rates for absorption ($\mathrm{d}W^{\mathrm{a}}/\mathrm{d}t$) and induced (stimulated) emission ($\mathrm{d}W^{\mathrm{i}}/\mathrm{d}t$) between the states of energy $\mathscr{E}_1$ and energy $\mathscr{E}_2$
where

$$N(\mathscr{E}_1) = D(\mathscr{E}_1)\,f_{\mathrm{V}}(\mathscr{E}_1), \quad N(\mathscr{E}_2) = D(\mathscr{E}_2)\,f_{\mathrm{C}}(\mathscr{E}_2)$$

and

$$P(\mathscr{E}_1) = D(\mathscr{E}_1)\left(1 - f_{\mathrm{V}}(\mathscr{E}_1)\right), \quad P(\mathscr{E}_2) = D(\mathscr{E}_2)\left(1 - f_{\mathrm{C}}(\mathscr{E}_2)\right)$$

are the occupation densities for electrons and holes, which depend on the state densities $D(\mathscr{E}_i)$ and the occupation probabilities for electrons in the valence and the conduction band $f_{\mathrm{V}}(\mathscr{E}_1)$ and $f_{\mathrm{C}}(\mathscr{E}_2)$, respectively (compare e.g. [2.11, 2.12]). The two relations above, i.e. (2.80a) and (2.80b), have the same proportionality factor. Hence the requirement

$$\frac{\mathrm{d}W_{21}^{\mathrm{i}}}{\mathrm{d}t} > \frac{\mathrm{d}W_{12}^{\mathrm{a}}}{\mathrm{d}t} \tag{2.81}$$

leads to

$$f_{\mathrm{C}}(\mathscr{E}_2)\left(1 - f_{\mathrm{V}}(\mathscr{E}_1)\right) > f_{\mathrm{V}}(\mathscr{E}_1)\left(1 - f_{\mathrm{C}}(\mathscr{E}_2)\right), \tag{2.82}$$

from which

$$f_{\mathrm{C}}(\mathscr{E}_2) > f_{\mathrm{V}}(\mathscr{E}_1) \tag{2.83}$$

follows. This means that the occupation probability of states at the higher energy $\mathscr{E}_2$ must be greater than of states at $\mathscr{E}_1$ in order to achieve a gain for light of the frequency $\omega_{21} = (\mathscr{E}_2 - \mathscr{E}_1)/\hbar$ in the semiconductor. Just as in isolated one-electron systems occupation conditions of this kind can only be reached and maintained by pumping. In general, inside of a band very effective radiationless relaxation processes occur. For this reason, in each band a quasi-equilibrium probability distribution

$$f_{\mathrm{V,C}}(\mathscr{E}) = \frac{1}{1 + \mathrm{e}^{(\mathscr{E} - F_{\mathrm{V,C}})/k_{\mathrm{B}}\mathscr{T}}} \tag{2.84}$$

which is characterized by a quasi-Fermi level, e.g. F_{V} or F_{C}, builds up very fast. If we substitute these expressions in (2.83), we obtain

$$F_{\mathrm{C}} - F_{\mathrm{V}} > \mathscr{E}_2 - \mathscr{E}_1 = \hbar\omega . \tag{2.85}$$

Table 2.3. Wavelength of emission and type of excitation of some semiconductor lasers

Type	λ/μm	Pumping by		
		Injection	e-Beam	Laserbeam
GaAs	0.84	×	×	×
GaSb	1.6	×	×	
InP	0.9	×		
InAs	3.1	×	×	
InSb	5.2	×	×	×
GaP_xAs_{1-x}	0.65–0.84	×		
$Ga_xIn_{1-x}As$	0.84–3.1	×		
InP_xAs_{1-x}	0.9–3.1	×		
$Al_xGa_{1-x}As$	0.6–1.0	×		
$In_xGa_{1-x}P$	0.56–1.0	×		
$In_xGa_{1-x}As$, $InAs_{1-x}P_x$	1–3.2	×		
CdS	0.5		×	
CdTe	0.8		×	
PbS	8.5	×		
PbTe	6.5			
PbS_xSe_{1-x}	5–9	×		
$Pb_{1-x}Sn_xSe$	9–20	×		
$Pb_{1-x}Sn_xTe$	7–20			

This means the pumping process must work such that the spacing of the quasi-Fermi levels of the two participating bands is greater than the photon energy of the radiation to be amplified. The excitation can occur by irradiation of the sample with light, by electron beams or by charge carrier injection (see Table 2.3). Under optical pumping the semiconductor must be irradiated with radiation whose photon energy is greater than the width of the energy gap. Radiation of such wavelengths is absorbed in a thin surface layer (see Fig. 2.20a). In this region electrons are raised from the valence band into the conduction band. As a result of the radiationless relaxation processes these electrons accumulate at the bottom of the conduction band and the respective holes at the top of the valence band. Using powerful irradiation the condition (2.85) can be satisfied, where the width of the energy gap is to be substituted for $(\mathscr{E}_2 - \mathscr{E}_1)$. Instead of using light, the material can also be excited by bombardment with fast electrons, where electron energies of $10^4 - 10^5$ eV are employed. Considerably higher energies are ruled out because they damage the material. The penetration depth as with optical irradiation is very small so that the excited region is very thin (several micrometers). The fast electrons generate electron-hole-pairs through collisional excitation and lose energy in these processes, in which about one third of the electron beam energy is converted into excitation energy.

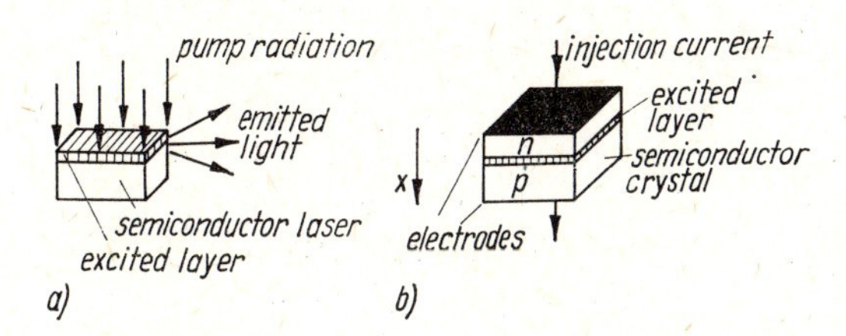

Fig. 2.20. Pumping of a semiconductor laser with light or an electron beam (a) and with charge carrier injection (b)

At the outset of this subsection we mentioned that the possibility of miniaturization and integration is a considerable advantage of semiconductor lasers. Both are possible only under application of the third excitation method, i.e. pumping by means of charge carrier injection. Let us consider a semiconductor diode that has a p,n-transitions Under suitable doping concentration the warping of the energy bands, which appears in the transition region, can be greater than the energy gap (see Fig. 2.21a). Let the

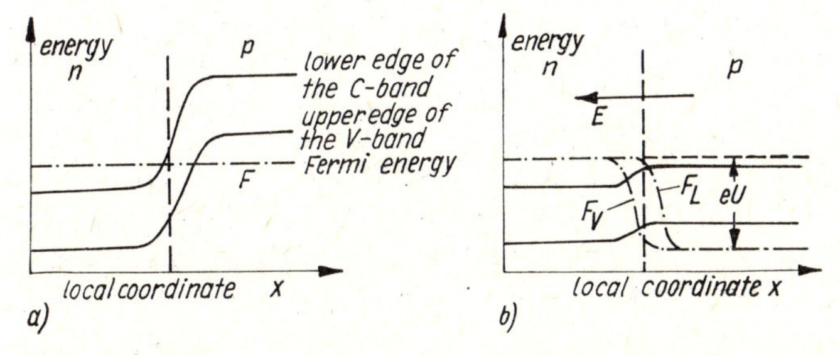

Fig. 2.21. Scheme of the bands in a p,n-transition versus the space coordinate x; a) without external voltage, b) with external voltage U in the direction of transmission

Fermi level be in the conduction band for the n-region and for the p-region in the valence band. If we apply a voltage U to the semiconductor diode, the charge carrier must run against an energy barrier eU at the p,n-transition; i.e., the Fermi level takes on two values on both sides of the transition, which differ by energy eU. Depending on the direction of the voltage applied the warping of the energy bands decreases or increases. In Fig. 2.21b the effect of a voltage in the forward direction, which causes a decrease in the warping of the energy bands, is depicted. An increased penetration of electrons and holes in the transition layer occurs; thus, charge carriers are injected into this region. If the recombination between the conduction band and the valence band by interband relaxation processes occurs more slowly than the supply of charge carriers by injection, then a non-equilibrium probability distribution of electrons and holes arises in the transition layer, which must again be characterized by quasi-Fermi levels for the individual bands as depicted in Fig. 2.21b. In a certain local range $(F_C - F_V)$ is greater than the band gap, and the amplification condition is satisfied in this layer of the semiconductor. The thickness of the laser active layers lies in the order of 10^{-1}—10^{2} μm, the injection current density is 10^{6}—10^{7} A/m². In order to increase the current density in the vicinity of the p,n-transition, special shapes of the semiconductor devices are usually chosen, which ensure a contraction of the current (electrical confinement) at the p,n-transition and a good transport of the dissipated energy. By such methods, e.g. with GaAs lasers, threshold currents of the order of 10^{-1} A can be attained.

Further advances become possible by applying single and multi-quantum well structures, where the electrons and holes are confined to certain spatial regions by potential wells produced by doping the material. Here the band structure does not only depend on the material properties but also on the extent of the wells. By phaselocking the radiation from several well structures on one semiconducting sample continuous wave powers as high as several Watts have been obtained (cf., e.g., [2.15], [2.16]).

The semiconductor laser differs from most other lasers by the small dimension of the active medium. The layer geometry of the amplifying region produces substantial diffraction of the emitted or amplified light (the beam divergence can be of the order of 1 rad). Due to the small length of the active material, one can construct extremely short resonators. In the simplest arrangements the end faces of the (highly refractive) semiconductors function as resonator mirrors. The small optical resonator lengths L lead to very great mode spacings $\delta\nu = c/2L$; at $L = 0{,}5$ mm we obtain $\delta\nu = 300$ GHz. This means that even with single mode lasers, where the duration τ_L of the shortest possible pulses is of the order of $1/\delta\nu$, picosecond pulses can be obtained (cf. [2.23]). If we want to put an external resonator into operation, we must carefully eliminate or at least strongly reduce the reflection of the semiconductor end faces.

Various methods of generating ultrashort light pulses can be used with semiconductor lasers. Most important is the method of active gain modulation with the injection laser, because one can easily modulate the injection current at high frequencies (compare 4.). Furthermore, the method of synchronous pumping of semiconductor lasers is employed, where one can use, for instance, modelocked dye lasers as pump sources (compare 5.). As with dye lasers and optically pumped solid state lasers the shortest pulses in the subpicosecond range are attained by means of passive modelocking (compare 6. and 7., especially 7.4).

2.5 *Q*-Switching of Lasers

We have until now only described stationary processes in lasers, the description of which thus allows us to neglect the time derivatives in the basic equations. Before we discuss in the next section the most important principle for the generation of ultrashort light pulses, the modelocking principle, we should discuss briefly another nonstationary laser regime, the Q-switching of lasers.

To begin with it should be mentioned that nonstationary regimes in lasers can evolve even without additional measures. In the calculation of the radiation power according to equation (2.15) all time derivatives were neglected from the start; this neglect is of course only possible after a certain time lapses once the pump radiation has been switched on, because the transient behavior of the laser material, which occurs until the steady state has been reached, cannot be described correctly. If we account for the time derivatives in the basic equations, it can be seen that the transient processes after the switch-on of the laser, even in the case of single mode operation, cannot be described by a monotonic evolution, but mostly in the form of temporally decaying, non-harmonic oscillations of the radiation field and occupation inversion, which eventually, after a certain time lapse, approach their steady-state values. These damped oscillations are called relaxation oscillations of lasers in single mode operation. In the consideration of multimode regimes the relationships become more complicated. Due to the time-space interferences of the various modes and their change during the transient oscillations the output radiation of the laser consists of irregular pulses having stochastically fluctuating amplitudes. It is essential here that the laser in general does not pass into the steady-state regime, but retains its unsteady nature even after a long period of time.

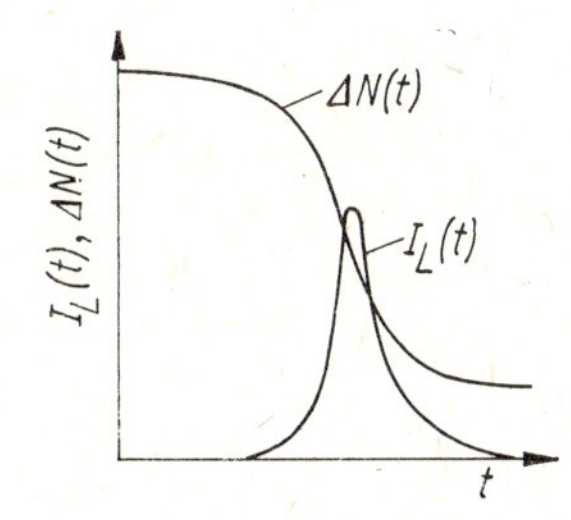

Fig. 2.22. Typical time dependence of the occupation number inversion $\Delta N(t)$ and the photon number flux density $I_L(t)$ in Q-switched lasers (from B. A. Lengyel [2.13])

A nonstationary laser regime which is produced intentionally in contrast to the mostly undesired relaxation oscillations can be obtained by changing as fast as possible the quality, i.e. the Q-factor, of the resonator or the gain with respect to time. The principle of Q-switching is as follows. An optical shutter is attached inside the laser resonator as an additional component. When the shutter is closed, at first the laser generation cannot begin, and because of the pumping of the active medium the occupation inversion rises well over the laser threshold, if this is defined for the resonator without the additional loss. If the shutter is opened within a time that is short compared to the resonator lifetime T_g, the gain in the laser considerably exceeds the losses and the accumulated energy will be emitted in the form of a short and intense pulse. The temporal evolution of laser radiation under the use of Q-switching can, for example, be recognized in Fig. 2.22. In this figure the solution of the rate equations for the temporal change of the occupation number difference $\Delta N(t)$ and the photon flux density $I_L(t)$, which have been numerically obtained, are plotted (see e.g. [2.13]). As can be seen from the figure, the photon flux density $I_L(t)$ increases rapidly after the shutter has been opened. As a result of the growing intensity in the resonator the occupation number inversion $\Delta N(t)$ is simultaneously diminished due to the depletion of the active medium by stimulated emission. Thus, the photon flux density reaches its maximum value and falls again quickly because of the steadily decreasing occupation number difference $\Delta N(t)$. The result of the fast opening of the shutter is thus a rapid increase in the radiation power which then sinks again rapidly due to the depletion of the gain, i.e., the radiation is emitted in the form of a short and intense pulse. The duration of the pulse is longer than the resonator lifetime $T_{\mathscr{E}}$, which in its turn is at least several resonator round trip times u. The minimum pulse durations thus last, e.g., some ten nanoseconds in Q-switched ruby or Nd:YAG lasers of typical resonator length ($L \simeq 1$ m). As mentioned in 2.4.4, with high-gain semiconductors and dyes it is possible to construct lasers of very short cavity length. Thus resonator round trip times below 1 ps become attainable, and as a consequence one can generate picosecond light pulses with semiconductor lasers and dye lasers by gain switching or loss switching.

Various devices can be used as optical shutters. Very fast electronically controlled shutters can, for example, be constructed using electrooptical and acoustooptical modulators, which are described in section 4.3.1 (for details, see e.g. [4]). Shutters can also be built up in a purely mechanical way by using rotating mirrors or prisms, where given a typical resonator length, a rotational frequency of some 100 Hz must be applied. Besides Q-switching, switching of the gain, which is especially applicable with semiconductor lasers, can be employed. By switching or modulating the injection current the electrically pumped gain undergoes a temporal change (compare 7.4).

2.6 The Principle for Generating Ultrashort Light Pulses: The Modelocking Technique

In our description of lasers we have up to now not considered the way in which ordinary lasers differ from lasers for ultrashort light pulses or by which mechanisms such extremely short pulses can be generated. Before we treat in detail various methods for generating ultrashort light pulses in chapters 4. to 7., in this section and the next section the general principle of their generation will be discussed.

The origin of the generation of ultrashort light pulses in laser resonators is connected with the previously mentioned fact that in laser materials having relatively large bandwidths of the laser transition very many longitudinal eigen modes can be simultaneously excited. The total field strength

$$E(t) = \sum_m \frac{1}{2} \hat{E}_\mathrm{m} \, \mathrm{e}^{\mathrm{i}\varphi_m + \mathrm{i}(\omega_0 + m\delta\omega)t} + \mathrm{C.C.} \tag{2.86}$$

of the laser radiation results as the superposition of the field strengths E_m of M single longitudinal modes, where the summation runs from $m = -(M-1)/2$ to $m = (M-1)/2$, and $\delta\nu = \delta\omega/2\pi = c/2L$ is the mode spacing frequency which we have here assumed to be constant across the total oscillation range. This assumption is always satisfied, if we can neglect the dispersion of the samples, which influences the optical resonator length L. Depending on the properties of the active material and the resonator the phases φ_m of the various modes can be statistically dependent or statistically independent of one another. First the case of statistical independence will be considered. (This case is met, when independent gain reservoirs are available to the individual modes, e.g. with inhomogeneously broadened transitions.) Then for the total intensity we obtain

$$I \sim \overline{E(t)^2} = \sum_m \frac{1}{2} |\hat{E}_m|^2 . \tag{2.87}$$

This means that in the case of statistically independent, fluctuating phases φ_m, the total intensity can be represented as the sum of the intensities of the individual modes. In Fig. 2.23, the temporal structure of one stochastical realization of such multimode radiation within the laser resonator is depicted. In the frequency domain, this radiation consists of a large number of discrete spectral lines separated by frequency intervals of the value $c/2L$. Each mode oscillates independently of the others, and their phases are distributed stochastically in the interval $-\pi$ to π. In the time domain, the field consists of an intensity distribution that has the characteristic properties of Gaussian noise.

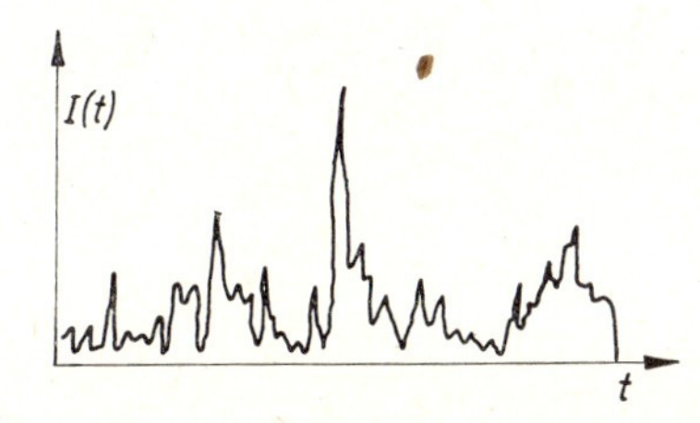

Fig. 2.23. Intensity of a multimode laser in the time domain

If, however, one succeeds in producing by a suitable mechanism steady relations among the phases of the various modes, then very interesting phenomena emerge, which qualitatively deviate from the above situation. The intensity of the output radiation depends then on time as a well-defined function. This operational regime of lasers is designated as modelocked.

A qualitative description of modelocked lasers is given by the following elementary consideration. Let us assume that all modes possess approximately the same amplitude E_0 and that between their phases the "synchronization condition" (or "phase locking condition")

$$\varphi_m - \varphi_{m-1} = \alpha = \text{const} \tag{2.88}$$

is fulfilled. Then the term $(m\alpha + \varphi_0)$ can be substituted in (2.86) for φ_m. The summation can be carried out analytically and we obtain

$$E(t) = \hat{E}_0 \frac{\sin\left[\frac{M}{2}(\delta\omega t + \alpha)\right]}{\sin\left[\frac{\delta\omega t + \alpha}{2}\right]} e^{i(\omega_0 t + \varphi_0)} + \text{C.C.} \tag{2.89}$$

In Fig. 2.24 the time dependence of the intensity of the output radiation is depicted for seven modes ($M = 7$). Due to the locking of the phases the modes in the resonator interfere so that the laser emits radiation in the form of short light pulses. The peaks

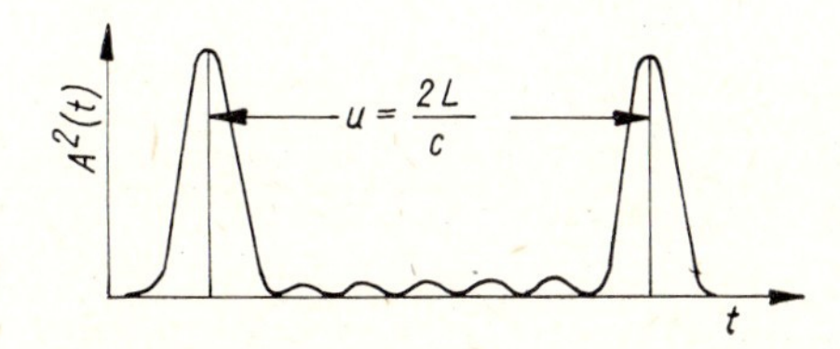

Fig. 2.24. Time dependence of the intensity under the generation of seven modes with synchronized phases and equal amplitudes

of the pulses appear at those moments in which the denominator in (2.89) vanishes, i.e., when $(\delta\omega t_q + \alpha)/2 = q$ holds, where q is an integer. The concept of modelocking is to be understood such that, at these times t_q, every mode provides a maximum contribution to the total field strength. The time interval u between two neighboring pulse peaks follows from (2.89) and is determined by

$$u = \frac{2\pi}{\delta\omega} = \frac{2L}{c}. \tag{2.90}$$

This is precisely the time that is necessary for a complete round trip in the resonator. Thus, inside the resonators of ideally modelocked lasers, only one pulse moving steadily forward and back is to be found. The pulse duration τ_L can likewise be evaluated from (2.89), providing us with

$$\tau_L = \frac{2\pi}{M\delta\omega} = \frac{2\pi}{\Delta\omega_{gen}}, \tag{2.91}$$

where $\Delta\omega_{gen}$ is the frequency interval in which laser modes are excited. Under intense pumping, $\Delta\omega_{gen}$ can attain values of the order of the linewidth of the laser transition $\Delta\omega_{21}$. Thus, the larger the spectral linewidth of the laser transition is and the more modes exceed the oscillation threshold, the shorter the light pulses that can be generated. A lower boundary, below which the pulse duration cannot significantly fall, is given by the reciprocal value of the linewidth. Thus it is evident that it is not possible with typical low-pressure gas lasers to generate pulse lengths in the picosecond range. With solid-state lasers, however, the minimum value of the pulse length lies in the order of 1 ps and with dye lasers even more than one order of magnitude below that value.

The pulses of modelocked lasers are distinguished not only by ultrashort pulse durations but also by high peak intensities. The maximum intensity according to (2.89) is proportional to $M^2 |\hat{E}_0|^2$, whereas for the case of a non-modelocked laser, according to (2.87), an expectation value for the peak intensity proportional to $M |\hat{E}_0|^2$ results. Thus, given in both cases the same number of modes, M, the peak intensity in the case of modelocking is larger by the factor M than the expectation value of the intensity under generation of laser radiation with stochastic relations between the phases of the individual modes.

The properties of ultrashort laser pulses are described by the relation (2.89) only in a very idealized form. In general the temporal structure of an optical pulse is completely determined by the modulus $|\hat{E}_L(t)|$ of the field strength amplitude (or the intensity $I_L(t)$) and its phase $\varphi_L(t)$. In most cases the temporal intensity profile $I_L(t)$ and in the frequency domain its spectral intensity $\tilde{I}_L(\omega)$ are measured, where due to the dependence of the field strength on the phase $\varphi_L(t)$ there is no unambiguous relation between these two quantities. Between the halfwidths τ_L and $\Delta\nu_L$, for instance, only an inequality

$$\Delta\nu_L \tau_L \geqq C_B \tag{2.92}$$

can be given, where C_B is a number factor of the order of one, whose precise value is determined by the temporal shape of the pulse under consideration. The shortest pulse, which can be obtained with given spectral halfwidth $\Delta\nu_L$ is said to be bandwidth-limited (or Fourier-limited), and its pulse duration is given by $\tau_L = C_B/\Delta\nu_L$.

Let us for example, consider a (general) Gaussian pulse with the field strength amplitude

$$\hat{E}(t) = \hat{E}_0 \exp(-\gamma t^2 + i\beta t^2) \tag{2.93}$$

that changes slowly with time.

The constant parameter γ describes here the envelope of the pulse and is connected to the temporal halfwidth of the radiation power of the pulse by the relation

$$\tau_L = \left(\frac{2 \ln 2}{\gamma}\right)^{1/2}. \tag{2.94}$$

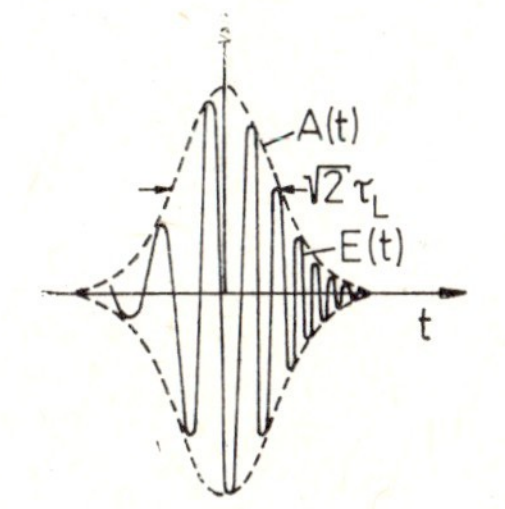

Fig. 2.25. Phase modulated pulse (chirped pulse)
The frequency in the leading edge of the pulse is larger than in the trailing edge — down chirp. The reverse behavior, i.e. increase of the frequency with time, is called up chirp.

The term $i\beta t^2$ describes a special kind of phase modulation, namely a linear frequency shift or frequency sweep across the pulse. The frequency sweep per unit time, i.e. the temporal ascent, which amounts here to 2β, is also called a chirp. Such a temporal change of the instantaneous frequency of a down chirped pulse is represented schematically in Fig. 2.25. Under actual experimental conditions, such phase modulation can arise due to dispersion and optical nonlinearity of matter (compare 8.3). Taking the Fourier transform $\underset{\sim}{E}(\omega)$ of $E(t)$ from (2.93) and calculating $\tilde{I}(\omega)$, which is proportional to $|\underset{\sim}{E}(\omega)|^2$, we obtain

$$\Delta\nu_L = \frac{1}{\pi}\left[2 \ln 2 \left(\frac{\gamma^2 + \beta^2}{\gamma}\right)\right]^{1/2} \tag{2.95}$$

for the halfwidth of the spectral profile $\tilde{I}_L(\omega)$ of the intensity. Hence, in this case, the pulse-duration-bandwidth-product is given by

$$\tau_L \Delta\nu_L = \frac{2 \ln 2}{\pi}\left[1 + \left(\frac{\beta}{\gamma}\right)^2\right]^{1/2}. \tag{2.96}$$

For the special case, where $\beta = 0$ (no phase modulation), $C_B = 0.441$ results for this product. From (2.96) it is evident that under fast phase modulation, i.e. for $\beta/\gamma \gg 1$, the pulse-duration-bandwidth-product in (2.92) can be considerably greater than 1.

For the temporal shape of pulse intensity that is represented by the squared secant-hyperbolic function $(\cosh 1.76t/\tau_L)^{-2}$ used sometimes to describe the pulse shape of modelocked dye lasers, the value 0.315 results for the parameter C_B.

2.7 Methods of Modelocking

To obtain experimentally optimal use from the effect of modelocking, one has to excite as many longitudinal modes with constant phase differences as possible in the resonator containing a laser-active medium with broad spectral gain profile. For this purpose, there are various methods that can be used. These methods are described in detail in chapters 4. to 7. In this section, we want only to mention and briefly characterize the most important methods used to modelock lasers.

2.7.1 *Active Modelocking*

The method of active modelocking using a periodic modulation of a resonator parameter is founded upon the fact that inside the resonator a modulator, which is controlled by an outer signal, alters the losses of the resonator (or another essential resonator parameter, e.g. the optical path length) periodically in time with a certain modulation frequency. If we choose the modulation frequency such that it is the same as the frequency of the spacing between adjacent longitudinal modes then, due to the modulation, for each mode side bands are generated whose frequency coincides with the frequencies of the two neighboring modes. As a result of the interaction, which thus arises between the modes, all modes will be synchronized at a sufficiently strong modulation. For active modelocking acoustooptical or electrooptical modulators are mostly used.

2.7.2 *Synchronous Pumping*

Modelocking can be achieved not only by periodic modulation of the losses, but also by periodic modulation of the gain. This can be achieved by pumping a laser with the pulse train of another, e.g. actively modelocked laser. The advantage of this method is that, with periodic pumping, we can obtain pulse lengths which are considerably shorter than those of the pump pulses. In the case of synchronously pumped dye lasers, we can furthermore continuously tune the frequency of the ultrashort light pulses in certain spectral ranges.

2.7.3 *Passive Modelocking*

An effective method of generating ultrashort light pulses is by so-called passive modelocking, in which a saturable absorber is placed inside the laser resonator in addition to the other laser components. A saturable absorber consists of matter that possesses an absorption transition whose cross section should be rather high at the laser frequency. Organic dyes in solution are particularly suited for such purposes. If a light pulse of the laser irradiates this kind of absorber, absorber molecules become excited and the incoming radiation field is partially absorbed. If we consider, for example, the change in the level occupation in a two-level system under the influence of a radiation field, we obtain with quasi-stationary excitation (i.e. $\tau_L \gg T_{21}$) from (1.22) and (1.23) the relation

$$\Delta N = \frac{N}{1 + I/I_S} \tag{2.97}$$

for the occupation inversion $\Delta N = N_1 - N_2$ where $I_S = 1/2\sigma_{21}T_{21}$ is the saturation intensity of the absorber. With increasing intensity, the occupation inversion ΔN, which according to (1.21) is responsible for the absorption of radiation, decreases monotonously. If the intensity becomes large compared to the saturation intensity I_S of the absorber, no substantial absorption can take place — the absorber is saturated, and thus the radiation field is not attenuated any more.

Similar appearances also occur under nonstationary excitation, in which the time derivatives in equations (1.22), (1.23) cannot be neglected. If we consider, for instance, the nonstationary case, in which the relaxation time T_{21} is large compared to the pulse duration $\tau_L (T_{12} \gg \tau_L)$, we obtain

$$\Delta N(t) = N \exp \left\{ -2\sigma \int_{-\infty}^{t} I_L(t')\, dt' \right\} \tag{2.98}$$

from (1.22) and (1.23). In this case the absorption is decreased with increasing pulse energy, from which an asymmetry in the time behavior follows. Whereas the leading edge of the pulse is markedly attenuated, since at these times the pulse energy is still small and consequently the absorption is not yet saturated, at later times the trailing edge can pass through the absorber nearly unattenuated because of the absorption saturation.[1])

For various purposes of application optically pumped solid state lasers and dye lasers have proved especially suitable for passive modelocking. These two laser types distinguish themselves considerably not only in reference to the properties of the generated pulses, but also in the mechanism of the generation processes. The passive modelocking of dye lasers is characterized by the fact that the relaxation time of the laser dye is of the order of the resonator round trip time and, likewise, the relaxation time of the absorber dye is large compared to the duration of the light pulses in the steady-state regime of the continuously pumped laser. These conditions enable the gain depletion to play an important part in the generation and formation of the pulses. Through the combined effect of the saturable absorber, which suppresses the leading edge of the pulses, and the amplifier, which suppresses the trailing edge of the pulses, a laser regime is possible in which the build-up of ultrashort light pulses occurs. The passive modelocking of typical optically pumped solid state lasers in contrast to that of dye lasers is characterized by the fact that the relaxation time of the amplifier is very large compared to the resonator round trip time. The build-up of ultrashort light pulses is possible under this condition owing to the effect of a fast relaxing saturable absorber that favours the most intense fluctuation peak with respect to the noise background. This fluctuation peak experiences net gain on every round trip, and it is the only one which survives and attains high peak power whereas all other, less intense fluctuation peaks are finally suppressed.

---

[1]) Actual saturable absorbers can be better described in general by a three-level system, where level 3 represents an excited vibrational level of the S_1 state (level 2) (see Fig. 6.4.). Due to the fast relaxation time T^{b}_{32} the occupation of level 3 can be neglected ($N_3 \approx 0$). By a simple calculation it can be shown that for this kind of absorber relations similar to (2.97) and (2.98) hold, where 2σ must be replaced solely by σ.

2.8 Distributed Feedback Dye Laser

Until now we have dealt with laser resonators in which the feedback of the radiation at the ends of the resonator is produced using mirrors. In contrast to these, distributed fedback dye lasers do not possess any external feedback components. Rather, the feedback is generated in the active medium itself.

This feedback principle can be employed in solids, liquids, gases and semi-conductors. For the generation of ultrashort light pulses as well as in regard to the possibility of continuous tunability of the wavelength, distributed feedback dye lasers — abbreviated DFDL — are of particular interest.

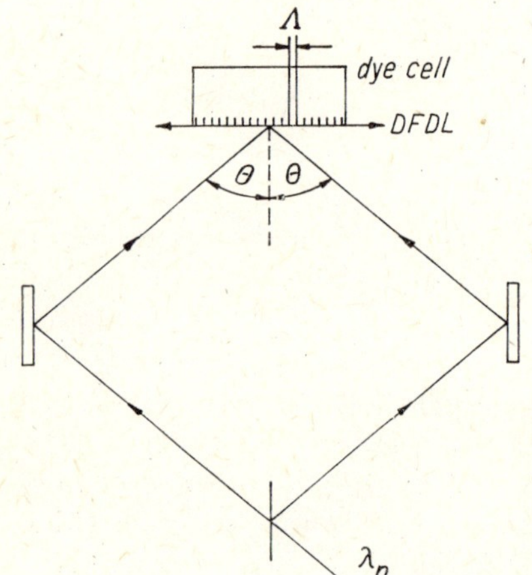

Fig. 2.26. Scheme of a distributed feedback dye laser (DFDL)

λ_P is the pump radiation, Λ is the fringe separation and DFDL represents the generated laser light.

Let us illustrate the principle of distributed feedback using Fig. 2.26. The coherent light of the pump laser is divided by a beam splitter into two partial beams that strike the dye cell at angle θ and interfere there. The interference fringe pattern of the pump light thus generated produces a spatially periodic modulation of the amplification and refractive index of the laser medium. As Kogelnik and Shank showed in 1971 [2.18], at this spatial structure a Bragg reflection of the light wave appears, which effects the feedback. The development of a standing light wave in this laser resonator during the duration of the pump pulse can be understood as follows: due to spontaneous emission two weak oppositely travelling waves arise. The waves are amplified in the inverted ranges of the interference pattern and coupled with one another by the Bragg reflection, so that through their superposition an increasing standing wave develops. The process depends strongly on the spatial direction, because the interference pattern generated from the pump light extends along the longitudinal axis of the laser medium. Due to the Bragg condition the feedback is highly wavelength-selective. The generated laser wavelength is given by

$$\lambda_D = 2 n_L \Lambda \tag{2.99}$$

where n_L is the refractive index of the solution at λ_D and Λ is the fringe separation. This fringe separation is

$$\Lambda = \frac{\lambda_P}{2 \sin \theta}; \tag{2.100}$$

λ_P is the pump wavelength and θ the angle at which the two partial beams superpose.

As pump laser for the DFDL the employment of nitrogen and excimer lasers is attractive, because these lasers are sturdy and particularly easy to handle. Tuning of the laser wavelength can be achieved by varying n_L (by adjusting the solvent mixture) or by changing the pressure in the dye solution. A broader tuning range is obtained by varying the interference fringe spacing Λ, which according to (2.100) can be achieved by changing the incident angle θ. This can be realized by rotating oppositely the two folding mirrors around vertical axes.

With the excitation of the DFDL laser with nano- and subnanosecond pulses a nonsteady behavior of the laser results, which is characterized by relaxation oscillations that lead to the formation of a certain number of short laser pulses. Theoretical and experimental investigations of the temporal behavior of DFDL lasers were conducted in particular in [2.19]—[2.22].

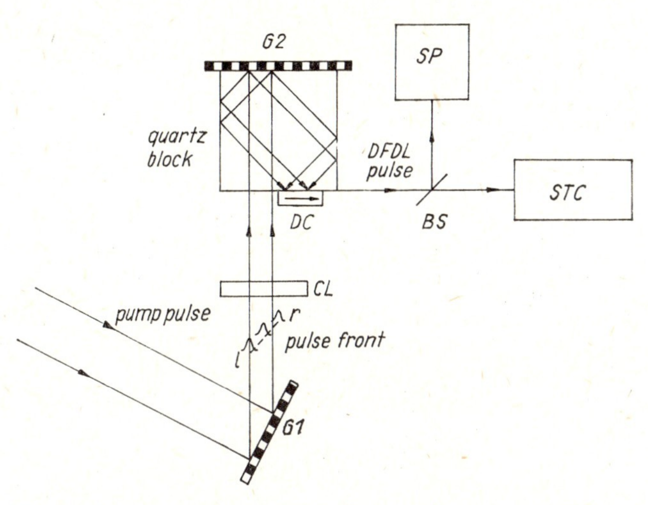

Fig. 2.27. Experimental set-up of a travelling-wave DFDL

A holographic grating (G 1) produces the delayed wave front which is focussed by means of a cylindrical lens (CL) into the dye cell (DC) after it is divided by the holographic grating G 2. After the beam splitter (BS) the DFDL-pulses can be measured simultaneously by a spectrograph (SP) and a 2ps streak camera (STC) (from [2.21]).

These investigations showed that between the laser threshold and the threshold for the occurrence of a second pulse a single short pulse develops. Just below the threshold of the second pulse, its duration is about 50 times shorter than that of the pump pulse. In this manner, DFDL single pulses of 70 ps duration can be generated, for instance of a nitrogen laser that produces pulses of 3.5 ns duration is employed as pump source. With the excitation of a DFDL by the 2. and 3. harmonic, respectively, of a modelocked

Nd : YAG laser (yielding pulses of 16 ps duration) 1.6 ps pulses could be generated [2.21]. Even shorter pulses can be achieved by means of travelling wave excitation of the DFDL using two holographic diffraction gratings [2.22]. If an ultrashort light pulse having a duration of several picoseconds is incident on such a grating, it is diffracted according to its wavelength. At every grating groove, a spatial delay of one wavelength arises with respect to the direction of maximum intensity. If N grating ranges are illuminated by the incident beam, then the temporal delay along the pulse front comes to $\Delta t = N \frac{\lambda}{c}$. If, before the diffraction at the grating, the pulse front was perpendicular to the wave vector, then afterwards they include an angle γ which is given by $\tan \gamma = \lambda \frac{\mathrm{d}\beta}{\mathrm{d}\lambda}$ ($\frac{\mathrm{d}\beta}{\mathrm{d}\lambda}$ is the angular dispersion of the grating). If a pulse front of this kind strikes the DFDL cell in the way depicted in Fig. 2.27, then the superposition of the two partial beams in the dye cell leads again to interference fringes as in the usual DFDL. Here, the fringes are stationary, but their envelopes move from left to right through the cell with the velocity $v = c/\tan \gamma$. This pump light travelling wave now produces a travelling wave in the DFDL as well, which passes through the dye solution with a velocity v'. If both waves travel synchronously (i.e. $v = v'$), then the angle must satisfy the condition $\tan \gamma = n_{\mathrm{L}}$. This can be achieved by changing the orientation of the delay grating or adjusting the refractive index by varying the solvent mixture. Experiments using this method provided pulses with minimum duration of about 1 ps, whereby single pulses were also generated far above the threshold.

A further pulse shortening was achieved in [2.24] using a new achromatic DFDL arrangement that led to improvements with respect to some characteristics of the pumped region, i.e. to perfect, high visibility interference fringes and to a small-size active volume. DFDL pulses as short as 350 fs were obtained at 610 nm with this setup where the laser was pumped with the 8 ps pulses from an excimer-laser-pumped cascade dye laser, in which gain depletion by stimulated emission is mainly responsible for pulse shaping [2.25].

3. Methods of Measurement

3.1 Basic Concepts for the Measurement of Fast Processes

As early as 1740 Segner determined that our eyes can only analyse events that are separated by more than 150 ms. Therefore, in high speed physics we must always make use of certain auxiliary techniques which allow signals that occur over short time intervals to be separated and recorded so that at a certain time during or after the measurement we can examine them simultaneously. With this aim, methods of measurement were developed long ago in physics; the fundamental principles of which can also be applied today in the measurement of ultrashort light pulses.

3.1.1 Streak Technique

With the streak technique, the temporal sequence of signals is transformed into a spatial one. Segner applied this principle in the measurement of the time resolution of the human eye. On this basis, Wheatstone built a mechanical streak camera with a rotating mirror (see Fig. 3.1) in 1834 to investigate spark discharges in the microsecond

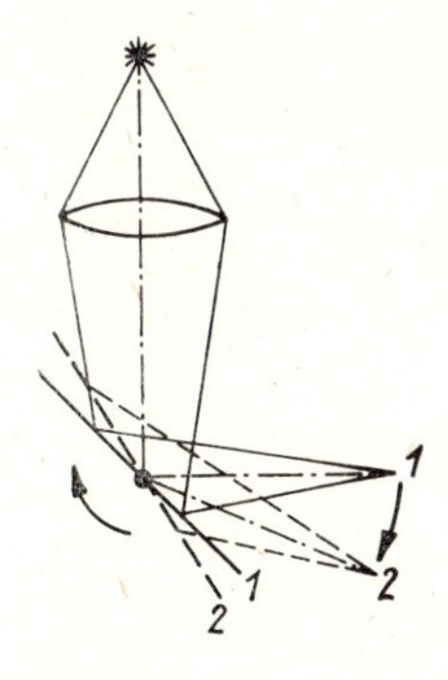

Fig. 3.1. Mechanical streak camera with rotating mirror
A spark is imaged at time t_1 (mirror position 1) at point 1, and at time t_2 (mirror position 2) at point 2.

range. The same fundamental principle is used in oscilloscopes, in which an electron beam is deflected linearly in time across the display screen and the signal is shown either as a deflection of the beam perpendicular to the time axis or by its intensity. The electro-optical streak camera functions similarly. Here a signal point (or a one-dimensional variety of signal points) is displayed on a photocathode. By means of electron optics photoelectrons transfer the signal to a display screen. If the electron beam is deflected very fast and linearly in time across the screen, we obtain on the screen the signal (or the signals) versus time.

3.1.2 *Stroboscopy, Sampling Technique*

Given a series of identical signals, the temporal profile of the signal can be measured with a "shutter" having a short opening time by varying the moment of opening with respect to the signal. Now the signal receiver, e.g. the human eye or a photodetector, is not required to have a very high time resolution, because it must only have "stored" the recorded value and be ready again for reception before the appearance of the next

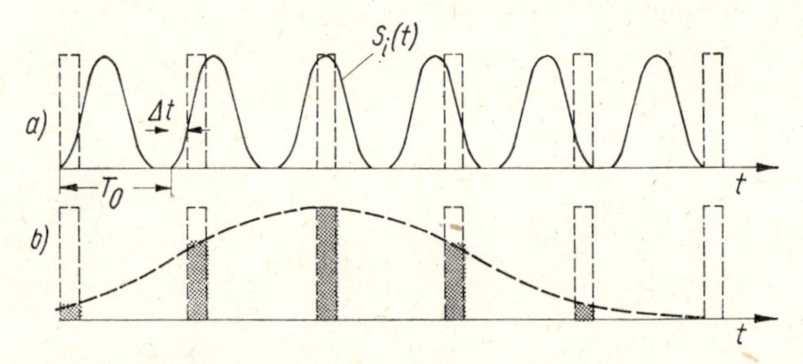

Fig. 3.2. Schematic representation of the sampling principle

a) The periodically occuring signal $S_i(t)$ with period T_0 is recorded with a shutter that has a short opening time and a subsequent detector that integrates over the shutter time. The moment of opening is (relative to the signal) changed by Δt from signal to signal.

b) The measured values appear with time interval $T_0 + \Delta t$ at the detector. Recorded with respect to time they produce a graph of the signal that appears to be expanded by a factor of $T_0/\Delta t$.

signal (see Fig. 3.2). The method of measurement designated as stroboscopy or sampling technique can also be applied in order to "photograph" a repeating process at various moments with light flashes of short duration, where now, of course, the use of a fast "shutter" in front of the receiver is not necessary. Besides mechanical shutters, in "high speed photography" electrooptical switches are employed. These switches make use of the Pockels effect and the Kerr effect, which are founded upon the field induced birefringence in crystals and liquids, respectively (see Fig. 3.3). Since these switches do not contain mechanically moving components and the switching time is determined only by the rise time of the field, higher time resolution can be attained than with mechanical shutters.

Extremely short switching times are obtained with devices in which the short field strength pulse is supplied by a light pulse that affects the orientation of the molecules in a Kerr cell. In this case, we speak of the light induced or optical Kerr effect (compare 3.3.4). In modified form the sampling technique can also be used when only one unrepeatable signal process (a "single-shot" signal) is under consideration. We can either produce from this single signal a series of signals at suitable intervals by means of a delay line or, if that is not possible, we can photograph the single event with several appropriately synchronized "shutters" or "light flashes", at various moments.

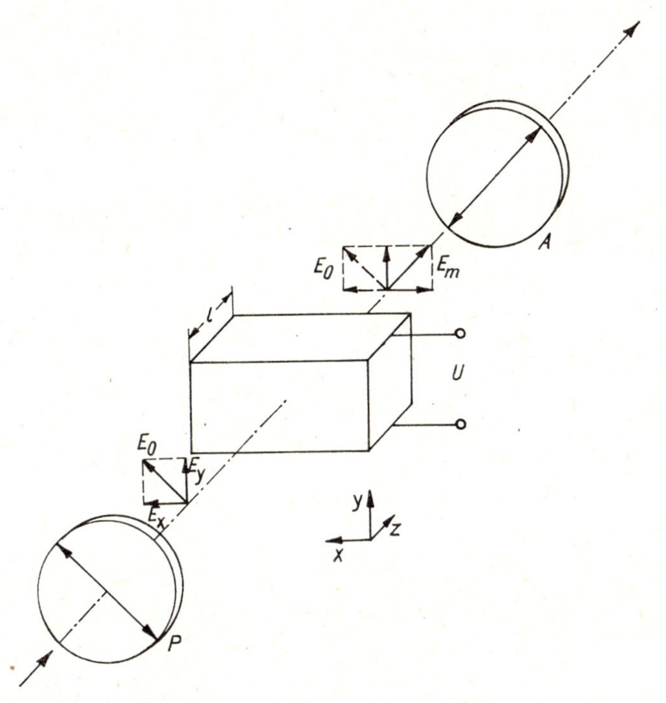

Fig. 3.3. Electrooptical switch

The polarization direction of the incident light is determined by a polarizer P. The preferred direction of the analyzer A lies perpendicular to that of the polarizer. If no voltage is applied to the electrooptical material, the material is optically isotropic and the switch allows no light to pass through. In an electric field (as a result of the applied voltage U) the material becomes birefringent; i.e. the light components polarized parallel and perpendicular to the field have different refractive indices and thus different propagation velocities whereby, in general, elliptically polarized light results which is not completely blocked by the analyzer. The transmission is given by

$$T(t) = \sin^2 \frac{1}{2} \varphi(t)$$

where $\varphi(t) = 2\pi \Delta n(t)\, l/\lambda_0$ ($\Delta n = n_\parallel - n_\perp$, λ_0 is the vacuum wavelength, l is the sample length). From this we can calculate the time integral over a transmitted test pulse of intensity $I_T(t + \tau)$ as a function of the input intensity $I_0(t + \tau)$. Thus, we have

$$\int_{-\infty}^{\infty} dt I_T(t + \tau) = \int_{-\infty}^{\infty} dt\, I_0(t + \tau) \sin^2 \frac{1}{2} \varphi(t),$$

which for $\varphi(t) \ll 1$ leads to

$$\int_{-\infty}^{\infty} dt I_T(t + \tau) \approx \left(\frac{\pi l}{\lambda_0}\right)^2 \int_{-\infty}^{\infty} dt I_0(t + \tau)\, [\Delta n(t)]^2$$

(The maximum transmission of the shutter is attained at $\Delta n l/\lambda_0 = 1/2$, whereby at the output of the electrooptical material the electric field E_m is again linearly polarized. Its polarization direction is, however, rotated 90° with respect to the input signal.) When the applied electric field alters as a pulse the device functions as an electrooptical shutter.

A special advantage of optical shutters in connection with the sampling technique is that by using such shutters we can trace the temporal evolution not only of one signal point but of a two-dimensional variety of signal points, i.e. of a complete picture.

3.1.3 *Measurement of Time Intervals as Spatial Displacements*

A very precise method of changing the opening time of "shutters" or the ignition of "illuminating flashes" relative to the signal uses the transformation of the temporal change Δt to a spatial displacement Δl, where the finite propagation velocity of the signal or a subsidiary radiation, e.g. the illuminating flash, is utilized. If the propagation velocity is given by the velocity of light in free space, then, for example, the following correspondences hold: 1 ns $\triangleq$ 300 mm, 1 ps $\triangleq$ 0.3 mm. We see that for processes on the nanosecond, picosecond and femtosecond time scale the method of light delay leads to easy to handle measuring ranges of displacements in the laboratory. Historically, this method has developed — in the sense of a reversal — out of the classical measurement techniques for the velocity of light according to Fizeau and Foucault.

3.1.4 *Signal Conversion*

Frequently with time resolving measurements the original signal is converted into another form at a very early stage of the measuring process, e.g. before the streak principle or the sampling technique is applied. Of particular importance is the conversion of radiation into an electrical signal in photoelectric receivers or at the cathode of an image converter. After suitable (broadband) amplification the signal from the photoelectric receiver may enter an oscilloscope, which then puts the streak technique into action, and which, if necessary, may also put into practice the sampling technique to record the signal. Instead of the oscilloscope, fast transient recorders can be used to store the photoelectric signal.

Besides the conversion into electrical signals, the conversion of the radiation signal into light at another wavelength by means of nonlinear optical methods also plays a part in the measurement technique of ultrashort light pulses (see 8.).

3.1.5 *Correlation Methods*

Correlation methods, particularly in electronics, have already been used successfully for many years. Here, the signal $S(t)$ is convoluted with itself or with a subsidiary signal $H(t)$ (see Fig. 3.4.). The corresponding auto- or cross-correlation function

$$G(\tau) = \int_{-\infty}^{\infty} \mathrm{d}t S(t)\, S(t-\tau)$$

or (3.1)

$$K(\tau) = \int_{-\infty}^{\infty} \mathrm{d}t S(t)\, H(t-\tau)$$

can be measured by temporal integration, where the integral depends on the mutual delay τ between the two signals. (In 3.1 reference was made to the correlation function of the lowest, i.e. the second order, in the signal quantities; where differentiation is necessary, we will also provide the order. Thus, for example, we may write $G^{(2)}(\tau)$ for the autocorrelation function in (3.1).) Under certain conditions we can calculate the temporal evolution of $S(t)$ from $G(\tau)$ or $K(\tau)$. If we know, for example, that the signal is Gaussian shaped according to

$$S(t) = S_0 \exp\{-4(\ln 2)\,(t/\tau_S)^2\} \tag{3.2}$$

we can determine the halfwidth τ_S of the signal from the halfwidth τ_A of the similar Gaussian shaped autocorrelation function. Thus, we have

$$\tau_S = \frac{1}{\sqrt{2}}\,\tau_A \tag{3.3}$$

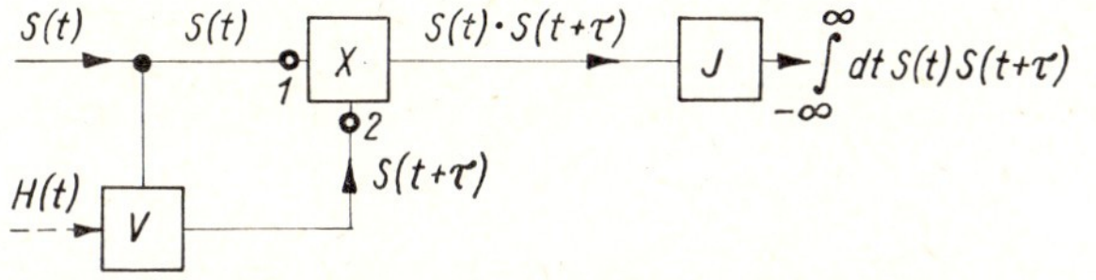

Fig. 3.4. Scheme of an arrangement for measuring a correlation function
The signal $S(t)$ is given at the input 1 of a multiplier, which represents a special nonlinear element. At the second input either the time delayed signal $S(t + \tau)$ or the auxilliary signal $H(t + \tau)$ is applied. After the multiplication of the two input quantities the product is temporally integrated.

The sampling technique described earlier (in 3.1.2) is obviously a special case of the recording of cross-correlation functions. In such measurements the temporally varying transmission of the fast shutter or the temporally varying intensity of the illuminating flash provide the subsidiary signal $H(t)$. If $H(t)$ corresponds to a very short pulse (and accordingly $H(t)$ can be approximately represented by a δ-function), then the relation $K(\tau) = S(\tau)$ follows from (3.1) under the normalization of the integral of $H(t)$ to 1.

A two-beam interferometer with the following photoelectric receiver represents the simplest optical autocorrelator (see Fig. 3.5). With this autocorrelator the field strength autocorrelation function $G_E{}^{(2)}(\tau)$ and, by applying an additional Fourier transformation to $G_E{}^{(2)}(\tau)$, the spectral power density, which is proportional to $|\underline{E}(\omega)|^2$, can be measured. From this measurement the bandwidth of the incident radiation can be specifically determined. Since in this process the phase information gets lost, just as in other spectrometers (we are dealing here with a Fourier spectrometer), the temporal evolution of $E(t)$, and in particular parameters such as the pulse duration characterizing the temporal profile of pulses, cannot be calculated unambiguously. Determination of the pulse duration from the spectral bandwidth is only then possible, when we already know that the light pulse is bandwidth-limited, or Fourier-limited, i.e., when the phase of the field strength does not change during the pulse, and accordingly the spectral width is only determined by the profile of the amplitude (compare 2.6.). With bandwidth-limited pulses

$$\frac{\Delta\omega}{2\pi}\,\tau_L = C_B\,, \tag{3.4}$$

follows from (2.92) for the halfwidth $\Delta\omega$ of the spectral power density and the pulse duration τ_L, which is the full width at half maximum (FWHM) of the pulse intensity in the time domain, where C_B is a constant that depends on the pulse shape. The quantity $\Delta\omega\tau_L/2\pi$ is designated as the bandwidth-pulse-duration-product. Generally, i.e. for pulses that are not bandwidth-limited, this product assumes values that are greater than C_B. If we measure pulse duration and bandwidth independently, we can therefore reciprocally apply this product to the assessment of the pulse quality, i.e., from the deviation of the product from C_B we infer the deviation of the pulse from the bandwidth limitation.

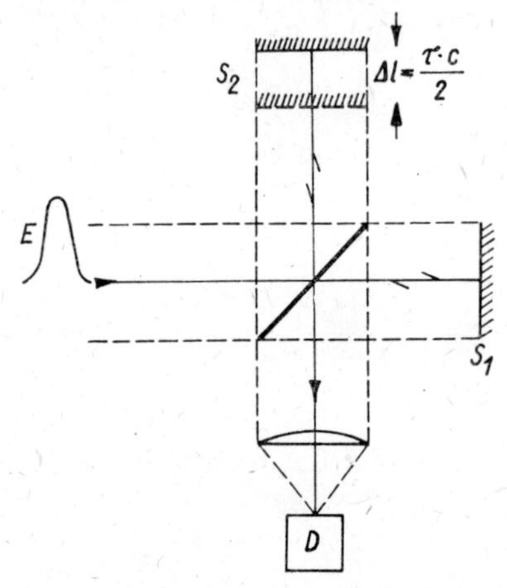

Fig. 3.5. Two beam interferometer of the Michelson type as an optical field strength correlator
At the detector D the field strengths $E_1(t) \propto E(t)$ and $E_2 \propto E(t + \tau)$ of the input pulses reflected on mirrors S_1 and S_2 respectively, are superimposed, whereby the delay τ from light pulse to light pulse can be controlled by a translation of one of the mirrors. If the integration time of D is sufficiently great, the output signal is proportional to the energy that enters the detector. This energy is proportional to

$$\int_{-\infty}^{\infty} \mathrm{d}t \, (E_1 + E_2)^2 \sim [G_E(0) + G_E(\tau)],$$

whereby

$$G_E(\tau) = \int_{-\infty}^{\infty} \mathrm{d}t \, E(t)E(t + \tau) = \frac{1}{2\pi} \int_{-\infty}^{\infty} \mathrm{d}\omega \, |E(\omega)|^2 \, \mathrm{e}^{-\mathrm{i}\omega\tau}$$

is the field strength autocorrelation function.

We can obtain further information about light pulses by measuring intensity correlation functions (see e.g. [3.1]). With this aim, if the requirements for time resolution are not very high, we can produce the correlation signal in an electronic multiplicator at whose inputs the signals of two photoreceivers are given according to the basic scheme in Fig. 3.4. The multiplication can be executed by analogue or digital devices; the second type offers itself particularly if photon counting techniques are applied.

The variable delay τ between the two input signals is obtained either by changing the optical path or by using electronic means to delay the electric signal behind the photoelectric receiver. After the signal passes through the multiplier and the integrator the intensity auto-correlation function

$$G^{(2)}(\tau) = \int_{-\infty}^{\infty} \mathrm{d}t I(t) \, I(t + \tau)$$

is obtained. This auto-correlation function attains high values only for those values of the delay τ, for which the intensities $I(t)$ and $I(t + \tau)$ are simultaneously high. Under certain conditions which we will discuss later, we can therefore calculate the light pulse duration τ_L from the width τ_A of $G^{(2)}(\tau)$; e.g. for bandwidth-limited, Gaussian shaped pulses equation (3.3) holds, where $C_B \approx 0.44$.

Until now, the electronic formation of correlation functions has been limited to the temporal range above 100 ps. If we want to produce intensity correlation functions with a time resolution in the picosecond range, we must execute the multiplication by optical means, thus turn to methods of nonlinear optics. The interactions between radiation and matter, which are mostly utilized here, can be considered inertialess even in the femtosecond range and ensure a correspondingly high time resolution (see 3.3).

It should be noted that we have restricted ourselves here to intensity correlation functions of the lowest order (i.e. of the second order in intensity or of the fourth order in field strength). Further information — for example, about the asymmetry in the temporal profile of the pulse — can be obtained from the correlation functions of higher order (see e.g. [3.2, 3.3]).

3.2 Limits for the Time Resolution

Through their inertia and the memory effect resulting from it, which are characterized by a certain rise time, the physical components limit the temporal resolving power of the experimental equipment. The output $\tilde{S}$ of a component in response to a signal S is, in general, not identical with S, but is given by the convolution

$$\tilde{S}(t) = \int_{-\infty}^{\infty} \mathrm{d}t' S(t') \, B(t' - t) \tag{3.5}$$

of the signal with the response function $B(t)$ of the component, if restricting to linear systems. With physically real systems, for reasons of causality, $B(t' - t) = 0$ holds for $t' > t$. (Fig. 3.6 shows as an example a response function with low-pass character.) From (3.5) we see that the output signal only coincides with the signal if $B(t)$ is equal to the delta function, thus if no memory is present, or if $B(t)$ differs from zero only in time intervals so small that the signal function does not noticeably change. The simple relation

$$\underset{\sim}{\tilde{S}}(\omega) = \underset{\sim}{S}(\omega) \, \underset{\sim}{B}(\omega) \tag{3.6}$$

follows from the convolution (3.5) for the Fourier transforms $\underset{\sim}{\tilde{S}}(\omega)$, $\underset{\sim}{S}(\omega)$ and $\underset{\sim}{B}(\omega)$. The upper limiting frequency of a device, which is a characteristic parameter in the frequency domain, corresponds to the finite response time of the device in the time domain. Accordingly, the cut-off of the high signal frequencies by the device corresponds to the smoothing of signals changing rapidly with time. It is evident that, in principle, we can calculate the input signal from the recorded quantity $\tilde{\underset{\sim}{S}}(\omega)$ or $\tilde{S}(t)$ in a simple manner from (3.6). This reverse calculation however is only possible without error using noise-free recordings. With actual measurements, in addition to the signal, noise is always processed in the component under consideration. This can lead to the fact that under the reduction of the systematic error through this deconvolution procedure, which we carry out up to a high frequency, the total error as a function of this upper frequency increases again after passage through a minimum. For this reason, we must in each case examine up to which frequencies a deconvolution makes sense.

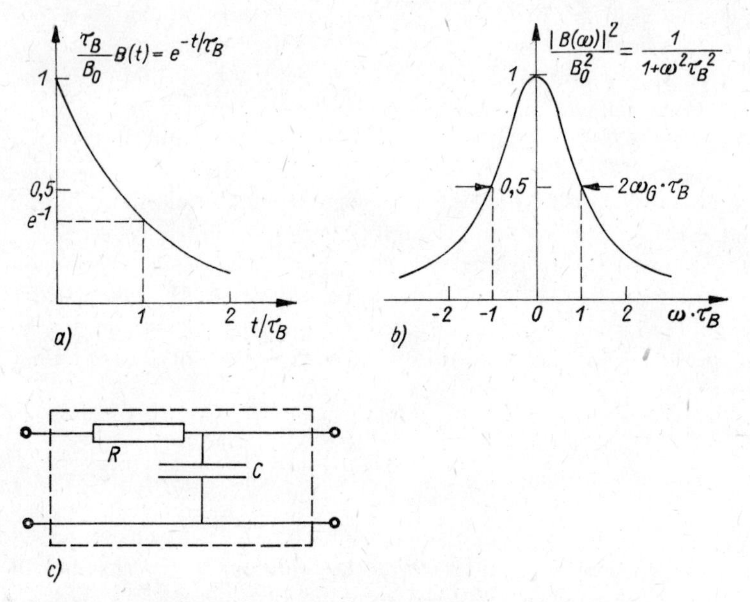

Fig. 3.6. Response function of a low-pass filter

a) Response B as a function of time

b) The square of the absolute value of the Fourier-transform of $B(t)$ (frequency response $\underset{\sim}{B}(\omega) = B_0(1 + i\omega\tau_B)^{-1}$)

c) RC-element as equivalent circuit diagram for a low-pass with an exponentially decreasing response function. The time constant of the low-pass is given by $\tau_B = RC$; the constant B_0 is equal to 1. Instead of by the time constant the component can also be characterized by the upper threshold frequency $\omega_G = 1/\tau_B$ at which $|\underset{\sim}{B}(\omega)|^2$ falls to one half of its maximum value.

The time resolution can, for example, be limited by the finite duration of the light pulse used, inaccuracies of the delay lines, dispersion effects, the finite deflection velocity in streak techniques, memory effects in light-matter interactions as well as by the response time of electronic and optoelectronic devices used for detecting, processing and storing the signal.

In the following, we will give some estimations for the time resolution of physical components.

3.2.1 *Rotating Mirrors*

As a simple example, let us investigate the mechanical streak camera with a rotating mirror arrangement which is outlined in Fig. 3.1. The minimum distance between distinguishable picture elements in the display surface is given by the diffraction of light at the rotating mirror.

With a distance r of the mirror edge from the axis, a distance L of the axis from the display surface and an angular velocity ω, we obtain $d \simeq \lambda L/r$ for this minimum distance and $\Delta t_A \simeq dr/vL \simeq \lambda/\omega r$ for the resolvable time interval, where $v = \omega r$ is the velocity of the mirror edge and λ the wavelength of light. Let us estimate the maximum possible velocity of the mirror edge, which is determined by the mechanical resistance of the material, such that at this limiting speed the additional kinetic energy $Mv^2/2$ of an atom resulting from the rotation achieves the value of the binding energy in solids which is given by $\mathscr{E}_B \simeq 1.6 \times 10^{-19}$ J $\triangleq$ 1 eV. Taking $M \simeq 5 \times 10^{-26}$ kg this estimation yields $v_{max} \approx 2.5 \times 10^3$ m/s.

A limiting resolution of the order of $\Delta t \simeq 10^{-9}$ s follows from the maximum possible velocity of the mirror edges and the minimum distance between distinguishable picture elements at wavelengths of about 1 μm. Thus, we cannot penetrate the picosecond range with mechanical streak cameras.

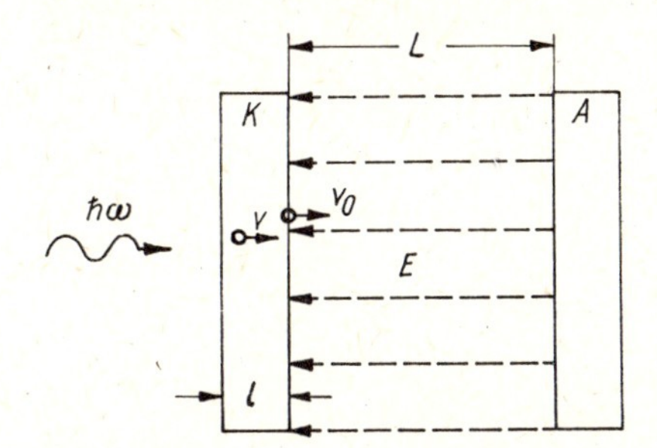

Fig. 3.7. Scheme of a photoelectric detector consisting of a cathode K and an anode A. v is the velocity of an electron within the cathode layer and v_0 the velocity of an electron at the exit surface.

The spread of the escape time of the electrons coming from the cathode of thickness l is of the order of $\Delta t_K = l/v$, where $v = \sqrt{m^{-1}\, 2\mathscr{E}_k}$ is a characteristic value for the velocity of the electrons within the cathode. This value is determined by the kinetic energy $\mathscr{E}_k$, which for typical cathode materials under excitation in the visible range is of the order of 1eV $\triangleq$ 1.6 $\times 10^{-19}$ J, from which the time spread can be estimated to be $\Delta t \simeq 10^{-14}$ s if $l \simeq 10$ nm. The transit time of an electron from the cathode to the anode is

$$t_L = \sqrt{2Lm/eE + v_0^2 m^2/e^2E^2} - v_0 m/eE,$$

which, with $mv_0^2/2eEL \ll 1$ leads to

$$\Delta t_L \simeq \Delta v_0 \cdot m/eE$$

for the transit time spread. This relation shows that the time resolution can be increased by strong accelerating fields between anode and cathode. (With complicated electrooptical arrangements it is a matter of accelerating the electrons particularly in the immediate vicinity of the cathode.) With $\Delta v_0 \simeq 10^5$ m/s and $E \simeq 10^6$ V/m we obtain a minimum transit-time spread of the order of $\Delta t_L \simeq 10^{-12}$ s.

3.2.2 *Photoelectric Detectors*

Let us discuss the time resolution of photoelectric detectors taking as examples light detectors based on the external photoelectric effect, which belong to the detectors producing a rather high time resolution (see Fig. 3.7). The elementary process of photoionization can be considered completely inertialess for picosecond investigations (the rise time amounts to $\simeq 10^{-14}$ s with typical band structures of the cathode material). The time resolution is primarily limited by the spread of the escape time from the cathode and the transit time spread of the electrons between cathode and anode resulting from a spread of the initial velocities. This electron-optically determined time reso-

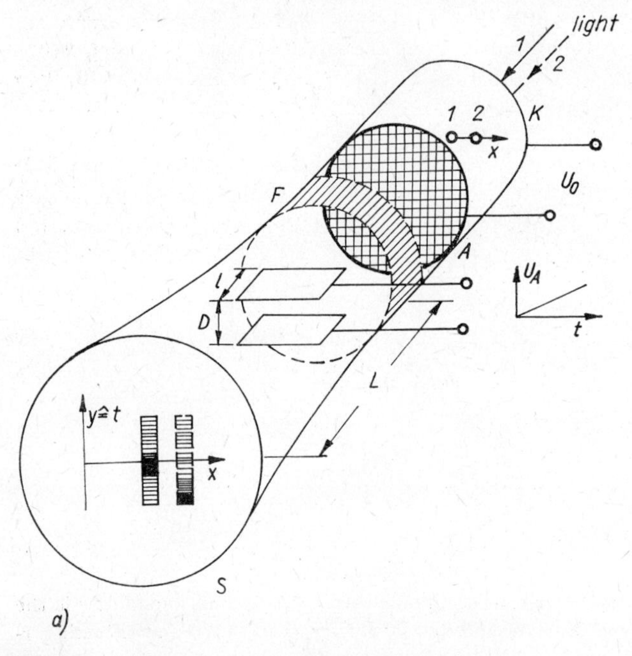

Fig. 3.8. Scheme of an electron-optical streak camera

a) Construction (K is the cathode, A is the anode, F is the focussing system, U_0 is the accelerating voltage, U_A is the deflecting voltage, l is the length of the deflecting plates, L is the distance of the deflecting plates from the display screen S and D is the distance between the deflecting plates) The "light spots" 1 and 2, which illuminate the cathode, appear as streaks blurred on the display screen. Different instants of time correspond to different spots of the streaks, where the variation in brightness along a streak in the y-direction represents the temporal variation of the signal intensity. Under a constant deflecting voltage U_A the electron beam with the initial velocity v_0 (which results from the accelerating voltage between cathode and anode) strikes the display screen at

$$y = elLU_A/(mv_0^2D) = \big(lL/(2U_0D)\big)\,U_A$$

where $l \ll L$ is assumed. From this the streak velocity at varying deflecting voltage can be written as

$$v_A = dy/dt = \big(lL/(2U_0D)\big)\,(dU_A/dt)\,.$$

Time resolutions of $\Delta t_A \simeq 10^{-12}$ s are achieved at $dU_A/dt \simeq 10^3$ V/10^{-9} s and with a screen resolution of $d \simeq 0.1$ mm.

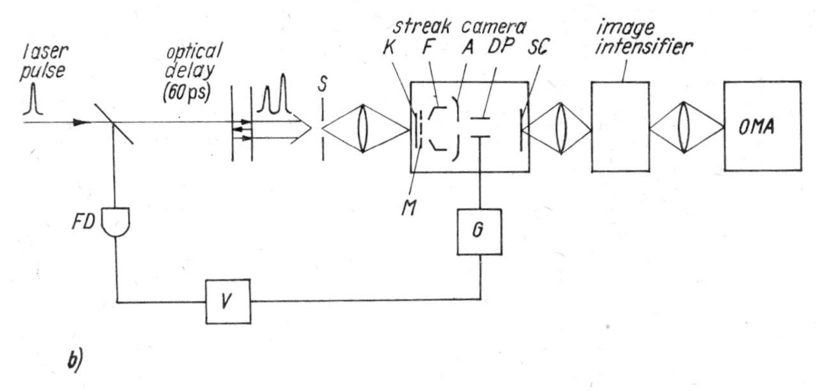

b) Arrangement for measuring the pulse duration of ps pulses with a streak camera (from Bradley, see [16])
The laser pulse is divided into two partial pulses which are separated by 60 ps in an optical delay line. This separation, which can be accurately measured, serves to calibrate the time. The two pulses reach the streak camera and their image is amplified and photographed in an image intensifier or recorded in an optical multichannel analyzer (OMA). The generator G for the deflecting voltage of the streak camera is triggered by an electrical pulse which the photodiode FD delivers at the striking of the laser pulse. The trigger moment can be set by means of an electrical delay line V. (S is the slit, K is the cathode, M is the accelerator grid near the cathode, F is the focussing cone, A is the anode, SC is the screen and DP shows the deflecting plates.)
The newest streak camera types (see [3.4, 3.5, 3.6]) contain a very high-performance microchannel plate amplifier and an optical fiber coupler instead of an imaging objective.

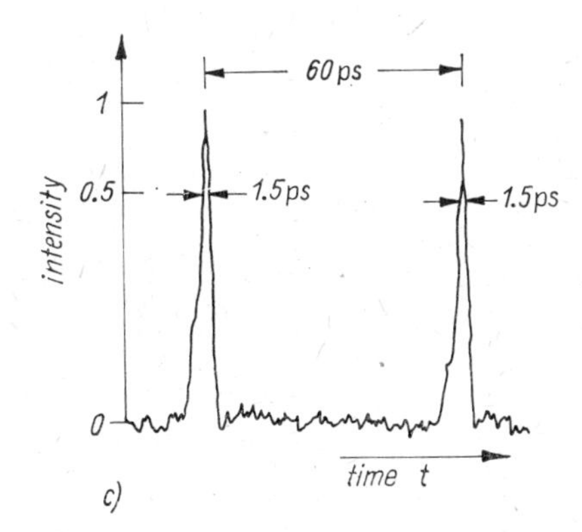

c) Microdensitogram of the streak camera recording of a pair of pulses recorded using the arrangement outlined in b) (from Bradley, see [16])
The pulse durations (flashlamp pumped, passively modelocked dye laser) are 1.5 ps.

lution can be reduced to a minimum of about 10^{-12} s, which until now had not been achieved in photoelectric detectors (though it had been in streak cameras, compare 3.2.3). At present, the best photoelectric detectors possess a time resolution of about 50 ps being limited by electron-optical as well as electronic effects. In photomultipliers the overall transit time spread through the longer paths of the electrons and the additional delays in the dynodes is in general greater ($\geq$ 100 ps); only by using special electron-optics does one obtain values below 100 ps. From the photoelectric detector, the electrical signal passes across appropriately matched transmission lines with large bandwidth ($\simeq 10^{11}$ Hz) and, if necessary, through broad-band amplifiers ($\simeq 10^{10}$ Hz) to the oscilloscope. The time resolution of oscilloscopes is in general not limited by the electron-optical process, which as described above would allow picosecond response times, but by the electronic components of the transmission lines and the electronic signal processing devices [9].

In single-photon counting, which distinguishes itself by a particularly favorable signal-to-noise ratio, the time resolution obtained thus far is about 100 ps. This limit is mainly caused by the jitter of the electrical pulses, which arises in the processes of pulse amplification and shaping (compare 9.1).

3.2.3 *Electron-Optical Streak Cameras*

The influence of the escape-time spread as well as of the transit-time spread of the electrons can — as discussed in section 3.2.2 — be reduced to about 10^{-12} s in optimized electron-optical devices.

Moreover, similar to the mechanical streak camera, the time resolution is limited by the time in which the beam with deflection velocity v_A passes the minimum resolvable distance d on the display screen (see Fig. 3.8). From this process limitations result that are also of the order of 10^{-12} s.

Time resolutions of the order of 1 ps are actually achieved with various streak camera types for the visible and near infrared spectral range [3.4, 3.5, 3.6] (compare 9.1).

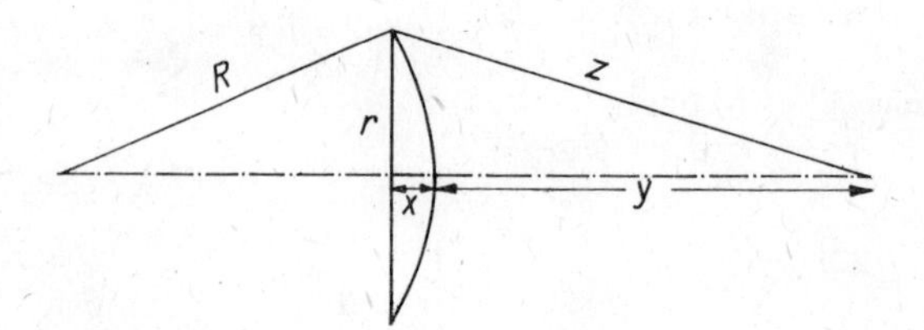

Fig. 3.9. Focussing with a plane-convex spherical lens (R is the radius of curvature, r is the lens radius, n is the refractive index, f is the focal length, $a = 2r/f$ is the aperture ratio)
The marginal ray has travelled the additional optical path

$$\Delta l = z - y - x \cdot n \approx \frac{1}{8}\, n^2(n-1)\frac{r^4}{R^3} = \frac{1}{128}\frac{n^2}{(n-1)^2}\, a^4 f$$

up to the intersection point with the optical axis. Assuming $n = 1.5$, $a = 1$ and $f = 100$ mm, then $\Delta l \approx 0.7$ mm follows. (It should be noted that this aberration can be somewhat reduced by reversing the lens.) Accordingly, in uncorrected systems with large apertures, transit-time spreads up to picoseconds may occur.

3.2.4 *Focussing Systems*

In the example of a focussing spherical lens we shall see how uncorrected optical systems positioned in the beam path of light pulses affect the time resolution (see Fig. 3.9). The transit-time differences in typical arrangements lie mostly in the subpicosecond range and reach several picoseconds only in extreme cases. This error can be held negligibly small with suitably corrected optical systems.

3.2.5 *Dispersing Systems*

Let us consider as an example the effect of plane-parallel plates having refractive index $n(\omega)$ on the time displacement between two pulses of frequencies ω_1 and ω_2 and group velocities v_1 and v_2, respectively, as well as on the temporal profile of a single pulse (see Fig. 3.10). The pulse with the greater group velocity v_1 reaches the point $z = l$ earlier by

$$\Delta t = l\left[\frac{1}{v_2} - \frac{1}{v_1}\right]. \tag{3.7}$$

The duration of an initially bandwidth-limited ultrashort light pulse is approximately doubled due to the different transit-times of its frequency constituents on a path of the order

$$l \simeq \tau_L^2 \bigg/ \left(\frac{d^2k}{d\omega^2}\right)_{\omega_L} \tag{3.8}$$

($(d^2k/d\omega^2)_{\omega_L}$ is the second derivative of the wave number with respect to the angular frequency ω at $\omega = \omega_L$.)

The order of magnitude of this length can be easily estimated in the following manner: imagine the bandwidth-limited pulse of duration τ_L and halfwidth $\Delta\omega$ composed of two parts having center frequencies ω_L and $\omega_L + \Delta\omega$, respectively. From (3.7) we obtain a transit time difference

$$\Delta t = l\left[\frac{1}{v(\omega_L + \Delta\omega)} - \frac{1}{v(\omega_L)}\right] \approx l\left[\frac{d^2k}{d\omega^2}\right]_{\omega_L} \Delta\omega$$

between the two pulse components after they have passed the length l. Substituting $\Delta t = \tau_L$ and $\Delta\omega = 4 \ln 2/\tau_L$, where this numerical factor holds for a Gaussian shaped pulse, this equation yields

$$l \approx 0.4\left[\left(\frac{d^2k}{d\omega^2}\right)_{\omega_L}\right]^{-1} \tau_L^2 .$$

The exact solution of the wave equation for Gaussian shaped pulses provides the same functional relationship for the length, over which the pulse doubles its duration, where the numerical factor is about 0.6 instead of 0.4.

From this relationship and the numerical example in Fig. 3.10 we see that pulse broadening due to dispersion must be taken into account with extremely short pulses and for great interaction lengths (e.g. in optical fibers) even with pulses of moderate duration (compare 8.3).

3.3 Nonlinear Optical Methods for Measuring Ultrashort Light Pulses

As we have seen in the previous sections, it is extremely difficult to measure temporal processes in the picosecond and all the more so in the subpicosecond ranges using only electronic and electron-optical devices. Nonlinear optics allows the fundamental principles mentioned earlier — especially that of correlation techniques — to be applied in the measurement of extremely short light pulses. Only in this way was it possible to measure the pulse duration of the first modelocked lasers shortly after their development [3.7, 3.8].

We will explain these techniques using as an example the generation of the second harmonic and the two photon fluorescence (for the theoretical basis of these effects see [11]). Furthermore, we will deal with optical switches on the basis of the laser induced Kerr effect.

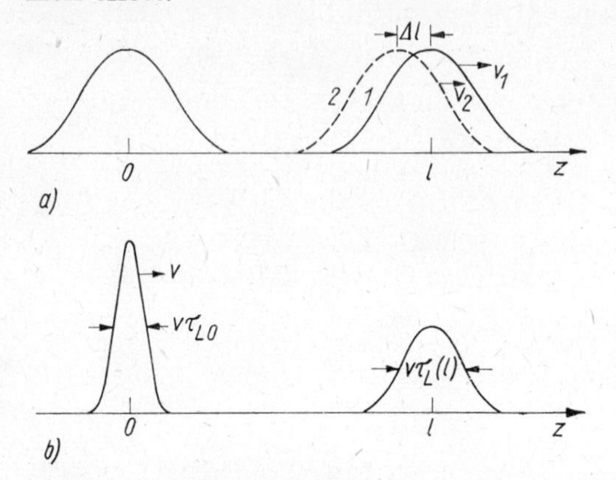

Fig. 3.10. Influence of a dispersing plate on the pulse propagation
a) The pulse with the greater group velocity $v_1 = (\mathrm{d}\omega/\mathrm{d}k)_{\omega_1}$ arrives, at $z = l$, $\Delta t = \Delta l/v_1$ earlier.
b) The initial duration of an ultrashort Gaussian shaped pulse doubles, if $l \approx 0.6\tau_{\mathrm{L0}}^2/(\mathrm{d}^2k/\mathrm{d}\omega^2)_{\omega_l}$. ($(\mathrm{d}^2k/\mathrm{d}\omega^2)_{\omega_{\mathrm{L}}}$ characterizes the dispersion of the reciprocal group velocity.) From the simple dispersion relationship $n = 1 + A[1 - (\omega/\omega_0)^2]^{-1}$ and with the representative parameters for transparent materials — e.g. glass — $A = 0.5$ and $\omega_0 = 3 \times 10^{16}\ \mathrm{s}^{-1}$ we obtain the following values at $\omega_1 = 3 \times 10^{15}\ \mathrm{s}^{-1}$ ($\lambda_1 = 628$ mm) and $\omega_2 = 5 \times 10^{15}\ \mathrm{s}^{-1}$ ($\lambda_2 = 377$ mm): $n_1 = 1.505$, $n_2 = 1.514$, $v_1 = 1.980 \times 10^8$ m/s, $v_2 = 1.956 \times 10^8$ m/s, $(\mathrm{d}^2k/\mathrm{d}\omega^2) \approx 6 \times 10^{-26}\ \mathrm{s}^2/\mathrm{m}$. With these values, $\Delta t/l \approx 70$ ps/m follows for the advance of the faster (red) pulse and the relationship $(l/\tau_{\mathrm{L0}}^2) \approx 10\ \mathrm{m/ps}^2$ determines the length in which an ultrashort pulse broadens to twice its initial duration τ_{L0}. This means that extremely short pulses of 0.1 ps duration broaden to double their initial duration in a 10 cm glass plate (compare 8.3.2).

3.3.1 *Measurement of the Intensity Correlation Function by Means of Second Harmonic Generation*

The arrangement outlined in Fig. 3.5 for recording the field strength autocorrelation function is modified in the following manner: a nonlinear optical crystal, which partially converts light of frequency ω (fundamental wave) into light of frequency 2ω (second harmonic wave) in an inertialess process, is placed in front of the detector. The remaining light at frequency ω is absorbed by a filter. Moreover, one can insert polarizers in the beam path in front of the two mirrors in such a way that the two reflected light waves are polarized perpendicular to one another. With a suitable choice of the crystal and its orientation (compare 8. and [11, 22])

$$|\hat{E}_2(t;\tau)|^2 = C\,|\hat{E}_1(t)|^2 \cdot |\hat{E}_1'(t+\tau)|^2$$

and (3.9)

$$I_2(t;\tau) = C' I_1(t)\, I_1'(t+\tau)$$

hold, where E_1 and E_1' are the electric field strengths of the linear polarized waves in the two arms of the arrangement, and I_1 is the light intensity at the entrance of the correlator, τ is the delay time which results from the displacement Δl of the mirror M_2. The intensity I_2 of the second harmonic, which is proportional to $|\hat{E}_2(t, \tau)|^2$, depends on the time t as well as on the time delay τ that acts as a parameter. The constants C and C' are determined by the crystal parameters and the experimental geometry. The nonlinear optical crystal acts thus as an optical multiplier. The detector temporally integrates the signal which is proportional to the power of the second harmonic. This temporally integrated output signal is accordingly proportional to the intensity correlation function of the second order (or to the field strength correlation function of the fourth order). In this special experimental set-up the polarizers can act in connection with a suitable crystal in such a manner that the second harmonic is generated only if radiation from both channels enters the crystal simultaneously. In this case the correlation function is thus recorded background-free. (Without the polarizers additional signals would occur — e.g. proportional to $\int \mathrm{d}t\, |\hat{E}_1(t)|^4$ — which produce a background.) Instead of using polarizers a background-free measurement can also be achieved by noncollinear irradiation of both pulses into the nonlinear optical crystal (see Fig. 3.11).

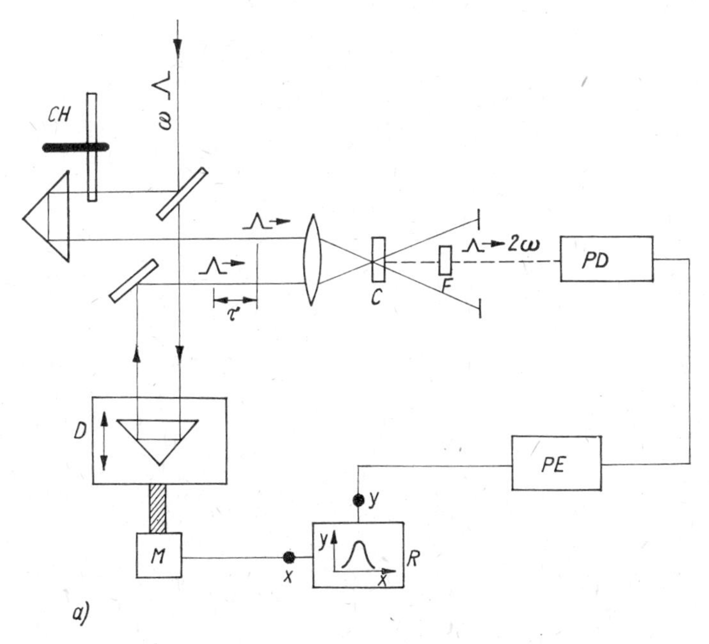

Fig. 3.11. Measurement of the second order intensity correlation function of light pulses of high repetition frequency by means of noncollinear generation of the second harmonic [3.9, 3.10, 3.11]

a) Scheme

CH is a mechanical chopper, which modulates the pulse train of high repetition frequency ($\simeq 10^8$ Hz) for the purpose of favorable electronic detection with a low frequency ($\simeq 10^2$ Hz). D is the optical delay path, which is driven by the motor M. The second harmonic of frequency 2ω is generated in the nonlinear optical crystal C under noncollinear irradiation. The remaining fundamental radiation of frequency ω is absorbed in the filter F. The second harmonic is detected in the photodetector PD where the electrical signal is amplified in an amplifier with phase-sensitive rectification PE, which is matched to the frequency and phase of the mechanical chopper. Finally, the signal is recorded by an x-y plotter R whose x and y axes correspond to the optical delay and the intensity of the second harmonic, respectively.

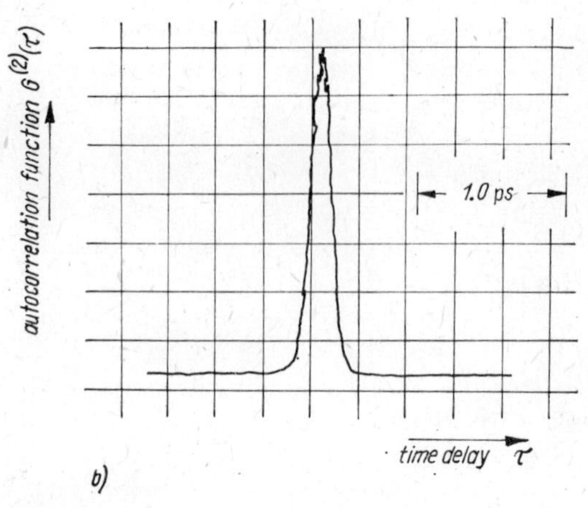

Fig. 3.11

b) Measured autocorrelation function of pulses from a passively modelocked dye laser (compare 6.) [3.11]

The autocorrelation function has a halfwidth of 0.15 ps, from which a pulse width of about 0.10 ps can be estimated.

With suitably cut nonlinear optical crystals second harmonic radiation with extraordinary polarization can be generated from two fundamental waves of ordinary polarization entering the crystal. In this process, the emission direction of the harmonic is determined by the relationship

$$\vec{k}^{e}(2\omega) = \vec{k}_1^{0}(\omega) + \vec{k}_2^{0}(\omega)$$

for the wave number vectors — the so-called phase matching relation (compare 8.). In this direction the inequalities $\vec{k}^{e}(2\omega) \neq 2\vec{k}_1^{0}(\omega)$, $2\vec{k}_2^{0}(\omega)$ hold and thus, under the exclusive irradiation of only one component, an effective frequency doubling is not possible. The techniques of measuring the intensity correlation function with the second harmonic that have been described lend themselves particularly to lasers having high pulse repetition frequencies, e.g. continuously pumped, modelocked lasers, where the repetition frequency usually lies in the order of 100 MHz (compare 4.—6.).

In some cases the background affected correlation function is deliberately recorded interferometrically by means of the generation of the second harmonic at high spatial resolution with respect to the delay, whereby with an accurate evaluation additional information about the signal can be obtained [3.12]. For this purpose collinear superposition of the two incident beams without polarizers in the beam paths is used. (It should be noted that in the usual (noninterferometric) correlation measurements we must accurately average over the fast oscillating contributions, which can otherwise lead to distortions — especially near the maximum of the autocorrelation function.)

With the generation of the third harmonic (either in a nonlinear optical material with a large nonlinear susceptibility of the third order or by generating the second harmonic and the subsequent production of radiation at frequency 3ω through sum frequency generation from light of frequencies 2ω and ω) we can plot the intensity correlation function of third order (compare 8.). Whereas the correlation function of the second order is always symmetrical in τ and therefore, e.g., no conclusions about the asymmetry of the pulses can be drawn, from an additional analysis of the correlation function of the third order we can also determine specific details of the pulse profile — such as asymmetries [3.13, 3.2, 3.3].

3.3.2 *Measurement of the Intensity Correlation Function by Means of Two Photon Fluorescence*

If we are interested in the temporal profile of non- or only slowly repeatable pulses, we must record the entire correlation function with one single pulse. For this purpose two photon fluorescence lends itself particularly as a suitable method of high time resolution and sensitivity. A typical experimental set-up is displayed in Fig. 3.12. The molecules are excited by simultaneous absorption of two photons, i.e. by two photon

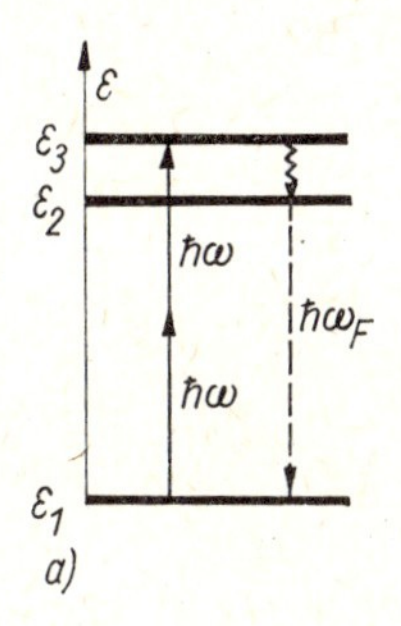

Fig. 3.12. Measurement of intensity correlation functions of the second order by means of two photon fluorescence (TPF)

a) Energy level scheme with radiative and radiationless transitions. The atomic systems are excited by the absorption of two photons having energy $\hbar\omega$ from the ground level into the level with energy $\mathscr{E}_3 \approx \mathscr{E}_1 + 2\hbar\omega$. By fast radiationless relaxation processes the molecules reach level 2 from which fluorescence quanta with energy $\hbar\omega_F \approx \mathscr{E}_2 - \mathscr{E}_1$ are emitted.

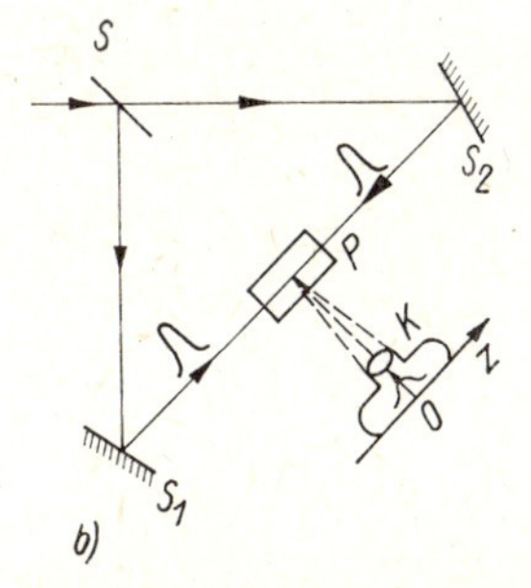

b) Triangular arrangement for the recording of the correlation function

The incoming pulse is split at a partially transmitting mirror and after reflection at S_1 and S_2 the partial pulses travel through the sample P from opposite directions. The pulse moving to the right has the field amplitude $\hat{E}_r(t, z) \sim (-A(t - z/v)\, e^{-ikz})$, and the one moving to the left, $\hat{E}_l(t, z/v) \sim A(t + z/v)\, e^{ikz}$. The fluorescence light, whose intensity depends on the number of excited particles at position z, is recorded with time integration using a camera K. Instead of a camera with photographic recording on a film we can use a set of photoreceivers connected to an electronic storage, a so-called optical multichannel analyzer (OMA).

absorption, and subsequently emit light due to fluorescence, which can have a shorter wavelength than the exciting light. The absorption process can (if the reciprocal value of the homogenenous linewidth is small compared to the pulse duration) be regarded as inertialess. In two photon absorption the transition rate is proportional to the square of the light intensity at the position of the molecule, and thus to the fourth power of the field strength. If we choose materials that have long lifetimes in the excited state compared to the pulse duration, we obtain for the occupation number density in the excited state 2 after the excitation by two photon absorption with respect to z

$$N_2(z) \propto \int_{-\infty}^{\infty} \mathrm{d}t\, |\hat{E}(t, z)|^4 \tag{3.10}$$

where

$$\hat{E}(t, z) \propto \left\{-A\left(t - \frac{z}{v}\right) \mathrm{e}^{-\mathrm{i}kz} + A\left(t + \frac{z}{v}\right) \mathrm{e}^{\mathrm{i}kz}\right\}.$$

(The symbols are explained in the description of Fig. 3.12. The negative sign in front of one of the amplitudes results from the reflection of the wave moving in the positive direction at the beam splitter.) The occupation number contains contributions which oscillate rapidly with respect to z (spatial frequency $2k$) as well as slowly in z varying contributions whose z-dependence is caused solely by the time-dependence of the pulse intensity in accordance with the fact that the space coordinate z is related to the delay time τ by $\tau = 2z/v$. The fluorescence signal that is photographed with spatial resolution is proportional to $N_2(z)$. If the photographic resolution interval is $\Delta z \gg \lambda$ (which e.g. is always satisfied in silver-halide photography), the fast oscillating terms from equation (3.10) are "averaged out". (The fast oscillating contributions only make an essential contribution to the recorded signal, if more complicated transition processes must be considered [3.14].) With these assumptions

$$F(\tau = 2z/v) \propto \int_{-\infty}^{\infty} \mathrm{d}t \{I^2(t - \tau/2) + I^2(t + \tau/2) + 4I(t - \tau/2)\, I(t + \tau/2)\}, \tag{3.11a}$$

which leads to

$$F(\tau = 2z/v) = 2F_0\{1 + 2G^{(2)}(\tau)/G^{(2)}(0)\} \tag{3.11b}$$

holds for the dependence of the recorded fluorescence signal on z. (F_0 is the fluorescence signal, which is measured after blocking one of the two partial beams. $2F_0 = F(\infty)$ holds for signals of finite duration). The z-dependence (or the τ-dependence) of the fluorescence signal is thus determined by the intensity correlation function (see Fig. 3.13) which, however, is here measured with the background. This background is caused because fluorescence light also appears if the pulses do not superpose in the area concerned, but pass it one after another. From the width of the autocorrelation signal we can determine the pulse duration. Typical cases are shown in Fig. 3.13. Therefore, it is important for the interpretation to take into consideration additional parameters of the intensity correlation function besides its width. As evident from Fig. 3.13, the so-called contrast ratio

$$K = F(0)/F(\infty)$$

represents an important quantity for the evaluation of the signal. For pulses of limited duration $K = 3$ follows from (3.11); for bandwidth-limited noise, however, $K = 1.5$. Furthermore, as already mentioned, we can determine additionally the pulse-duration-bandwidth-product (compare (3.4)).

Intensity correlation functions of higher order can be measured by means of multiple photon fluorescence, i.e. of n-photon absorption ($n > 2$) with subsequent fluorescence [16].

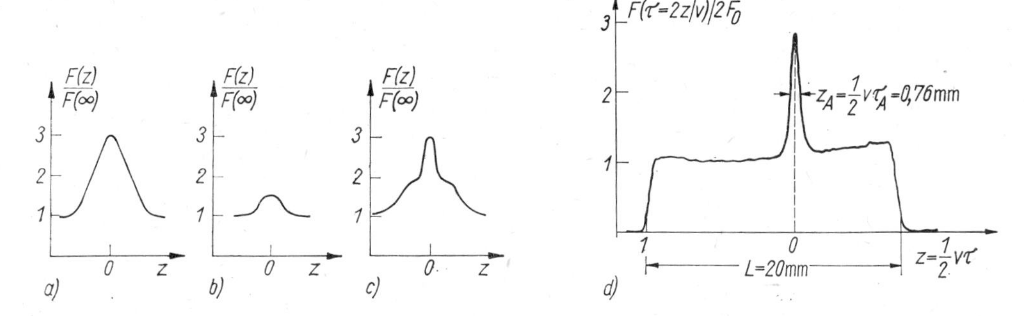

Fig. 3.13. Fluorescence signal dependent on $\tau = 2z/v$ for different representative input signals
a) Single pulse
b) Narrow-band Gaussian noise of long duration
c) "Cleft pulse" (Noise within a short time interval)
d) Measured two photon fluorescence [3.15] (Nd: glass laser pulse, two photon fluorescence in rhodamine 6G dissolved in ethanol, recording with an optical multichannel analyzer)
The signal corresponds to a bandwidth-limited pulse resulting from good modelocking (contrast ratio $K = 2.95 \pm 0.1$, $\tau_A = 6.1$ ps, $\Delta\omega\tau_L/(2\pi) = 0.4 \pm 0.1$). From these values a negligible phase modulation can be concluded. The shape of the autocorrelation function justifies the assumption of pulses with a Gaussian profile, having a duration of $\tau_L = \tau_A/\sqrt{2} \approx 4$ ps. (The origin of the z-axis is in the maximum of the autocorrelation function, which here does not lie exactly in the center of the dye cell. The asymmetry in the shape of the "shoulders" results from a local variation of the sensitivity of the vidicon.)

3.3.3 *Measurement of Intensity Cross Correlation Functions*

The techniques of measuring intensity autocorrelation functions described in sections 3.3.1 and 3.3.2 can also be employed with minor modifications to determine cross correlation functions.

If the two signals have center frequencies ω_1, ω_2 and intensities $I_1(t)$, $I_2(t + \tau)$, respectively, in a suitably cut crystal, light of the sum frequency $\omega_1 + \omega_2$ whose intensity is proportional to $I_1(t)\, I_2(t + \tau)$ can be generated, whereby the relative delay can be controlled by an optical delay line. A time integrating receiver provides an output signal that is proportional to the cross correlation function

$$K_{12}^{(2)}(\tau) = \int_{-\infty}^{\infty} \mathrm{d}t I_1(t)\, I_2(t + \tau)\,. \tag{3.12}$$

An interesting special case results — as already mentioned in section 3.1.5 — if one of the input pulses, e.g. $I_1(t)$, is short compared to the other, because then $K(\tau) \sim I_2(\tau)$

holds. This measuring technique was for instance used [3.16] to sample the light pulses of a modelocked argon ion laser ($\tau_{L1} \simeq 100$ ps) with the considerably shorter pulses of a dye laser ($\tau_{12} \simeq 1$ ps), whereby the dye laser was excited by the argon ion laser using the method of synchronous pumping (compare 5.). Such a measurement allows one to determine accurately not only the temporal profile of the argon laser pulse but also the relative position between the argon laser pulses and the dye laser pulses.

Besides sum frequency generation and two photon fluorescence other nonlinear optical effects are also successfully used to measure intensity cross correlation functions, e.g. parametric amplification which lends itself particularly to the measurement of weak signals (compare section 8.2.3).

3.3.4 Laser-Controlled Optical Kerr Gates

As described in section 3.1.2, we can construct optical gates or shutters with Kerr cells whose time resolution reaches up to the picosecond range, if instead of an external electric field, in the sense of an optical correlation technique, we employ an ultrashort light pulse.

The optical Kerr effect rests on the fact that, under the influence of a changing field, molecules with an anisotropic polarizability α_{ij} align according to the Boltzmann distribution, preferably in a direction such that the potential energy $\mathscr{E}_p = -\frac{1}{2} \sum_i \alpha_{ii} E_{Li} E_{Li}$ has a minimum. This molecular distribution leads to a difference between the refractive indices $n_\parallel$ and $n_\perp$ for signal light that is polarized parallel or perpendicular to the orientating field $\vec{E}_L$ (see [16]).

Since, in general, the relationship $\mathscr{E}_p \ll k_B \mathscr{T}$ also holds with rather strong laser fields, it suffices to develop the Boltzmann distribution up to the first-order term in $\mathscr{E}_p$. In this way we obtain (under constant irradiation with light) a refractive index difference $\Delta n = n_\parallel - n_\perp$ proportional to the square of the field strength or the intensity of the laser light,

$$\Delta n = n_2 I_L, \tag{3.13}$$

whereby n_2 is the nonlinear optical birefringence coefficient. If the intensity of the laser light is dependent on time, the refractive index difference Δn also varies temporally. However, due to "friction processes" in liquids the "response" does not result instantaneously, but a memory effect occurs. Using a simple relaxation model $\mathrm{d}\Delta n(t)/\mathrm{d}t = -\Delta n(t)/\tau_0 + n_2 I_L(t)/\tau_0$ where τ_0 is the orientational relaxation time, we obtain

$$\Delta n(t) = \int_{-\infty}^{t} \mathrm{d}t' I_L(t') \, \frac{n_2}{\tau_0} \, \mathrm{e}^{-(t-t')/\tau_0}. \tag{3.14}$$

In a favorable geometrical arrangement (see Fig. 3.14a) the time resolution of the optical gate is determined only by the duration of the switching pulse and the orientational relaxation time. A suitable substance for Kerr switches is, for instance, carbon sulphide which is distinguished by its large nonlinear coefficient and a short orientational relaxation time ($\tau_0 \approx 2$ ps) [16].

In a precise measurement using a fs-laser Tang [28] found out that the refractive index change of CS_2 decayed according to two superposed exponential functions with time constants of 2.1 ps and 0.36 ps.

If the pulse duration is great compared to the orientational relaxation time then (3.13) results from (3.14) as a special case and for the range of weak switching ($\varphi \ll 1$) the time integrated intensity of the transmitted signal pulse $I_T(t + \tau)$ can be calculated from the relationships given below figure 3.3,

$$\int_{-\infty}^{\infty} dt I_T(t + \tau) \propto K_{0LL}^{(3)}(\tau, 0) \equiv \int_{-\infty}^{\infty} dt I_0(t + \tau) I_L^2(t). \tag{3.15}$$

The time integrated output signal is thus a measure of the intensity cross correlation function of third order explained in (3.15). If the switching pulse is short compared to the signal pulse, $K^{(3)}(\tau, 0) \propto I_0(\tau)$ follows, whereby the temporal profile of the signal pulse can be measured directly using methods analogous to the sampling technique with many recordings at various delay times.

The special advantage of the optical gate is that not only a single signal but also a two dimensional variety of signal points $I_0(t, x, y)$ (i.e. a complete picture) and its temporal variation can be investigated (see Fig. 3.14a). If we want to observe only a one dimensional signal distribution, e.g. a spectrum as a function of time, this can be done with a single recording using the method of crossed pulses (see Fig. 3.14b).

Optical gates whose switching time is not limited by the orientational relaxation time or another relaxation time, can be obtained on the basis of sum frequency generation, parametric amplification as well as nonlinear optical phase conjugation (see [3.17, 3.18]).

Besides using optical gates for analyzing fast emission processes (compare 9.1), we can employ them for "optical radar" (gated picture ranging). In [3.19] this made it possible to photograph an object through a strongly scattering material (see Fig. 3.15). For this purpose the pulse of a mode locked Nd:glass laser is divided into an illuminating pulse and a switching pulse. The light reflected from the object reaches the optical gate which can be opened by the switching pulse at definite moment. Due to the short opening time of the gate, the film records only optical information that comes from a dis-

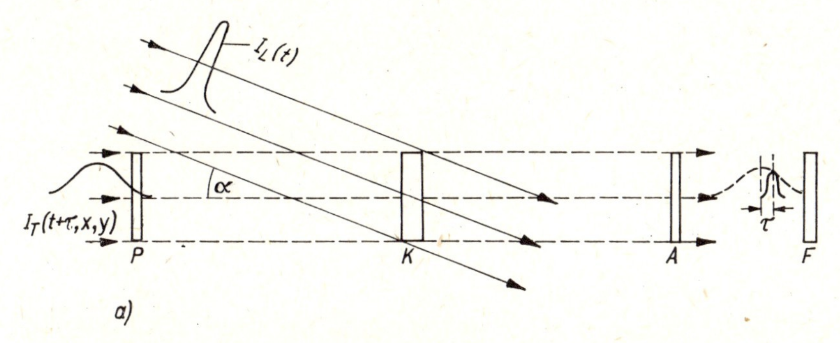

Fig. 3.14. Typical arrangements for the measurement of temporal processes with an optical gate which consists of polarizer P, analyzer A, Kerr cell K and recorder F (e.g. film)

a) Configuration for the recording of two-dimensional pictures according to the sampling technique with several recordings. ($I_T(t + \tau, x, y)$ is the signal intensity with respect to time $(t + \tau)$ and the image coordinates x, y.) The time delay relative to the switching pulse can be varied with an optical delay line which is in the path of one of the two pulses. The angle α between exciting light and signal light must be kept as small as possible, in order that the time resolution is not reduced by geometric influences of the order of $(d/c) \tan \alpha$, where d is the beam diameter.

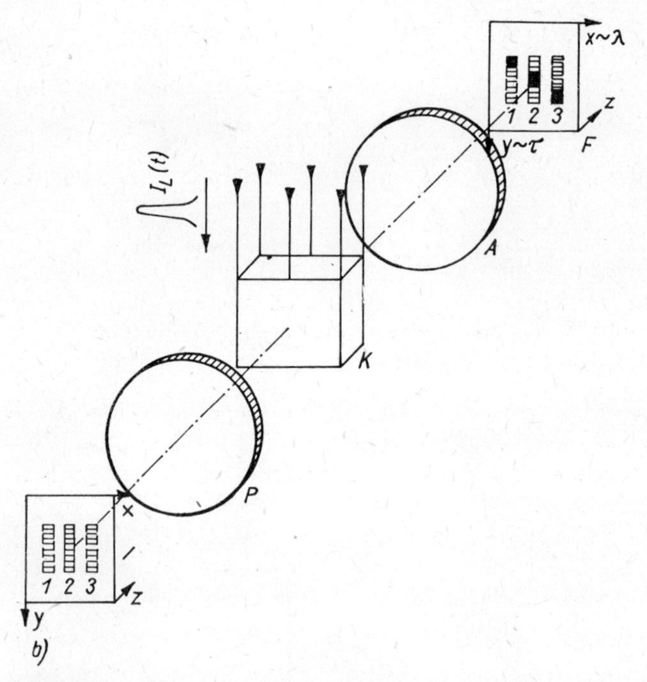

Fig. 3.14

b) Configuration with crossed beam paths for the temporal analysis of a one-dimensional "image" (e.g. of a spectrum) with a single recording

The switching pulse passes through the Kerr cell perpendicular to the signal pulse. Thus, the cell is opened at points with different y for different delay times $\tau(y = v \cdot \tau)$. At a fixed point x, the signal intensity of, for example, a spectral line is recorded as a function of the delay time in the film plane F along the x-axis. The x-dimension can be used to depict, e.g., a spectrum as a function of time.

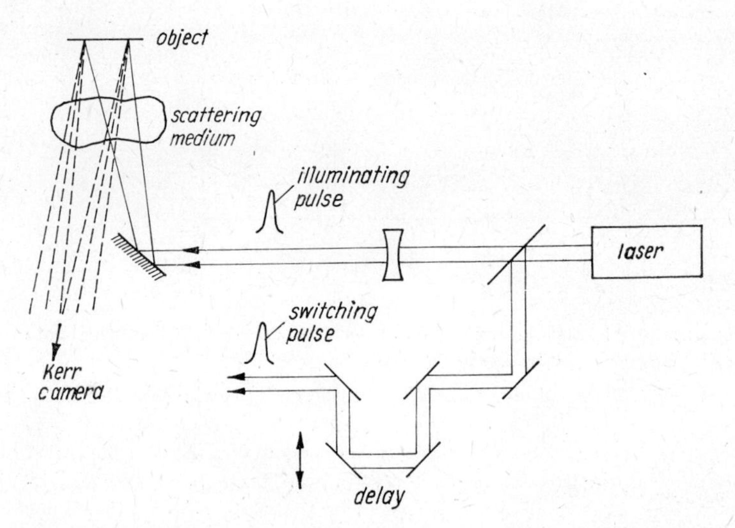

Fig. 3.15. "Optical radar" with ps-resolution on the basis of a Kerr shutter (according to [3.19])

tance from the camera between x_0 and $x_0 + \Delta x$ where $\Delta x \approx \frac{1}{2} \tau_A c$; τ_A is the opening time of the gate and x_0 is given by the delay time between the illuminating and switching pulses. In the example given the ultrashort pulses of a Nd:glass laser ($\tau_L \approx 7$ ps) and a CS_2-Kerr effect switch ($\tau_0 \approx 2$ ps) were used, and accordingly, the recording interval Δx was about 1 mm. With a suitable choice of x_0 only the light reflected from the desired object, and not scattered light from scattering materials in front of the object, is recorded. In [3.19] this technique was applied satisfactorily by photographing a flat object that was located 10 mm behind a diffusing screen and could not be imaged by usual photographic techniques.

Using laser pulses with a pulse duration of $\tau_L \simeq 0.1$ ps and inertialess optical gates — for example, on the basis of generating the second harmonic or of parametric effects — a spatial resolution of $\Delta x \simeq 10$ μm can be attained.

3.4 Laser-Controlled Optoelectronic Shutter

Ultrashort laser pulses can be employed for fast switching of optoelectronic components. The principle of operation is here similar to that of the optical gate in which, instead of light an electrical voltage or microwaves are switched. Thereby, the shortest electronic signals are generated and measured. The first component of this kind was devel-

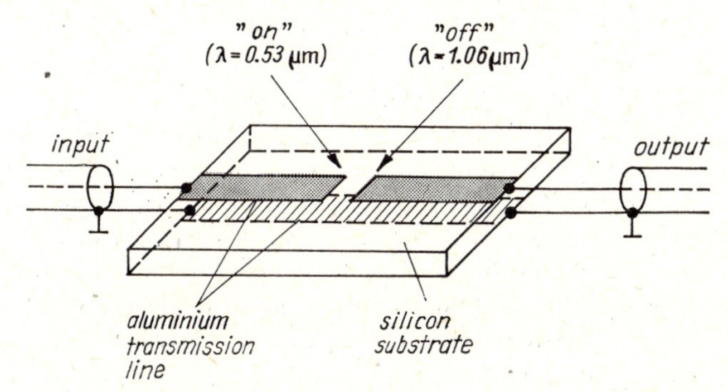

Fig. 3.16. Laser-controlled optoelectronic shutter from [3.20]. Turn-on by a pulse with $\lambda = 0.53$ μm, cut-off by a pulse with $\lambda = 1.06$ μm

oped in 1975 by Auston [3.20] (see Fig. 3.16). It consists of a wafer made of highly resistive silicon ($\simeq 10^4$ Ω/cm), which is coated on its underside and top face with aluminum microstrips where the upper conductor track is broken by a slit. On this slit a light pulse is focussed that is absorbed by the silicon and thereby generates a large number of electron hole pairs. (In [3.20] the second harmonic of an Nd:glass laser ($\lambda = 0{,}53$μm) was used for switching.) In this way, a highly conductive connection across the slit is created. The component may be connected on the left side to a coaxial cable to which a steady voltage is applied; then, at the cable located to the right, this voltage is switched on by the ultrashort light pulse in a time comparable with the pulse duration. For the switch to function well, it is important that the dark resistance of the slit is large compared to the impedance Z_0 of the cable ($Z_0 \simeq 50\ \Omega$) and that the resistance of the irradiated slit is small compared to Z_0. This resistance $\mathcal{R}_{\mathrm{Sp}}$ follows from the number of charge carrier pairs $\mathcal{N}_{\mathrm{p}}$ generated with homogeneous distribution in the slit region, from the slit width l and the effective charge carrier mobility $\mu_{\mathrm{eff}} = \mu_- + \mu_+$; thus we have

$$\mathcal{R}_{\mathrm{Sp}} = \left[\mu_{\mathrm{eff}} \frac{1}{l^2} e \mathcal{N}_{\mathrm{p}}\right]^{-1}.$$

Under the assumption that all photons penetrating into the substrate are absorbed near the surface and generate charge carrier pairs with the quantum efficiency η, and that the light pulse of energy $\mathcal{E}_{\mathrm{L}}$ is short compared to the lifetime of the charge carriers,

$$\mathcal{N}_{\mathrm{p}} = \eta(1 - R_{\mathrm{Si}})\, \mathcal{E}_{\mathrm{L}}/\hbar\omega_{\mathrm{L}}$$

holds, with which we obtain

$$\mathcal{R}_{\mathrm{Sp}}^{-1} = \eta(1 - R_{\mathrm{Si}})\, \mu_{\mathrm{eff}} \frac{1}{l^2} e \frac{\mathcal{E}_{\mathrm{L}}}{\hbar\omega_{\mathrm{L}}}.$$

(Here, the reflectivity of the silicon plate is denoted with R_{Si} to avoid confusion with the electrical resistance $\mathcal{R}_{\mathrm{Sp}}$; for silicon R_{Si} [$\lambda = 0.53\ \mu\mathrm{m}$] ≈ 0.65 holds.) Obviously, for $\eta\mu_{\mathrm{eff}} \simeq 1\ \mathrm{V}^{-1}\ \mathrm{cm}^2\mathrm{s}^{-1}$ and a slit width of 0.2 mm, pulse energies in the order of 1 mJ are sufficient to achieve a slit resistance of the order of 1 Ω and with that to fulfil the condition $\mathcal{R}_{\mathrm{Sp}} \ll Z_0$. After the passage of the optical switching pulse the conductivity decreases due to relaxation processes, which are caused in crystalline silicon solely by the electron hole recombination and which proceed slowly ($T_{\mathrm{Recomb}} >$ ns). Thus, the electrical pulse has a sharp leading edge, but falls off after long lapses of time. Therefore, Auston used a second light pulse of the fundamental wavelength of the Nd laser which switched off the signal connection in the picosecond range. In contrast to light of the second harmonic which, due to the very high absorption coefficients ($\simeq 10^4\ \mathrm{cm}^{-1}$), only excites layers of silicon near the surface, the fundamental wave has a relatively large penetration depth ($\simeq 1$ mm) and thus generates electron hole pairs over the entire plate thickness up to the lower circuit strips. Between the top and bottom microstrip this charge carrier plasma produces a short circuit that breaks the connection to the right so that all electrical signals arriving from the left side are reflected. In place of a switch-off light pulse of different wavelength we can also use a light pulse with the same wavelength as the single pulse that we focus on a second circuit path disconnection whose bridging produces a short to the grounded base board (see Fig. 3.17). Overall, however, the operation with two optical switching pulses is experimentally very complicated. Therefore, a course has been adopted in which semiconductors with very fast relaxation of the conductivity are used for the switch. For this reason GaAs and highly defective silicon layers particularly are employed with success [3.21–3.24]. In silicon the lifetime can be significantly shortened by introducing deep impurities. Huge impurity concentrations are attained by fast evaporation of dielectric substrates or by ion implantations. As opposed to the crystal, many bonds in the layer are broken, and so-called "dangling bonds" occur. They cause the non-localized conduction charge carriers to be trapped very quickly and thus localized, without a recombination having to take place. In nearly amorphous layers the conductivity decreases with a relaxation time of $\lesssim 4$ ps [3.23]. It should be noted, though, that under the amorphization the charge carrier mobility, and thus the sensitivity of the components, diminishes.

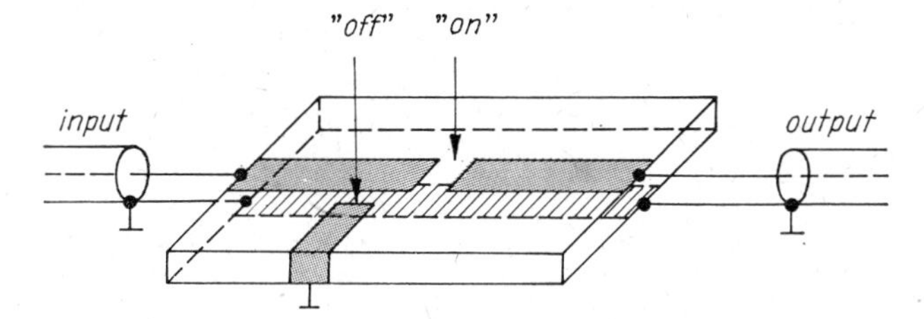

Fig. 3.17. Laser-controlled optoelectronic shutter
Turn on by one pulse and turn off by another pulse where $\lambda = 0.53$ µm for each pulse.

If we want to measure accurately the temporal variation of the switching process, we must again make use of suitable optical ultrashort measuring techniques. For this purpose, Auston has already used in [3.20] the optical correlation technique depicted in Fig. 3.18. The electronic signal at the output of the component is measured with respect to the delay τ between which the pulses strike the first and the second slit. The second switch operates here as a sampling detector (compare 3.1.2).

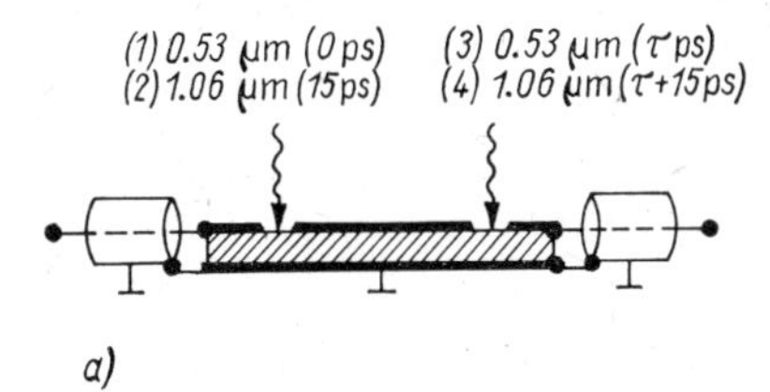

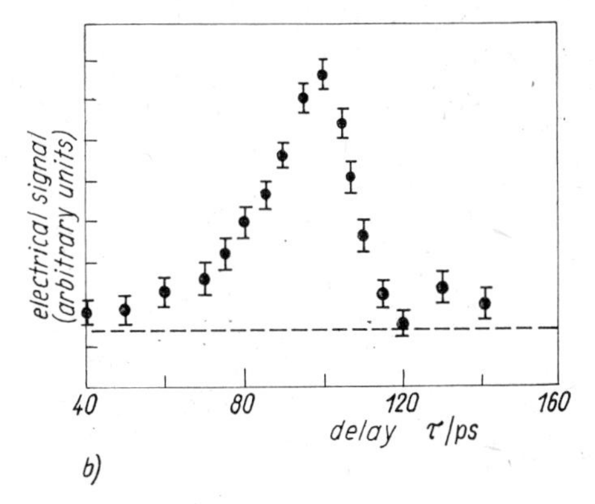

Fig. 3.18. Measurement of the switching time by an optoelectronic correlation method
a) Scheme: two gates switched in sequence
b) Time integrated electrical signal at the output of the device with respect to the delay time τ

The optoelectronic switches described are ready to be employed for various purposes. It is possible to drive and test electronic components in the ultrashort time range by means of extremely short electrical pulses. Here, it has also been possible to obtain high-voltage pulses in the kV range with steep rise edges and very high reproducibility of their temporal characteristics [3.25, 3.26]. With such pulses, for example, it has also been possible to drive Pockels cells and streak cameras with high precision. The high bandwidths (up to 10^{12} Hz) of the components used furthermore made a switching and modulation of electromagnetic radiation in the microwave range possible [3.27, 3.28].

In [3.29] the electrooptical switch was used to investigate signals from a vacuum photodiode with a fast sampling technique.

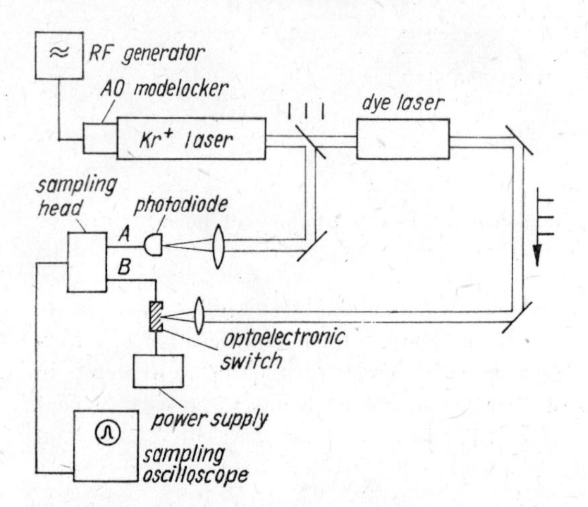

Fig. 3.19. Scheme of an experimental set-up for switching with the pulses of a synchronously pumped dye laser (according to [3.24]), compare 6.

At the waveguide structure a d.c. voltage of about 100 volts was applied. The optically induced electrical signal at the slit was transmitted over short coaxial lines directly to the input B of a sampling head (HP 1430C) that had a rise time of $\lessapprox$ 20 ps. The sampling head was triggered externally by the signal of an avalanche diode connected to the input A. The diode is radiated by a fraction of the light of the likewise modelocked pump laser (krypton ion laser) with a pulse repetition frequency of 76 MHz. The dye laser pulses (peak power 100—150 W, pulse duration 5—10 ps, repetition frequency 76 MHz) were focussed by means of a lens ($f = 40$ mm) on the active detector face (0.45×0.03 mm^2). In this arrangement the optoelectronic switch can also be used as a fast photoreceiver. Its sensitivity lies at about 1 mV per 1 mW of the average laser power.

Most of the laser-switched electrooptical gates described thus far were switched with light pulse energies in the order of 1 mJ. However, through a drastic decrease of the slit width (10—30 μm) it has been possible to produce components which can be driven with the pulses from modelocked dye lasers (compare 5., 6.) the energy of which lies in the nJ range [3.22, 3.24]. Simultaneously, with the extremely short pulses of these lasers ($\tau_L \lessapprox 1$ ps) it seems to be possible to reduce further the switching times, whereby about 1 ps is considered possible. Fig. 3.19 shows the experimental set-up used in [3.24]. By means of such an arrangement Auston [28] obtained an optoelectronic sampling oscilloscope with which the pulse response function of a GaAs-FET was measured with an accuracy of about 1 ps.

A sub-picosecond electrooptical sampling technique was developed by Valdmanis, Mourou and Gabel [3.30] and improved by Meyer and Mourou [3.31]. Here, the voltage which is switched by means of a laser-driven optoelectronic switch, is applied directly to an electrooptical crystal in which the polarization properties of the test pulse are influenced. Fig. 3.20a) shows schematically one such sampler which is composed of a Cr-doped GaAs crystal and a $LiTaO_3$ crystal. On the surface of both crystals a coplanar strip transmission line is arranged, which is interrupted at one point. The switching pulses that originate from a fs-laser ($\tau_L \simeq 100$ fs) are focussed on the slit in the transmission line. In this way, a connection to a d.c. power supply of 50 V is produced and a switching signal of about 30 mV develops. This voltage generates an electric field in the electrooptical crystal, whereby the penetration depth of the field is of the order of the slit width, which like the strip width comes to only 50 µm. The test pulse is influenced now nearly inertialessly by the electrooptical effect, whereby as in Figure 3.14

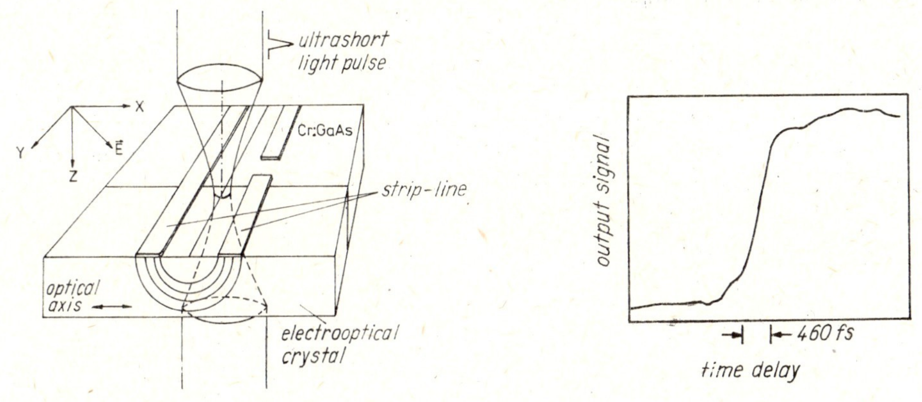

Fig. 3.20. Electrooptical sampling technique (from [3.30])

a) Experimental arrangement with electrooptical crystal and coplanar strip-line geometry

b) Electrical response of a Cr:GaAs shutter

The rise time of the fastest component is 460 fs (measured between 10% and 90% of the maximum signal)

the crystal is arranged between two crossed polarizers. This means that in the zero-field case no test signal is measured at the output; the test signal can only be observed if the output signal has switched the Pockels cell. Again, the sampling technique rests on the fact that the experiment is repeated often, whereby the temporal delay between the exciting and test pulses is changed while all other parameters are held constant. Fig. 3.20b) shows the output signal with respect to the delay between the exciting and the test pulse. We see that the response of the device is characterized by a very fast rise; the rise time (measured between 10% and 90% of the maximum signal) comes to 460 fs. This fast rise results from the compact, miniaturized geometry which, in particular, broadens the bandwidth of the coplanar strip line. The temporal resolution depends essentially on the transit time τ_W of the electric signal passing through the test beam waist and the transit time τ_0 of the test pulse passing through the field region. The transit time τ_W can be decreased by reducing the focus diameter of the test beam. The length of the field region to be passed through decreases with the electrode distance and the electrode width. By such means and using extremely short light pulses ($\tau_L \simeq 20$ fs) it should be possible in the future to achieve rise times below 100 fs.

4. Active Modelocking

4.1 Principles of Operation

The basic design of an actively modelocked laser differs from that of a common laser in that an amplitude or phase modulator whose modulation frequency coincides with the mode spacing frequency of the axial modes $\delta\nu_\mu = c/2L$ is mounted near one of the mirrors (see Fig. 4.1). An amplitude modulation can thereby be produced by an acousto-optical or electrooptical modulator, whereas a phase modulation can only be achieved using an electrooptical modulator.

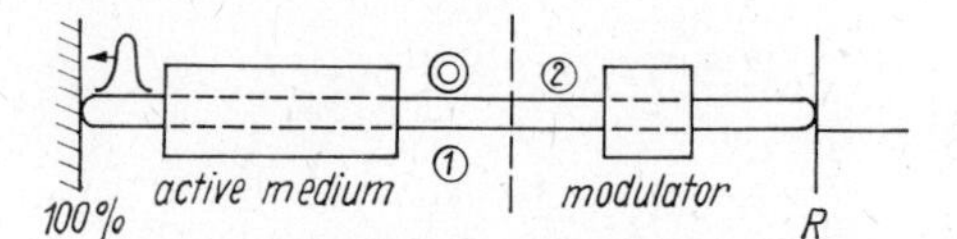

Fig. 4.1. Resonator configuration of an actively modelocked laser

4.1.1 *Amplitude Modulation*

Due to the time-varying transmission of the amplitude modulator, sidebands that overlap with the adjacent axial modes are excited for every oscillating mode. The occurrence of the modelocking can be explained as follows: If we excite the molecules of the laser medium by a pumping process, the threshold is reached first for that frequency ν_0 whose maximum lies next to the maximum of the amplification profile. The field of this mode is amplitude-modulated by the amplitude modulator with frequency $\delta\nu$, whereby sidebands are generated having a defined amplitude and phase at the frequency $\nu_0 \pm \delta\nu$.

Since both of these sidebands are eigenfrequencies of the resonator and lie within the amplification line of the laser transition, their fields are further amplified and modulated again in their turn with the modulation frequency $\delta\nu$, at which sidebands of the frequency $\nu_0 \pm 2\delta\nu$ develop. This process continues until all axial modes within the oscillation area are coupled or locked with one another.

We can also explain the same mechanism in the time domain where it should be taken into account that the modulation period is equal to the resonator round trip time u. For this reason, the laser radiation propagating forward and backward in the resonator always passes through the modulator at the same moment of its modulation period. Thus, during each resonator round trip, the radiation field is attenuated except for the portion that passes through the modulator at the moment in which almost no loss takes place (Fig. 4.2a). Accordingly, the radiation will concentrate itself in a short pulse within the time range of very small modulation losses.

4.1.2 *Phase Modulation*

The mechanism of modelocking by actively modulating the phase is similar to that of the amplitude modulation. Due to the temporal modulation of the phases, a number of sidebands whose phases are locked are generated for every mode within the amplification profile. In the time domain, this means that the radiation is concentrated on the time ranges in which the phase modulation $\Phi(t)$ is only slightly changed, that is, at one of the two extreme values within a modulation cycle where $\mathrm{d}\Phi/\mathrm{d}t = 0$ (see Fig. 4.2b). Outside these time ranges the field strength possesses a very fast temporal

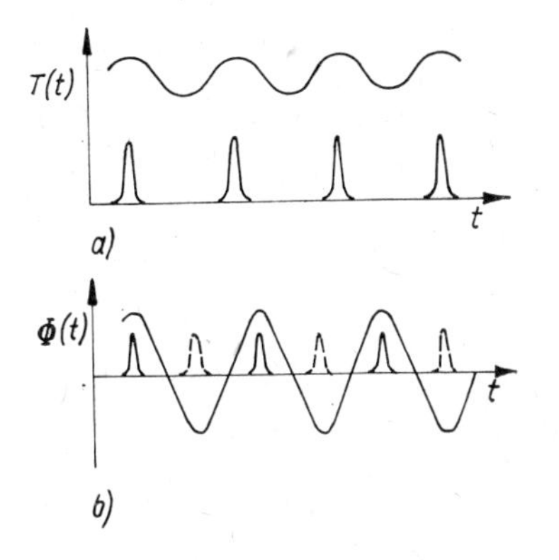

Fig. 4.2. Position of the optical pulse in relation to the modulation cycle for the case of amplitude modulation (a) and phase modulation (b)

modulation, which means the radiation obtains a frequency shift that is proportional to $\mathrm{d}\Phi/\mathrm{d}t$. During the round trips, these frequency shifts are added up so that the frequency of this portion of the radiation falls out of the bandwidth of the amplifying medium. Since two extreme values of the phase exist for every round trip an uncertainty results about the position of the generated pulse, because the pulse may originate with equal probability at each of these extreme points.

4.1.3 *Homogeneously and Inhomogeneously Broadened Laser Transitions*

The mechanisms of active modelocking of homogeneously and inhomogeneously broadened laser transitions differ greatly from each other.

In inhomogeneously broadened lasers, a large number of longitudinal modes whose phases are statistically distributed oscillate under sufficiently strong excitation, if there is no modulation signal. The locking of these modes is relatively simple to achieve, because only a weak sideband has to be generated by the modulator. This sideband serves as start signal for the adjacent mode and is amplified in a well-defined manner by the amplifier. Active phase locking was realized experimentally for the first time by Harris and Targ using an He—Ne laser of wavelength 0.633 μm with an acoustooptical loss modulator [4.1]. They obtained a periodic pulse train with pulse widths of about 2.5 ns. Detailed calculations of the active modelocking of inhomogeneously broadened lasers have been provided by Harris and McDuff [4.2] who started with the coupled equations of the individual modes in the frequency domain and solved them by assuming that the individual modes were saturated independently.

In homogeneously broadened lasers the mechanism of modelocking is very different and can be more favorably analyzed in the time domain. The modelocking can be described in the time domain with the generation of a short light pulse which moves back and forth in the resonator. Under continuous, steady pumping conditions (cw-regime), a situation of particular interest occurs: After a certain number of resonator round trips, the combined action of the amplifier and the modulator leads to the generation of a short pulse that reproduces itself after each round trip and no longer undergoes a change in its characteristics. The gain from the active medium is thus compensated by the losses in the modulator and at the mirrors, while the permanent pulse shortening is neutralized by the bandwidth limitation of the spectral amplification line or by a frequency selective element in the resonator. Accordingly, a continuous train of short pulses that remain unchanged in their characteristics leaves the resonator.

In the following, we want to restrict ourselves essentially to the case of homogeneously broadened lasers which, in particular, applies to the frequently used Nd:YAG lasers and the homogeneously broadened high pressure-CO_2-lasers. Active modelocking of the Nd:YAG laser was first achieved experimentally by Di Domenico et al. [4.3] using an amplitude modulator and by Osternik and Forster [4.4] using a phase modulator.

4.2 Theory

The theoretical description of a homogeneously broadened laser with active modelocking was first accomplished by Kuizenga and Siegman. In the following, we will for the most part base our discussion on this treatment.[1])

A complete theoretical analysis of this pulse-shaping process would have to consider the buildup of radiation from noise and the gradual formation of a pulse that achieves a cw-regime after a certain number of round trips and then ceases to evolve. As we shall see, however, the previous history is of little importance for the characteristics of cw-pulses. The requirement that a single pulse reproduces itself after passing through the active medium and the modulator determines its final shape which is independent from the preceding stages of the amplification process. We can find a self-consistent solution of this kind for the pulse profile by examining the distortion of a pulse in passing through the resonator elements as shown in the schematic arrangement in Fig. 4.1.

For the amplifier we assume a medium that can be described by a four-level system and assume further that we have steady pumping conditions. Since the finite spectral width of the laser transition is important in determining the pulse shape that appears in the cw-regime (in the event that no other optical elements with a smaller spectral bandwidth are present in the resonator), we cannot use the approximation of rate equations. A complete analysis of the four-level system using the fundamental equations (1.50) and (1.56) leads immediately to complicated mathematical problems that we can again simplify, however, under real conditions, because only the two levels 2 and 3 of

---

1) In our treatment we use a method different from the one in [4.5]. In particular, as opposed to [4.5] we do not need to assume a special pulse shape (Gaussian pulse). Rather, we obtain the shape of the pulse by solving the basic equations. The final results, however, are identical with those presented in [4.5].

the four in Fig. 2.2b) play an essential role. The other two levels under condition (2.17) cause a specific modification with regard only to the calculation of the occupation numbers. Taking into account the two laser levels, we therefore obtain for the field amplitude and the off-diagonal element ϱ_{23} from (1.50), (1.63) and (1.65)

$$\frac{1}{v}\frac{\partial A_L}{\partial t} + \frac{\partial A_L}{\partial z} = -\frac{i\mu_0\omega_L^2}{k_L}\mu_{32}\tilde{\varrho}_{23}, \tag{4.1}$$

$$\frac{\partial\tilde{\varrho}_{23}}{\partial t} + \left[\frac{1}{\tau_{32}} + i(\omega_L - \omega_{32})\right]\tilde{\varrho}_{23} = -\frac{1}{2i\hbar N}\mu_{23}N_3A_L. \tag{4.2}$$

In the determination of the occupation number density N_3 of the upper laser level, we can neglect the finite width of the laser transition which here causes only minor corrections, and consequently proceed from the rate equation (2.19) to obtain

$$\frac{\partial N_3}{\partial t} + \frac{1}{T_{31}}N_3 = -\sigma_{32}I_LN_3 + \sigma_{14}I_P(N - N_3). \tag{4.3}$$

In practice T_{32} is often relatively large, as for example in the case of actively mode-locked solid state lasers (e.g. Nd : YAG), so that the occupation of level 3 changes only little during one passage. We can therefore insert $N_3 \approx \overline{N}_3 = \frac{1}{u}\int_0^u d\eta N_3(\eta)$ in (4.2)

Equation (4.2) can then be solved and from (4.1) we obtain with $\omega_L \approx \omega_{32}$

$$\frac{\partial A_L}{\partial z} = \frac{\sigma_{32}\overline{N}_3}{2\tau_{32}}\int_{-\infty}^{\eta} A_L(\eta')\exp\left(-(\eta - \eta')/\tau_{32}\right)d\eta' \tag{4.4}$$

where $\eta = t - z/v$. As we shall see in the end, the pulse duration τ_L of the laser pulse is large compared to the relaxation time τ_{32}. In the frequency domain this means that not all axial modes within the laser transition can be generated because due to a mode selection process a large number of modes that experience smaller gain fall out. The main contribution in the integrand in (4.4) is supplied at the point $\eta = \eta'$. We can therefore develop $A_L(\eta')$ at point η up to the second order and obtain

$$\frac{\partial A_L}{\partial z} = \frac{\sigma_{32}\overline{N}_3}{2}\left(A_L - \tau_{32}\frac{\partial A_L}{\partial \eta} + \tau_{32}^2\frac{\partial^2 A_L}{\partial \eta^2}\right) \tag{4.5}$$

after carrying out the integration. When the boundary conditions are known, the solution (4.5) can be easily provided using the Poisson integral in accordance with the general methods of solving linear partial differential equations. If we still take into account that as shown in Fig. 4.1, the laser pulse goes through the amplifier a second time after it is reflected at the mirror, thus doubling the effective amplification length, then the pulse amplitude at point (1) in figure 4.1. is given by

$$A_L(1,\eta) = \frac{\exp\left(\frac{3}{4}g\right)}{2\sqrt{\pi}\,\tau_{32}\sqrt{g}}\int_{-\infty}^{\infty} d\eta' A_L(0,\eta')\exp\left(\frac{\eta - \eta'}{2\tau_{32}} - \frac{(\eta - \eta')^2}{4\tau_{32}^2 g}\right) \tag{4.6}$$

where $g = L^a \sigma_{32} \overline{N}_3$. In (4.6) we neglected that the mean occupation number $\overline{N}_3$ still depends slightly on $A_L(\eta)$ due to the bleaching of the amplifier. We can calculate this dependence from (4.3), in which we can neglect N_3/T_{32} according to our assumption, and obtain

$$\overline{N}_3 = \frac{\sigma_{14} I_P N}{1 + \sigma_{32} I_L + \sigma_{14} I_P}. \tag{4.7}$$

We now consider the change of the pulse as it passes through the amplitude modulator. For an acoustooptical modulator as well as for an electrooptical amplitude modulator the transmission function that describes a single passage is given by

$$\mathsf{T}_A(\eta) = \exp\left[-\delta_{AM} \sin^2\left(\omega_m(\eta - \eta_0) + \theta\right)\right] \tag{4.8}$$

in which δ_{AM} is the modulation index and ω_m the modulator frequency. η_0 is the moment at which the pulse maximum goes through the modulator. As we have already mentioned in section 4.1., in the case of modelocking a short pulse forms in the resonator in the time ranges where the losses are minimal. According to (4.8) this formation occurs twice in every modulation period. Thus $T_m = 2\pi/\omega_m = 2u$ must hold. In general the pulse does not pass through the modulator precisely within the transmission maximum, but is displaced from the maximum by a certain phase angle θ. Because $\omega_m \tau_L \ll 1$, (4.8) may approximately be written as

$$\mathsf{T}_A(\eta) = \exp\left\{-\delta_{AM}[\sin^2\theta + \sin(2\theta)\,\omega_m(\eta - \eta_0) + \omega_m^2(\eta - \eta_0)^2 \cos 2\theta]\right\}. \tag{4.9}$$

Taking into account the losses at the mirror and the modulator, we thus obtain for the pulse amplitude at position (2) of Fig. 4.1

$$\begin{aligned} A_L(2, \eta) = \sqrt{R} \exp\{-\delta_{Am}[\sin^2\theta + (\eta - \eta_0)\,\omega_m \sin 2\theta \\ + \omega_m^2(\eta - \eta_0)^2 \cos 2\theta]\}\, A_L(1, \eta), \end{aligned} \tag{4.10}$$

where R signifies the reflectivity of the output mirror. With (4.10) and (4.6) we have calculated the change of the pulse after a complete resonator round trip. In the cw-operation, the field strength of the pulse must reproduce itself after each round trip, whereby we obtain

$$A_L(2, \eta)\, e^{i\omega_L \eta} = A_L(0, \eta + h)\, e^{i\omega_L(\eta + h)}\, e^{-i\varphi} \tag{4.11}$$

as the self-consistency condition. The phase angle φ here includes a possible phaseshift during the round trip, while h signifies a time shift of the pulse maximum, which is caused by the amplification and the frequency selective properties of the active medium. Taking these properties into account, the effective resonator round trip time u of the pulse thus deviates from the resonator round trip time in the unpumped resonator u_0 by h: $u = u_0 - h$. Furthermore, it should be noted here that u_0 already differs from the value $2L/c$ (L — optical resonator length), since the pulses do not propagate with the phase velocity, but with the group velocity. If we insert the relations (4.10) and (4.6) in (4.11), we obtain a linear integral equation for the unknown amplitude function

$A_L(0, \eta)$. Thus we have

$$\frac{\sqrt{R}\exp\left(\frac{3}{4}g\right)}{2\sqrt{\pi}\,\tau_{32}\sqrt{g}}\exp\{-\delta_{AM}[\sin^2\theta+\omega_m(\eta-\eta_0)\sin 2\theta+\omega_m{}^2(\eta-\eta_0)^2\cos 2\theta]\}$$

$$\times\int_{-\infty}^{\infty}d\eta' A_L(\eta')\exp\left[\frac{\eta-\eta'}{2\tau_{32}}-\frac{(\eta-\eta')^2}{4\tau_{32}^2 g}\right]=A_L(\eta+h)\exp(i\omega_L h-i\varphi). \tag{4.12}$$

Since only Gaussian functions appear in (4.12) as integrand and as factors, we can evidently write the solution of (4.12) as

$$A_L(\eta)=A_0\,e^{-\gamma\eta^2}. \tag{4.13}$$

If we insert (4.13) in (4.6) we obtain

$$A_L(1,\eta)=\frac{A_0}{\sqrt{1+4g\tau_{32}^2\gamma}}\exp\left\{\frac{3}{4}g+\left[-\gamma\eta^2+2\gamma g\tau_{32}\eta+\frac{g}{4}\right]\frac{1}{1+4\gamma g\tau_{32}^2}\right\}. \tag{4.14}$$

The maximum of (4.14) is located at $\eta=g\tau_{32}$. The pulse is thus delayed by the amplifier by $\eta_V=g\tau_{32}$. Accordingly, we have to determine the time η_0 at which the pulse maximum passes through the modulator using this time delay: $\eta_0=g\tau_{32}$. If we insert (4.14) in (4.12), we obtain the following relations,

$$h=-g\tau_{32}+\frac{\omega_m\delta_{AM}\sin 2\theta}{2\gamma}, \tag{4.15}$$

$$\delta_{AM}\omega_m{}^2\cos 2\theta=\frac{g\gamma^2 4\tau_{32}^2}{1+4g\tau_{32}^2\gamma}, \tag{4.16}$$

$$\sqrt{\frac{R}{1+4g\gamma\tau_{32}^2}}\exp\left\{g-\delta_{AM}\sin^2\theta+\frac{\omega_m{}^2\delta_{AM}^2}{4\gamma}\sin^2 2\theta\right\}=1 \tag{4.17}$$

as conditions for the solution of the integral equation (4.12). With $h=u_0-\pi/\omega_m$ (4.15) to (4.17) represent a coupled system of transcendental equations from which the values γ, g and θ are to be determined as functions of the laser parameters ω_m, u_0, δ_{AM} and τ_{32}. The shortest and most intense pulses are obviously generated if the pulse maximum passes through the modulator in each resonator round trip at the moment of maximum transmission that is, when $\theta=0$. For this case the modulator frequency possesses a definite value $\omega_m=\omega_{m0}$, which is determined from the relations $h=-g\tau_{32}$ and (4.15) as

$$\omega_{m0}=\pi[u_0+\tau_{32}g]^{-1}. \tag{4.18}$$

The solutions of equations (4.15) to (4.17) allow deviations of the modulator frequency from this value by a certain amount, by which a stability range of the laser parameters is defined. For values of the laser parameters outside this stability range there exist no self-consistent solutions, that means there is no stable pulse regime.

Let us discuss in more detail the self-consistently determined laser parameters for the optimal case $\omega_m = \omega_{m0}$. After solving for γ equation (4.16) yields

$$\gamma = \frac{\delta_{AM}\omega_{m0}^2}{2} + \frac{1}{2}\sqrt{\delta_{AM}^2\omega_{m0}^4 + \frac{\delta_{AM}\omega_{m0}^2}{g\tau_{32}^2}}. \tag{4.19}$$

For typical values of the parameter we can assume $\omega_{m0}^2\delta_{AM} \ll 1$, because $\omega_{m0} \ll 1/\tau_{32}g\tau_{32}^2$. For the pulse duration $\tau_L = \sqrt{(2\ln 2)/\gamma}$ we thus obtain

$$\tau_L = \frac{2\sqrt{\ln 2}}{\sqrt{\omega_m}}\sqrt{\tau_{32}}\sqrt[4]{\frac{g}{\delta_{AM}}}. \tag{4.20}$$

The gain coefficient g is determined self-consistently as

$$g = \frac{1}{2}\ln\frac{1}{R} + \frac{1}{2}\ln(1 + 4g\gamma\tau_{32}^2), \tag{4.21}$$

according to (4.17), whereby the second term yields only a small correction compared to the first. With the definition of $g = L^a\overline{N}_3\sigma_{32}$ and relation (4.7), this means a self-consistent determination for the pulse energy. For this purpose, a more accurate value of g can be calculated by means of successive approximation; i.e. after inserting $g \approx (\ln 1/R)/2$ in (4.18) and (4.20) and in the last term of (4.21).

Let us now briefly discuss the case of a phase modulator. In this case the transmission function (4.8) must be replaced by

$$\begin{aligned} T_{PM}(\eta) &= \exp\{-i\delta_{PM}\cos(\omega_m(\eta - \eta_0) + \theta)\} \\ &\approx \exp\left\{\mp i\delta_{PM}\left[\cos\theta - \omega_m(\eta - \eta_0)\sin\theta - \frac{1}{2}\omega_m^2(\eta - \eta_0)^2\cos^2\theta\right]\right\}. \end{aligned} \tag{4.22}$$

The double meaning of the sign in (4.22) means that two possible solutions exist for the case of the phase modulation, one for every extreme value of the phase variation. Furthermore, it should be noted that in the case of phase modulation the modulator frequency coincides with the pulse repetition frequency $\left(\omega_m = \frac{2\pi}{u}\right)$, whereas $\omega_m = \frac{\pi}{u}$ holds under the amplitude modulation.

First, let us again consider the ideal case in which the pulse passes through the modulator precisely at one of the extreme values of the phase variation ($\theta = 0$). The modulator frequency in this case is given by $\omega_{m0} = 2\pi[u_0 + \tau_{32}g]^{-1}$. The above calculations can be carried out analogously, whereby δ_{AM} has to be replaced everywhere by $i\delta_{PM}$. Accordingly, we obtain

$$\gamma_{PM} = \frac{1}{2}\sqrt{\frac{\pm\delta_{PM}i\omega_m^2}{g\tau_{32}^2}} = (1 \mp i)\frac{\omega_m}{2\tau_{32}}\sqrt{\frac{\delta_{PM}}{g}} \tag{4.23}$$

from (4.19) for the factor γ in the exponent of (4.13). The developing signal thus possesses a Gaussian-shaped amplitude function and a time-varying phase which is proportional to η^2. The pulse duration follows from (4.10), whereby δ_{AM} must be replaced by δ_{PM}.

We can thus see that in both cases the pulse duration τ_L is proportional to $\sqrt{\tau_{32}/\omega_m}$. An enhancement of the modulation frequency and a corresponding change in the resonator length lead to a decrease in the pulse duration.

If we consider as an example a Nd : YAG laser with reflectivity of the mirrors $R = 0.9$, resonator length $L = 0.6$ m, linewidth $\Delta\omega/2\pi = 120$ GHz and modulation index $\delta_{PM} = 0.3$ rad, we obtain a pulse duration of $\tau_L \approx 70$ ps.

The case of a small deviation of the modulator frequency ω_m from ω_{m0} can be treated similarly to the previous calculations, whereby the term in the exponent in (4.22), which is linearly dependent on time, causes a frequency shift. Without going into the details of the calculation Fig. 4.3. shows the pulse widths that result from phase locking as a function of the detuning $\delta\omega_m = \omega_{m0} - \omega_m$ [4.5]. The smallest pulse duration occurs at a small negative detuning while it increases monotonically for a positive detuning. In comparison, in the case of amplitude modulation the minimum of the pulse width was at $\delta\omega = 0$, and the pulse duration increased monotonically with positive and negative detuning.

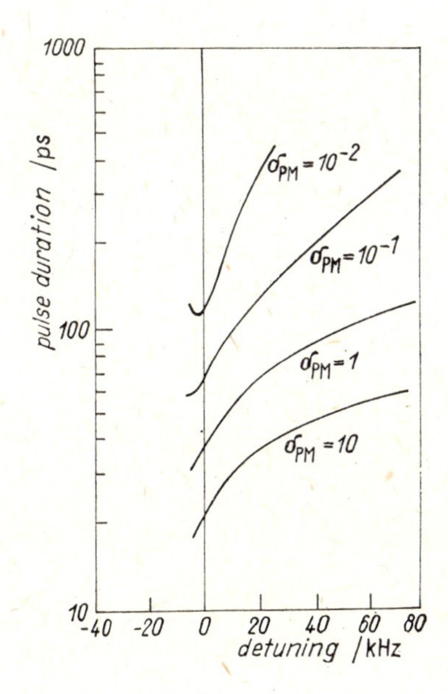

Fig. 4.3. The pulse duration as a function of the detuning for the case of modelocking by means of active phase modulation (from Kuizenga and Siegmann [4.5])

A modification of the results given in (4.20) and (4.23) is of practical relevance, if the spectral bandwidth limitation in the resonator is not given by the amplification line, but if, for example, a Fabry-Perot etalon considerably reduces the effective bandwidth. In this case the pulse width does not depend on the spectral bandwidth of the amplifier, but on the linewidth of the Fabry-Perot etalon $\Delta\omega_{FP}$. The pulse width in this case is given by

$$\tau_L = \frac{\sqrt{\ln 2}}{\sqrt[4]{2\delta_{AM}}\sqrt{\omega_m \Delta\omega_{FP}}}. \tag{4.24}$$

4.3 Experimental Investigations

The scheme of a possible experimental arrangement is given in figure 4.4, as it was used, for example in [4.6] for the modelocking of a Nd : YAG laser. It consists of a continuously (cw) pumped Nd : YAG laser rod, a modulator situated inside the resonator, and components for measuring the pulse spectrum and pulse duration. The essential element here for the modelocking is the modulator.

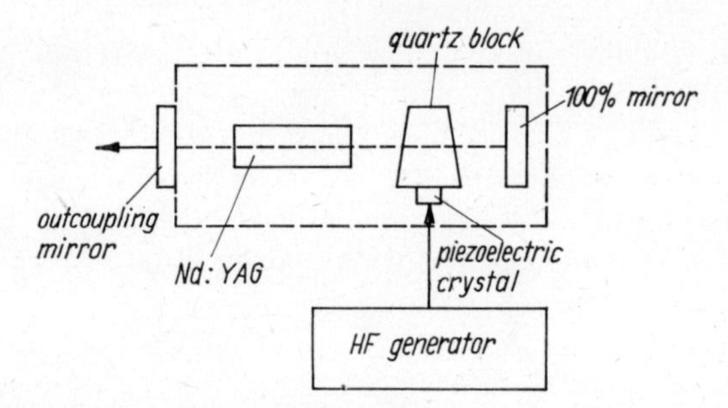

Fig. 4.4. Experimental arrangement for active modelocking of a Nd: YAG laser

4.3.1 Modulators

The amplitude modulation is usually achieved by using an acoustooptical modulator. In such a modulator an ultrasonic wave is sent through an optically transparent medium such as quartz (see Fig. 4.5). An ultrasonic wave which passes through a transparent material or a corresponding standing wave field causes a phase grating. This results

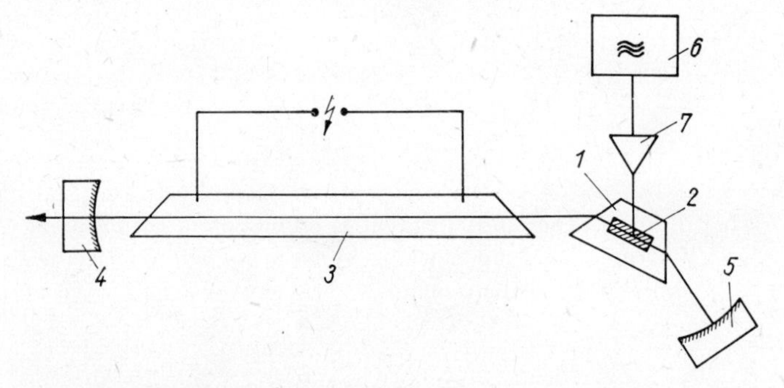

Fig. 4.5. Actively modelocked noble gas ion laser
1 is the mode-locker prism made of fused quartz (SQ 1) with Brewster surfaces; 2 is the piezoelectric oscillator made of $LiNbO_3$; 3 is the gas discharge tube; 4 is the output mirror; 5 is the end mirror; 6 is the frequency synthesizer; 7 is the power amplifier (from [4.10]).

from the photoelastic effect, through which the varying stress field of the ultrasonic wave is coupled with the refractive index. The grating that develops is, in modulators with standing ultrasonic waves as they are used for active modelocking, periodically time dependent. The spatial period is given by the half acoustic wavelength ($\Lambda/2$), whereas the amplitude of the modulation is proportional to the intensity of the sonic wave. If a light wave falls on this kind of grating, a portion of the intensity is deflected in another direction. This portion of the radiation can be directed out of the resonator with suitable laser parameters, whereby an energy loss results which is modulated periodically with time. The ultrasonic wave is fed into the optically transparent medium using a piezoelectric transformer which transforms a high frequency electrical signal into ultrasonic energy. (More detailed investigations of acoustic modulation — particularly of different kinds of deflections at thin or thick gratings — are found in [4, 4.3]).

A phase and amplitude modulation can be achieved by using an electrooptical modulator. An electrooptical modulator is based on the following principle of operation: due to the electrooptical effect or Kerr effect certain crystals or liquids under the influence of an external electric field can become birefringent. This change in the refractive index after applying an external electric field can be used to change the polarization characteristics of the light that propagates in the medium. This change in the polarization or in the refractive index for one polarizing direction can be used in various ways to achieve an optical modulation.

Let us consider somewhat more closely the performance of an electrooptical modulator using the example of lithium-niobate ($LiNbO_3$). We choose the coordinates such that the light propagates in the x-direction, while the modulating field is applied perpendicular to it in the z-direction along the optical axis of the crystal. Due to the influence of the external modulating field E_z the refractive index becomes dependent on the field strength of the external field. Depending on whether the light is polarized in the y- or the z-direction, a change in the refractive index occurs with respect to this direction, which is given by

$$\Delta n_y = -\frac{1}{2} r_{13} n_0{}^3 E_z, \tag{4.25}$$

$$\Delta n_z = -\frac{1}{2} r_{33} n_e{}^3 E_z; \tag{4.26}$$

n_0 and n_e are the ordinary and extraordinary refractive indices, and r_{13} and r_{33} the electrooptical coefficients of $LiNbO_3$. From a refractive index difference Δn of this kind an optical phase difference $\delta\varphi = 2\pi \Delta n a/\lambda_0$ relative to the zero-field case is produced in a crystal of length a with respect to a propagation of light in the x-direction. If we introduce the electrical voltage $V_z = E_z d = V_0 \cdot \cos \omega_m t$ in place of the field strength, we obtain for the phase difference

$$\delta\varphi_z = -\frac{n_e{}^3 \pi r_{33}}{\lambda_0 d} a V_0 \cos \omega_m t; \tag{4.27}$$

d is here the length of the crystal in the z-direction. Due to the temporal modulation of the external field a corresponding modulation of the phase having the same modulation frequency ω_m thus results. If the incident light is polarized in the y-direction, the term $r_{33} n_e{}^3$ in (4.27) must be replaced by $r_{13} n_0{}^3$. For $LiNbO_3$, these parameters at $\lambda_0 = 1.06$ µm are given by the following values: $n_0 = 2.24$ and $n_e = 2.16$, $r_{13} = 8.6 \times 10^{-12}$ m/V and $r_{33} = 30.8 \times 10^{-12}$ m/V. For a potential of $V_0 = 300$ V and a 5 mm $\times$ 5 mm $\times$ 20 mm crystal there follows accordingly a maximum phase difference of $\delta\varphi_{max} = 1$ rad.

4.3.2 Modelocked Noble Gas Ion Lasers

The active modelocking of the argon ion laser was first achieved in [4.7] and [4.8] using an amplitude modulator and in [4.9] with a phase modulator. Like the krypton ion laser, this type of laser has gained particular importance in recent years from its use as a pump pulse source for synchronously pumped dye lasers, which are described in chapter 5. Actively modelocked noble gas ion lasers of this kind are in use today in many laboratories, whereby industrially fabricated lasers that have already been mentioned in section 2.4.2 and which are completed with the appropriate modulators, are often employed. In section 2.4.2 noble gas ion lasers have already been discussed. In this section, we shall briefly discuss the particulars of their active modelocking. In Fig. 4.5 an example of a resonator arrangement for an actively modelocked argon ion laser is given (according to [4.10]). The modelocking of an argon ion laser ILA 120 from VEB Carl Zeiss Jena (3) was achieved here using a prismatic modulator (1). The main part of this prismatic standing wave modulator consists of a fused quartz block (1) with a piezoelectric generator (2). The prismatic quartz block may simultaneously be used for the wavelength selection in the laser resonator. The facets of the prism are cut such that the laser beam enters the modulator at the Brewster angle. In this manner the resonator losses are kept to a minimum and the appearance of subresonances is avoided completely. The modulator block can be calibrated and stabilized to a certain temperature using a heating element with an electronic regulation, in order that the required modulation frequency always coincides with an acoustic resonance of the standing wave modulator. By this means a maximum degree of modulation is achieved. The modulator is matched to the cable of the radio-frequency control electronics via an LC element and is supplied by a frequency synthesizer (6) at the required resonance frequency (which here amounts to 62 MHz). The modulator and the laser mirror (5) may be attached adjustably to a common optical carrier. With this set-up a continuous pulse train can be generated with pulses having a duration of 110 ps, a pulse power of 50 W, a pulse repetition frequency of 124 MHz and a mean output power of 0.65 W. With longer and more powerful gas discharge systems, lower pulse repetition frequencies, mean output powers of about 1.5 W and pulse powers of 200 to 300 W are achieved for pulses that have the same or somewhat shorter durations (up to about 70 ps).

4.3.3 Experimental Investigations of Actively Modelocked Nd : YAG Lasers

Kuizenga and Siegman investigated the modelocking of a Nd:YAG laser and verified many aspects of the theory described in section 4.2 with the results of their experiments. For the modelocking of their Nd:YAG laser they used a $LiNbO_3$ crystal as an electro-optical phase modulator with a modulation frequency of 264 MHz. The laser bandwidth $\Delta\nu_L$ was measured using a Fabry-Perot interferometer and the pulse length τ_L with a fast photodiode. Shorter pulses were measured using a correlation technique on the basis of second harmonic generation (SHG) (compare 3.). Depending on the modulation index δ_{PM} chosen, they observed pulses with a length of 40 ps to 200 ps having a mean output power of about 300 mW. Without using any additional elements the crystal block of the modulator functioned as a Fabry-Perot etalon, which limits the bandwidth of the laser radiation. In order to obtain the shortest pulses, these mode selection effects must be eliminated by avoiding all interfering reflections in the modulator crystal (e.g.

in that the end faces of the crystal are inclined to the optical axis at the Brewster angle) or by inclining the crystal and thus widely deflecting the reflected beams. The bandwidth and the duration of the pulses were measured as a function of the modulation index δ_{PM}. The results are plotted in Fig. 4.6. The solid lines confirm the relations $\tau_L \sim (\delta_{PM})^{-1/4}$ and $\Delta\nu_L \sim \delta_{PM}^{1/4}$ that the theory predicted. As we can see, the results show relatively good agreement between theory and experiment. If we assume a loss of 10% per round trip, a line width of 120 GHz and a modulation frequency of 264 MHz, then (4.20) (where $\delta_{AM} \to \delta_{PM}$) provides a theoretical value of $\tau_L = 37.9(\delta_{PM})^{-1/4}$. The spectral halfwidth of a Gaussian pulse (4.13) with the value for γ calculated from (4.23) with the same parameters is given by $\Delta\nu_L = 16.5(\delta_{PM})^{1/4}$. On the other hand, the experimental results yielded $\tau_L = 46(\delta_{PM})^{-1/4}$ and $\Delta\nu_L = 17.6(\delta_{PM})^{1/4}$. Thus, the deviation of the number factors from the theoretical values is likewise minimal. The theoretically predicted Gaussian shaped pulse profile was also verified by experiment.

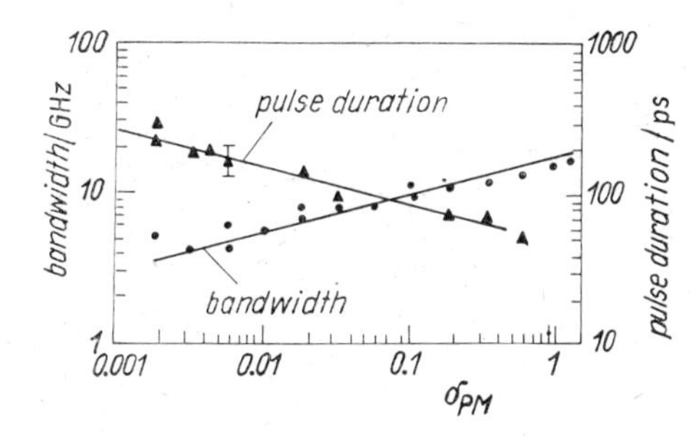

Fig. 4.6. Pulse width and spectral bandwidth of a phase locked Nd:YAG laser versus the modulation index δ_{PM} (from Kuizenga and Siegmann [4.6])

In comparison with the modelocked noble gas ion lasers, the modelocked continuously pumped Nd:YAG laser possesses some substantial advantages. Among these we should mention particularly the shorter pulse durations (about 50 ps) and the higher mean power (about 10 W) [4.11]. Furthermore, the fluctuations of the pulse parameters within the pulse train — particularly in the range above 100 kHz — are smaller, because the slowly relaxing active medium exercises a "stabilizing" effect. For the reasons mentioned, the Nd:YAG laser is frequently employed beside the noble gas ion laser as a pump source for synchronous pumping (compare 5.). The fundamental wave of the Nd:YAG laser is suited for the excitation of color center crystals, which emit in the range from 0.8 to 3.8 μm [2.14, 4.13], as well as for the excitation of special dyes that emit up to a wavelength of 1.45 μm [4.12]. For the pumping of common dye lasers operating in the visible spectral range the second or third harmonic ($\lambda = 0.53$ μm or $\lambda = 0.355$ μm) of the Nd:YAG laser radiation (compare 8.) is used.

Through a combination of Q-switching (compare 2.5) and active modelocking, it is possible to generate ps-pulses of increased power (to about 1 MW) with a continuously pumped Nd:YAG laser. In [4.14] the Q-modulation was achieved using an acoustooptical travelling wave modulator and the modelocking by means of an acoustooptical standing wave modulator. Here, groups of about 30 ultrashort light pulses were generated, whereby the length of one group was about 300 ns, and the group repetition frequency was in the kHz-range.

5. Synchronously Pumped Lasers

5.1 Principle of Operation

In the last section we described the principle of modelocking by modulating inner losses or the optical path length. Modelocking can, however, also be achieved through modulation of the gain. This can occur by pumping a laser with the continuous pulse train of another laser that is already modelocked (see Fig. 5.8). If the resonator length of the laser is nearly equal to that of the pump laser or an integral multiple of it, then under certain conditions the gain is temporally modulated with a modulation period that is equal to the resonator round trip time. As in the modulation of inner losses, similarly in this case a short pulse in the time range of the maximum gain builds up, whose pulse length under optimal conditions can be two to three orders of magnitude smaller than that of the pump pulses. The method of synchronous pumping is of special practical interest for dye lasers, since these are mostly optically excited and possess a wide gain profile ($\Delta\omega_{32}/2\pi \approx 10^{13}$ Hz to 10^{14} Hz). By using a frequency selective optical filter that considerably narrows the bandwidth of the laser radiation in the resonator and makes it possible to change the frequency of the resulting gain maximum (such as, for example, a Fabry-Perot etalon, a Lyot filter or a prism), the frequency in a dye laser can be tuned over a certain range. The spectral width of the filter profile, however, must not be too small since otherwise the pulses will be considerably lengthened. For the reasons mentioned synchronously pumped dye lasers have gained increasing importance in the generation of tunable picosecond and subpicosecond pulses in the last years. As opposed to passively modelocked lasers, which are dealt with in the next chapter, this method has the advantage that nearly the total spectral width of the laser transition can be used for tuning, whereas in passively modelocked lasers the tuning range is limited additionally by the spectral profile of the absorption band of the saturable absorber.

The modelocking of dye lasers by synchronous pumping had already been used early on, whereby the dye laser was pumped by the pulse train of a modelocked ruby laser [5.1] or by the second harmonic of a Nd:glass laser [5.2, 5.3]. The pulse durations achieved in these experiments with pulsed pump sources lay, however, only in the order of the pump pulses. Only when actively modelocked argon ion or krypton ion lasers were used as pump sources was it possible to satisfy more accurately certain experimental conditions necessary for a steady-state regime (cw) [5.4 to 5.7]. By this means synchronous pumping of dye lasers developed into a favorable method of generating ultrashort pulses up to the subpicosecond region [5.8, 5.10].

Let us illustrate in greater detail the mechanism of gain modulation in synchronous pumping using the graphs in Fig. 5.1. The dye laser is assumed to be pumped by a continuous pulse train of an actively modelocked laser. The pulse duration of these pump pulses in, for example, the frequently used actively modelocked argon ion lasers comes to about 100 to 200 picoseconds. The relaxation time of the upper laser level T_{32}, which lies in the nanosecond range in typical lasers (eg. $T_{32} = 5$ ns holds for rhodamine 6G), is thus large compared to the pulse duration of the pump and the generated laser pulses; it is, however, smaller than the resonator round trip time $u_0 \approx 2L/c$. Hence

$$\tau_L, \tau_P \ll T_{32}, T_{32} < u_0 . \quad (5.1)$$

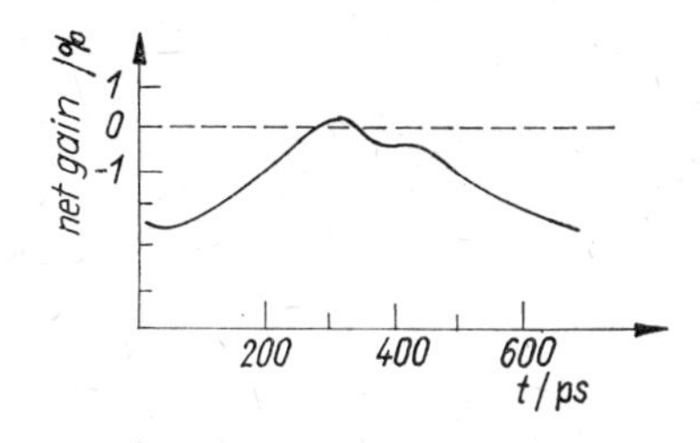

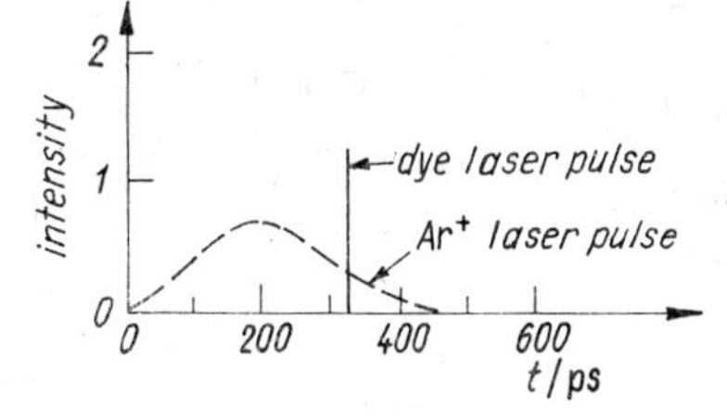

Fig. 5.1. Temporal evolution of the net gain and the intensity of the pump laser and dye laser in a synchronously pumped laser

Under these conditions, the occupation inversion of the active medium depends only on the pump energy supplied up to this moment. As depicted in Fig. 5.1, the gain coefficient increases gradually due to the pumping, until it exceeds the level of the losses, that is until the laser threshold is reached. As of this moment the laser begins to oscillate. The energy of the laser pulse increases rapidly thereafter and reaches the "saturation energy" of the active medium in the steady-state laser regime, that is, after the start-up processes of the amplification are finished. As a result, the occupation inversion of the dye is reduced by stimulated emission alone, whereby the gain again drops quickly to a value below the level of the losses.

Thus, a positive net gain occurs only within a small time interval during the pump pulse duration, so that the generated laser radiation is concentrated in this time range. The depletion of the gain caused by the generated laser pulse itself is here essential for a successful modelocking, because in this way the steepening of the pulse at the trailing edge occurs. It is important that the laser pulse in the continuous regime always travels exactly synchronously to the pump pulse through the active medium. This requires a relatively accurate matching of the resonator lengths of both lasers up to several micrometers. As in active modelocking, the effective limiting of the spectral width of the laser radiation also plays an important part, since in this manner the shortest achievable pulses are determined.

Color center lasers can be synchronously pumped in a very similar way to the dye lasers. Various color centers in alkali and alkaline earth halide crystals with wide fluorescence bands in the range between 0.8 μm and 3.8 μm also allow in this spectral range pulses of high repetition rate and very short durations to be generated, whereby either ion gas lasers, dye lasers or cw pumped Nd:YAG lasers are used for pumping (see e.g. [2.14, 4.13]). The pulse formation occurs in the same way in these lasers as in dye lasers, for which reason all explanations about dye lasers in the following sections also apply to the color center lasers.

The method of synchronous pumping is applicable and of particular interest in semiconductor lasers, as well. On the one hand, it is possible to use optical synchronous pumping for the excitation (e.g. with the help of dye lasers) whereas, on the other hand, these lasers can also be synchronously pumped without optical means by modulating the injection current [18].

5.2. Theory

5.2.1 Basic Equations

As in the active modelocking and passive modelocking of dye lasers, the pulse formation process in synchronously pumped lasers can be divided into three different phases. In the first gain phase a pulse builds up from noise whose energy increases with each round trip, and whose duration decreases. After a certain number of round trips the pulse reaches its maximum energy, where in this second phase it is shortened even more until it obtains its final duration. In the last phase the pulse ultimately reaches its final position relative to the pump pulse, and from this moment on the pulse duration, pulse energy and relative position of pump and laser pulse remain unchanged. The properties of the pulse in this steady state regime are relatively independent of the preceding development, assuming that the laser parameters satisfy certain conditions of the modelocking. Therefore, we shall again restrict ourselves first of all to the theoretical description of this regime (cw-regime). Here, we shall follow mainly the works of Herrmann and Motschmann ([5.11, 5.12][1]).

Consider a linear resonator arrangement that contains a dye as active medium and a frequency selective element for frequency tuning (see Fig. 5.2). The active medium is considered here as a four-level system (compare Fig. 2.2b) that is pumped by the pulse train of another, actively modelocked laser. For the description of the laser field and the atomic system we can again begin with equations (4.1) to (4.3). Furthermore, we assume the restricting condition (5.1) and that the fluorescence bandwidth of the dye is great compared to the bandwidths of the frequency selective element and the laser pulse. The bandwidth limiting due to the active medium which is caused by the frequency

---

[1]) Earlier theoretical investigations of synchronously pumped lasers are found in [5.13–5.18]. In [5.13, 5.14] the pulse evolution was studied by numerically solving the rate equations where a bandwidth limitation of the laser radiation was not taken into consideration. An approach of this kind fails to describe a steady state regime; the pulse duration approaches zero with increasing round trip number. Therefore conclusions about the pulse duration and other parameters cannot be drawn. Simple theoretical models were discussed in [5.15–5.18] where the bandwidth limitation of the laser radiation was accounted for which allows the solution of the basic equations to be found analytically. However, unrealistic approximations were used which strongly restricts the range of validity of the results obtained.

dependence of the emission cross section σ_{32} can then be neglected with regard to that of the optical filter, and therefore we obtain from equations (4.1) and (4.2) a similar equation to (1.71) (where σ_{21} and T_{21} are to be replaced by σ_{32} and T_{32} respectively). In place of the complex equation (1.71) we obtain

$$\frac{\partial I_L}{\partial z}(z, \eta) = \sigma_{32}(\omega_L) N_3 I_L(z, \eta) \tag{5.2a}$$

and

$$\frac{\partial \varphi_L}{\partial z}(z, \eta) = -\frac{\Delta_{32}}{2} \sigma_{32}(\omega_L) N_3 \tag{5.2b}$$

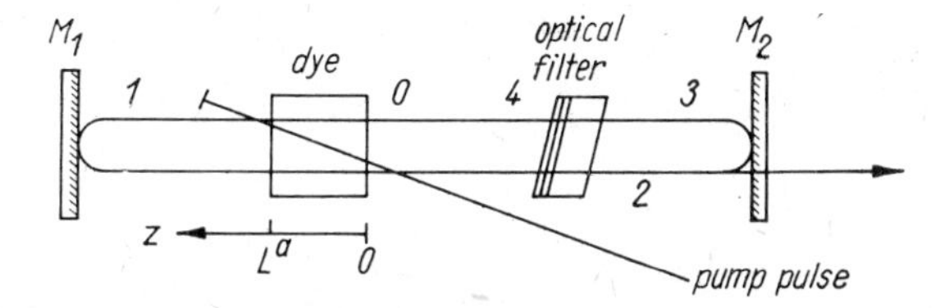

Fig. 5.2. Scheme of the resonator of a synchronously pumped laser
M_1 is the 100% mirror; M_2 is the output mirror with reflectivity R; 0, 1, ..., 4 designate the respective positions of the pulse during a round trip.

for the intensity I_L and the phase φ_L of the laser wave. From (4.3) under the condition (5.1) we further obtain

$$\frac{\partial N_3}{\partial \eta} = \sigma_{14}(\omega_P) I_P N_1 - \sigma_{32}(\omega_L) N_3 I_L, \quad N_3 + N_1 = N, \tag{5.2c}$$

and for the intensity of the pump radiation

$$\frac{\partial I_P}{\partial z} = -\sigma_{14}(\omega_P) N_1 I_P. \tag{5.2d}$$

In (5.2b), a possible detuning of the laser center frequency ω_L (which is determined by the adjustment of the optical filter) compared with the frequency at the maximum of the fluorescence profile ω_{32} was taken into account by the parameter $\Delta_{32} = \dfrac{\omega_L - \omega_{32}}{\tau_{32}}$.
In the system of equations (5.2), the same group velocity was assumed for the pump pulse and laser pulse ($v_P = v_L = v$), for which reason only one local time coordinate $\eta = t - z/v$ appears. This is justified for small lengths of the amplifier, typical disperion parameters and for pulse durations $\tau_L \gtrapprox 0.1$ ps.

We introduce a new function $\gamma(z, \eta)$

$$N_3 = -\frac{1}{\sigma_{32}} \frac{\partial}{\partial z} \ln \gamma(z, \eta), \tag{5.3}$$

which we insert into (5.2a) and (5.2d) and integrate with respect to z. In this manner we obtain

$$I_{\mathrm{L}} = f_{\mathrm{L}}(\eta)/\gamma\,, \tag{5.4}$$

$$I_{\mathrm{P}} = \mathrm{e}^{-\sigma_{14}Nz} f_{\mathrm{P}}(\eta)\, \gamma^{-\sigma_{14}/\sigma_{32}}; \tag{5.5}$$

$f_{\mathrm{L}}(\eta)$ and $f_{\mathrm{P}}(\eta)$ here are functions still unknown, which are determined by the boundary condition at $z = 0$. If we insert (5.3) into (5.2c), account for (5.2a) and (5.2d) and integrate with respect to z, then for $\gamma(z, \eta)$ we can derive the equation

$$\frac{\partial}{\partial \eta} \ln \gamma(z, \eta) = \sigma_{32}(I_{\mathrm{L}} + I_{\mathrm{P}})\,. \tag{5.6}$$

Equation (5.6) and the boundary conditions $I_{\mathrm{L}}(z = 0, \eta) = I_{\mathrm{L0}}(\eta)$, $I_{\mathrm{P}}(z = 0, \eta) = I_{\mathrm{P0}}(\eta)$ yield for the function $\gamma(z = 0) = \gamma_0$ the relation

$$\gamma_0 = \exp\left\{\sigma_{32} \int_{-\infty}^{\eta} \mathrm{d}\eta' \big(I_{\mathrm{L0}}(\eta') + I_{\mathrm{P0}}(\eta')\big)\right\}. \tag{5.7}$$

If we insert the relations (5.4) and (5.5) in (5.6) and eliminate $f_{\mathrm{L}}(\eta) = I_{\mathrm{L0}}\gamma_0$, $f_{\mathrm{P}}(\eta) = I_{\mathrm{P0}}\gamma_0{}^{\sigma_{14}/\sigma_{32}}$, then

$$\begin{aligned}\frac{\partial \gamma}{\partial \eta} &= \sigma_{32} I_{\mathrm{L}}(0, \eta) \exp\left\{\sigma_{32} \int_{-\infty}^{\eta} [I_{\mathrm{L}}(0, \eta') + I_{\mathrm{P}}(0, \eta')]\, \mathrm{d}\eta'\right\} \\ &\quad + \sigma_{32} I_{\mathrm{P}}(0, \eta)\, \gamma^{1-\sigma_{14}/\sigma_{32}} \exp\left\{\sigma_{14} \int_{-\infty}^{\eta} [I_{\mathrm{L}}(0, \eta') + I_{\mathrm{P}}(0, \eta')]\, \mathrm{d}\eta' - \sigma_{14} N z\right\}\end{aligned} \tag{5.8}$$

follows. Let us solve equation (5.8) for two special cases that are sufficient for our analysis of the pulse shaping process.

In order to achieve the highest possible conversion of pump energy into laser energy, the extinction $\sigma_{14}NL^{\mathrm{a}}$ should be as great as possible. For typical values $\sigma_{14} = 2 \times 10^{-20}\,\mathrm{m}^2$, $N = 2 \times 10^{-3}$ mol/l, $z = L^{\mathrm{a}} = 0.2$ mm, for example, we obtain $\sigma_{14}NL^{\mathrm{a}} \approx 5$. With that we can neglect the second term in (5.8) and obtain

$$\gamma(L^{\mathrm{a}}, \eta) = 1 + \sigma_{32} \int_{-\infty}^{\eta} \mathrm{d}\eta' I_{\mathrm{L}}(0, \eta') \exp\left\{\sigma_{32} \int_{-\infty}^{\eta'} [I_{\mathrm{L}}(0, \eta'') + I_{\mathrm{P}}(0, \eta'')]\, \mathrm{d}\eta''\right\}. \tag{5.9}$$

A special solution can be found for arbitrary values of $\sigma_{14}NL^{\mathrm{a}}$, if $\sigma_{32} = \sigma_{14}$. Hence,

$$\begin{aligned}\gamma(L^{\mathrm{a}}, \eta) &= 1 + \sigma_{32} \int_{-\infty}^{\eta} \mathrm{d}\eta' [I_{\mathrm{L}}(0, \eta') + I_{\mathrm{P}}(0, \eta') \exp(-\sigma_{14}NL^{\mathrm{a}})] \\ &\quad \times \exp\left\{\sigma_{32} \int_{-\infty}^{\eta'} [I_{\mathrm{L}}(0, \eta'') + I_{\mathrm{P}}(0, \eta'')]\, \mathrm{d}\eta''\right\}.\end{aligned} \tag{5.10}$$

In the following, we will refer mainly to the case of (5.9) and use (5.10) only to check the results at small values of $\sigma_{14}NL^{\mathrm{a}}$.

At position ① in figure 5.2 the intensity is

$$I_L(1, \eta) = \frac{1}{R} G(\eta) I_L(0, \eta), \tag{5.11a}$$

and the phase of the electric field is

$$\varphi_L(1, \eta) = -\frac{\Delta_{32}}{2} \ln\left(\frac{1}{R} G(\eta)\right) + \varphi_L(0, \eta) \tag{5.11b}$$

where we have introduced the effective gain G which already takes into account the outcoupling at the mirror with reflectivity R.

$$G(\eta) = \left\{\exp\left[\sigma_{32}\int_{-\infty}^{\eta}\left(I_L(0, \eta') + I_P(0, \eta')\right) d\eta'\right]\right\} [\gamma(L^a, \eta)]^{-1} \cdot R \tag{5.12}$$

The passage through the frequency selective element can be favorably described in the frequency domain. Hence,

$$A_L(2, \omega - \omega_L) = \mathsf{T}_F(\omega - \omega_L) A_L(1, \omega - \omega_L). \tag{5.13}$$

The transmission function T_F — as already mentioned — is given by a Lorentzian profile

$$\mathsf{T}_F(\omega - \omega_L) = \left[1 + 2i\frac{\omega - \omega_L}{\Delta\omega}\right]^{-1} \tag{5.14}$$

which applies to many practical cases, such as a Fabry-Perot etalon near the transmission maximum at ω_L. The bandwidth here is determined according to (2.40) by $\Delta\omega = (c/d)[(1 - R_{FP}) R_{FP}^{-1/2}]$ where R_{FP} is the reflectivity and d the thickness of the Fabry-Perot etalon. After the backtransforming of (5.13) and (5.14) into the time domain, we obtain for the intensity and phase at position ②

$$I_L(2, \eta) = \frac{\Delta\omega^2}{4R}\left\{\left[\int_{-\infty}^{\eta} e^{-\frac{\Delta\omega}{2}(\eta-\eta')}\sqrt{G(\eta') I_L(0, \eta')}\cos\varphi_L(0, \eta')\, d\eta'\right]^2 + \left[\int_{-\infty}^{\eta} e^{-\frac{\Delta\omega}{2}(\eta-\eta')}\sqrt{G(\eta') I_L(0, \eta')}\sin\varphi_L(0, \eta')\, d\eta'\right]^2\right\} \tag{5.15a}$$

and

$$\varphi_L(2, \eta) = \arctan\left\{\frac{\int_{-\infty}^{\eta} e^{-\frac{\Delta\omega}{2}(\eta-\eta')}\sqrt{G(\eta') I_L(0, \eta')}\sin\varphi_L(0, \eta')\, d\eta'}{\int_{-\infty}^{\eta} e^{-\frac{\Delta\omega}{2}(\eta-\eta')}\sqrt{G(\eta') I_L(0, \eta')}\cos\varphi_L(0, \eta')\, d\eta'}\right\}. \tag{5.15b}$$

After reflection at the mirror and the passage through the frequency selective element once more we obtain

$$I_L(4, \eta) = \left(\frac{\Delta\omega}{2}\right)^2 \left\{ \left[\int_{-\infty}^{\eta} d\eta' (\eta - \eta') \exp\left(-\frac{\Delta\omega}{2}(\eta - \eta')\right) \right.\right.$$
$$\left. \times \sqrt{G(\eta') I_L(0, \eta')} \cos\left(\varphi_L(0, \eta') - \frac{\Delta_{32}}{2} \ln\left(\frac{G(\eta')}{R}\right)\right)\right]^2$$
$$+ \left[\int_{-\infty}^{\eta} d\eta' (\eta - \eta') \exp\left(-\frac{\Delta\omega}{2}(\eta - \eta')\right) \right.$$
$$\left.\left. \times \sqrt{G(\eta') I_L(0, \eta')} \sin\left(\varphi_L(0, \eta') - \frac{\Delta_{32}}{2} \ln\left(\frac{G(\eta')}{R}\right)\right)\right]^2 \right\} \tag{5.16a}$$

and

$$\varphi_L(4, \eta) = \arctan \left\{ \frac{\int_{-\infty}^{\eta} d\eta' (\eta - \eta')\, e^{-\frac{\Delta\omega}{2}(\eta-\eta')} \sqrt{G(\eta') I_L(0, \eta')} \sin\left(\varphi_L(0, \eta') - \frac{\Delta_{32}}{2} \ln\left(\frac{G(\eta')}{R}\right)\right)}{\int_{-\infty}^{\eta} d\eta' (\eta - \eta')\, e^{-\frac{\Delta\omega}{2}(\eta-\eta')} \sqrt{G(\eta') I_L(0, \eta')} \cos\left(\varphi_L(0, \eta') - \frac{\Delta_{32}}{2} \ln\left(\frac{G(\eta')}{R}\right)\right)} \right\}. \tag{5.16b}$$

A steady-state regime has been achieved, if the pulse reproduces itself after each resonator round trip thus if the condition

$$I_L(4, \eta) = I_L(0, \eta + h) \tag{5.17}$$

is satisfied.

If we insert (5.16) into (5.17), we obtain a system of integro-differential equations for determining the steady-state pulse parameters. The value h — the so-called time mismatch — takes into account here a possible displacement of the maximum — caused by the amplification process and the frequency selective element. The maximum is shifted forward ($h_A > 0$) by the amplifier, whereas the frequency selective element delays the pulse ($h_F < 0$), where for the resulting time mismatch $h = h_A + h_F$ holds. As in active modelocking, the effective round trip time u thus deviates from the round trip time of the unpumped resonator u_0 by $-h$, ($u = u_0 - h$) due to the action of the amplifier and the frequency selective element. Thus, for the modelocking a condition between the spacing of the pump pulses u_P (which in actively modelocked lasers of modulator frequency ω_m is given by $u_P = \pi/\omega_m$ in the case of amplitude modulation and by $u_P = 2\pi/\omega_m$ in the case of phase modulation) and the effective round trip time u must be satisfied: $u_P = u = u_0 - h$.

Frequently, in place of the time mismatch h the equivalent length difference of the laser resonator $\delta L = \frac{hc}{2}$ which we term the resonator detuning is also used. Let us investigate first the case of small detuning of the laser mid-frequency ω_L in relation to the position of the fluorescence maximum $\omega_{32}(\Delta_{32} \approx 0)$. From (5.16 b) we obtain for the solution a constant phase $\varphi_L(\eta) = \text{const}$, that is the pulses do not have a phase modulation. In 5.2.4 let us consider separately the case of chirp generation under the condition $\Delta_{32} \neq 0$. In order to simplify the integral equation (5.17) we can assume that the bandwidth of the pulse is small compared to the bandwidth of the frequency selective element and that the time mismatch h is small compared to the pulse duration. Taking this assumption into consideration we obtain the following simplified equation from (5.16) and (5.17):

$$I_L(0, \eta) + h \frac{\partial I_L(0, \eta)}{\partial \eta} + \frac{h^2}{2} \frac{\partial^2 I_L(0, \eta)}{\partial \eta^2}$$
$$= I_L(0, \eta)\, G(\eta) - \frac{4}{\Delta\omega} \frac{\partial}{\partial \eta} [I_L(0, \eta)\, G(\eta)] + \frac{12}{\Delta\omega^2} \frac{\partial^2}{\partial \eta^2} [I_L(0, \eta)\, G(\eta)]$$
$$- \frac{2 \left[\frac{\partial}{\partial \eta} \left(I_L(0, \eta)\, G(\eta) \right) \right]^2}{\Delta\omega^2 I_L(0, \eta)\, G(\eta)}. \tag{5.18}$$

An exact analytical solution of this equation is not possible. Therefore, we develop $I_L(0, \eta)$ near the maximum into a power series. At the pulse wings the gain $G(\eta)$ is nearly constant, because in these ranges it depends only on the slowly varying pump pulse ($\tau_P \gg \tau_L$). Thus the solution of (5.18), where $G(\eta) = \text{const}$, yields an exponential function. Therefore, we choose the following ansatz for the pulse profile:

$$I_L(0, \eta) = I_0 \cdot \begin{cases} \frac{1}{2} \exp\,[\varrho_i(\eta - \eta_i)/\theta] & \text{for } -\infty < \eta < \eta_i, \\ [1 - (\eta/\theta)^2 + \mu(\eta/\theta)^3] & \text{for } \eta_i < \eta < \eta_f, \\ \frac{1}{2} \exp\,[-\varrho_f(\eta - \eta_f)/\theta] & \text{for } \eta_f < \eta < \infty. \end{cases} \tag{5.19}$$

The three time intervals are linked together at points η_i and η_f, where η_i and η_f follow from the solution of the equation $1 - (\eta/\theta)^2 + \mu(\eta/\theta)^3 = 1/2$. θ and μ are measures for the pulse duration and the pulse asymmetry, respectively. I_0 is the photon flux density in the maximum of the pulse. The pulse duration is given by $\tau_L = \eta_f - \eta_i$. For nearly symmetrical pulses ($\mu \leq 0.3$), $\tau_L \approx \sqrt{2}\,\theta$ holds. As pump pulse we assume a Gaussian shaped pulse (according to the results in active modelocking from the last section), whose maximum is displaced by a certain amount η_0 compared to the position of the maximum of the laser pulse. Thus, we have

$$I_P(0, \eta) = I_{P0} \exp\{-(\eta - \eta_0)^2/\tau_P^2\}. \tag{5.20}$$

We insert the ansatz (5.19) in equation (5.18) and compare the coefficients of the powers η^0 and η^1. Since the total energy $\mathscr{E}_{St}$ and the energy up to the pulse maximum $\mathscr{E}_0$ are included in the equations obtained in that manner, we integrate the equation (5.18) first from $\eta = -\infty$ to $\eta = \infty$, and secondly from $\eta = -\infty$ to $\eta = 0$ and obtain two more equations. Additionally, the integration of $I_L(\eta)$ (5.19) over the corresponding time intervals produces two relations linking the laser parameters to $\mathscr{E}_{St}$ and $\mathscr{E}_0$. With that we obtain six coupled transcendental equations for the following six unknown quantities $\tilde{\mathscr{E}}_{St} = \sigma_{23} \int_{-\infty}^{\infty} I_L(0, \eta) \, d\eta$ (normalized steady-state pulse energy), $\tilde{\mathscr{E}}_0 = \sigma_{23} \int_{-\infty}^{\infty} I_L(0, \eta) \, d\eta$ (normalized energy up to the pulse maximum), $\tilde{I}_0 = \sigma_{23} I_0/\Delta\omega$ (normalized intensity), μ (asymmetry parameter), θ (pulse duration) and η_0 (time interval between the two pulse maxima) as a function of h (time mismatch), R (reflectivity of the outcoupling mirror), $\tilde{\mathscr{E}}_P^{tot} = \sigma_{14} \int_{-\infty}^{\infty} I_P(0, \eta) \, d\eta$ (normalized total pump energy) and τ_P (pump pulse duration). These six transcendental equations have been solved numerically, and the results are shown in figures 5.3 and 5.4. We will discuss each of these solutions individually.

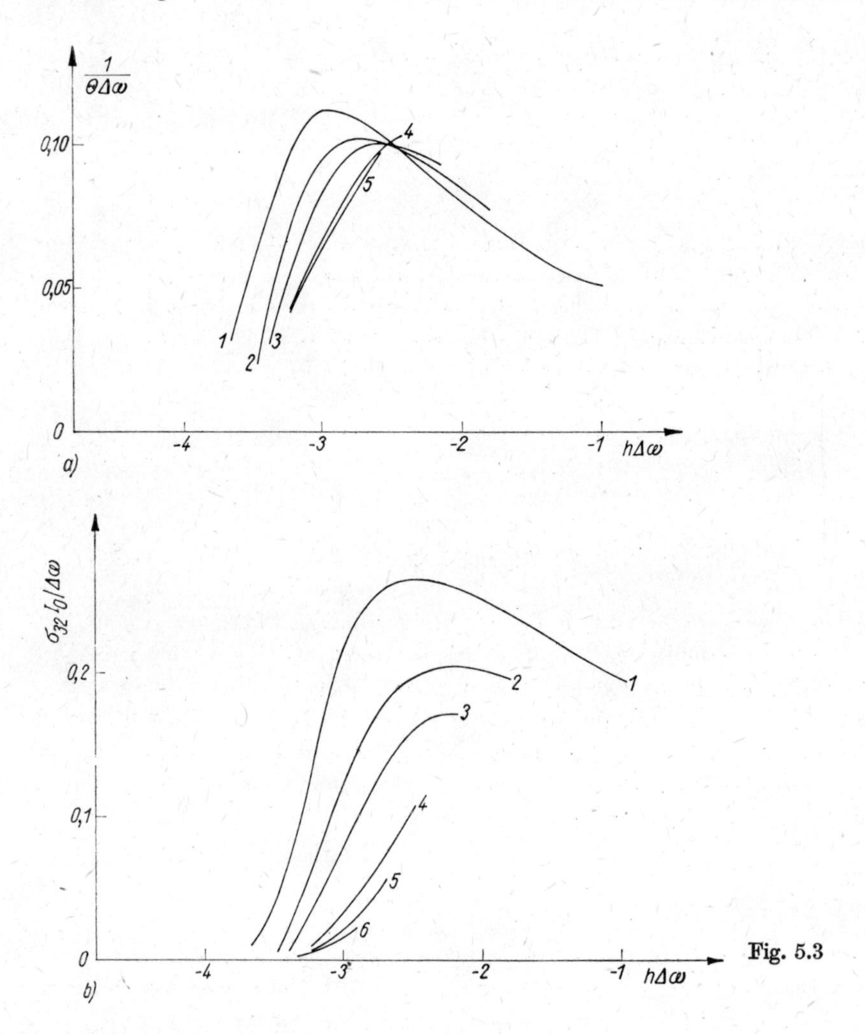

Fig. 5.3

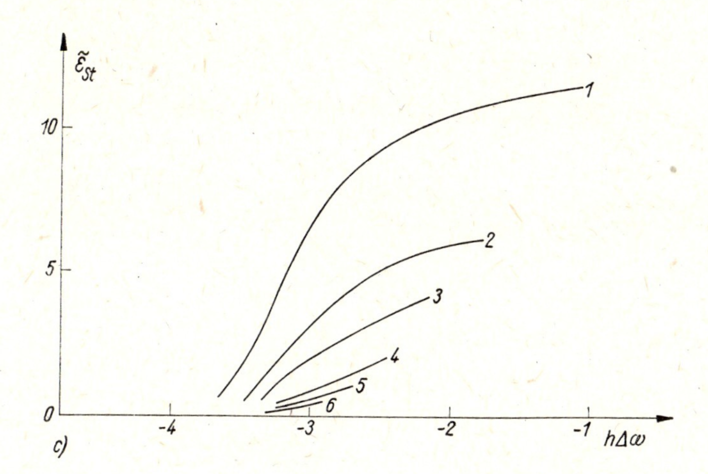
$\tilde{\varepsilon}_{st}$
10
5
0
-4
-3
-2
-1
$h\Delta\omega$
1
2
3
4
5
6
c)

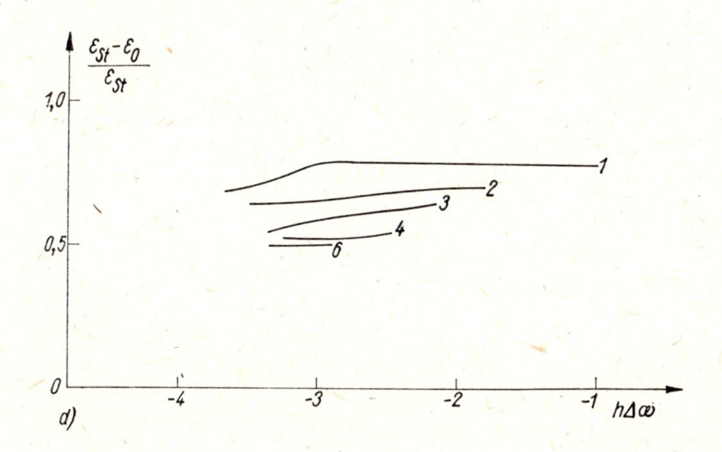
$\frac{\varepsilon_{st}-\varepsilon_0}{\varepsilon_{st}}$
1,0
0,5
0
-4
-3
-2
-1
$h\Delta\omega$
1
2
3
4
6
d)

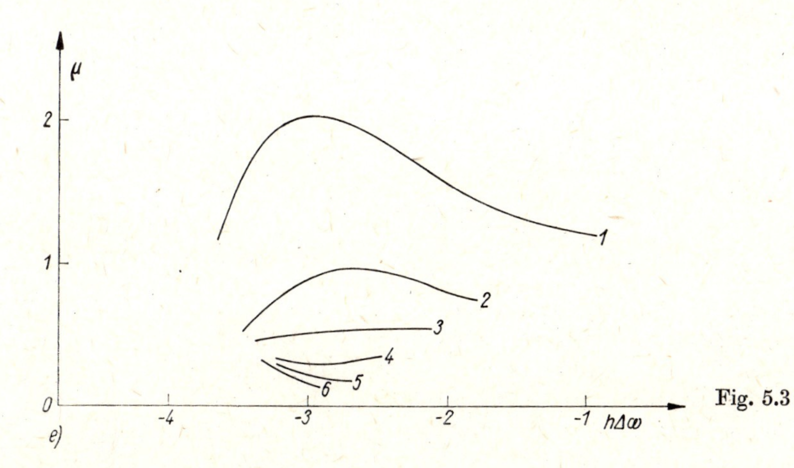
μ
2
1
0
-4
-3
-2
-1
$h\Delta\omega$
1
2
3
4
5
6
e)

Fig. 5.3

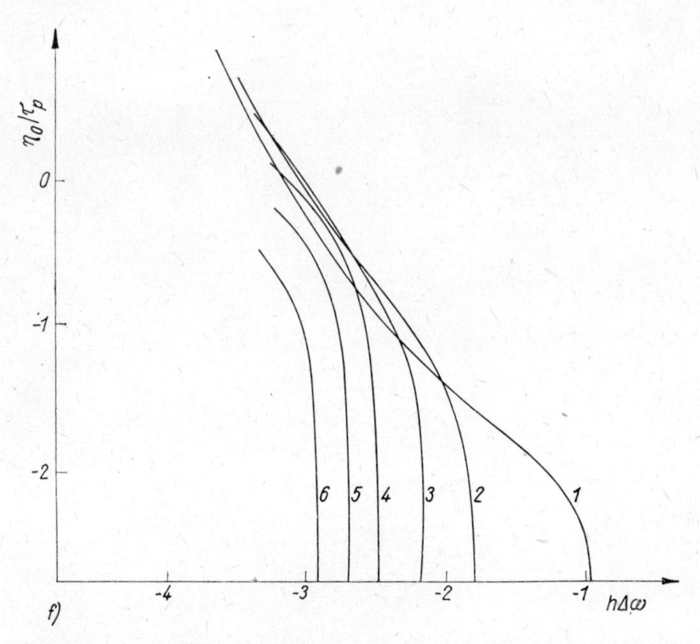

Fig. 5.3. a) to e) Pulse parameters, and f) time delay between the pump and laser pulse as functions of the resonator detuning ($h = 2\delta L/c$) and the reflectivity R

Parameters: curve (1): $R = 0.93$, curve (2): $R = 0.88$, curve (3): $R = 0.83$, curve (4): $R = 0.71$, curve (5): $R = 0.60$, curve (6): $R = 0.50$; $\tilde{\mathscr{E}}_{\mathrm{P}}{}^{\mathrm{tot}} = 1$, $\tau_{\mathrm{P}}\Delta\omega = 1000$.

(from Herrmann and Motschmann [5.11])

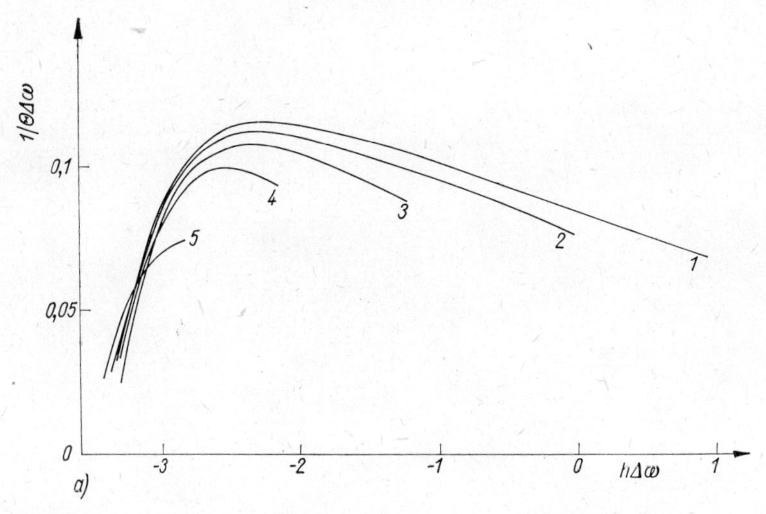

Fig. 5.4. a) to e) Pulse parameters and f) time delay between the pump pulse and laser pulse with respect to the resonator detuning ($h = 2\delta L/c$) and the pump energy

Parameters: curve (1): $\tilde{\mathscr{E}}_{\mathrm{P}}{}^{\mathrm{tot}} = 2.5$, curve (2): $\tilde{\mathscr{E}}_{\mathrm{P}}{}^{\mathrm{tot}} = 2.0$, curve (3): $\tilde{\mathscr{E}}_{\mathrm{P}}{}^{\mathrm{tot}} = 1.45$, curve (4): $\tilde{\mathscr{E}}_{\mathrm{P}}{}^{\mathrm{tot}} = 1.0$, curve (5): $\tilde{\mathscr{E}}_{\mathrm{P}}{}^{\mathrm{tot}} = 0.54$; R = 0.83, $\tau_{\mathrm{P}}\Delta\omega = 1000$.

(from Herrmann and Motschmann [5.11])

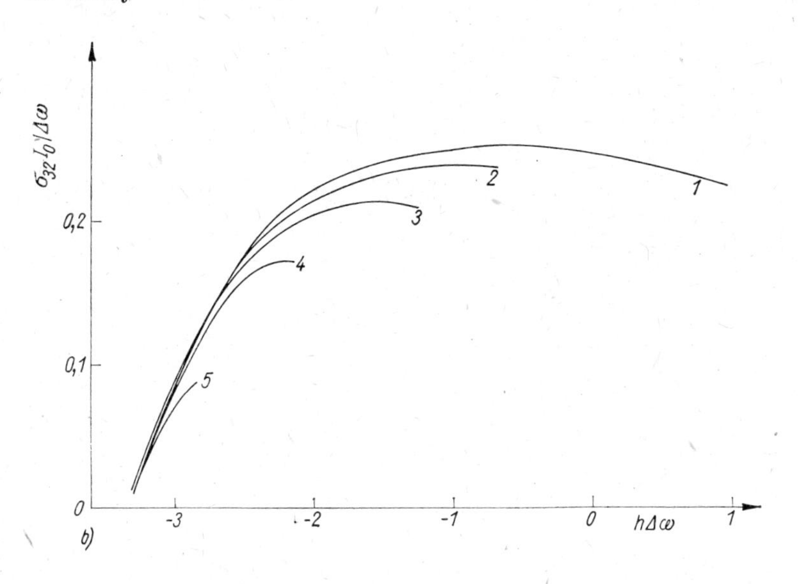
0,2
0,1
0
-3
-2
-1
0
1
hΔω
1
2
3
4
5
b)

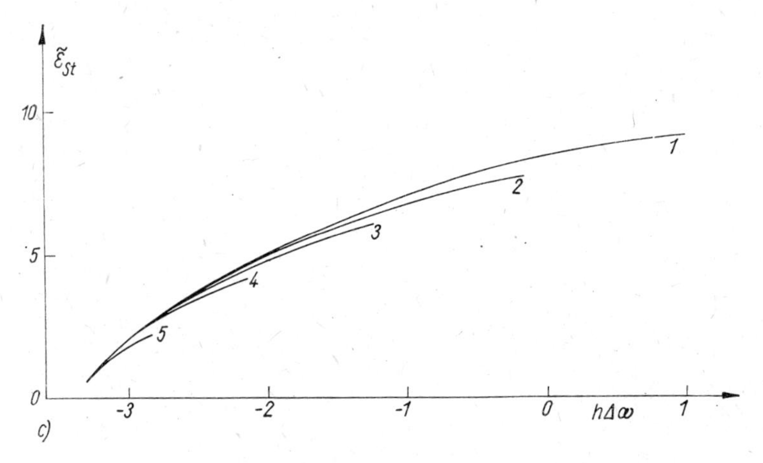
ε̃St
10
5
0
-3
-2
-1
0
1
hΔω
1
2
3
4
5
c)

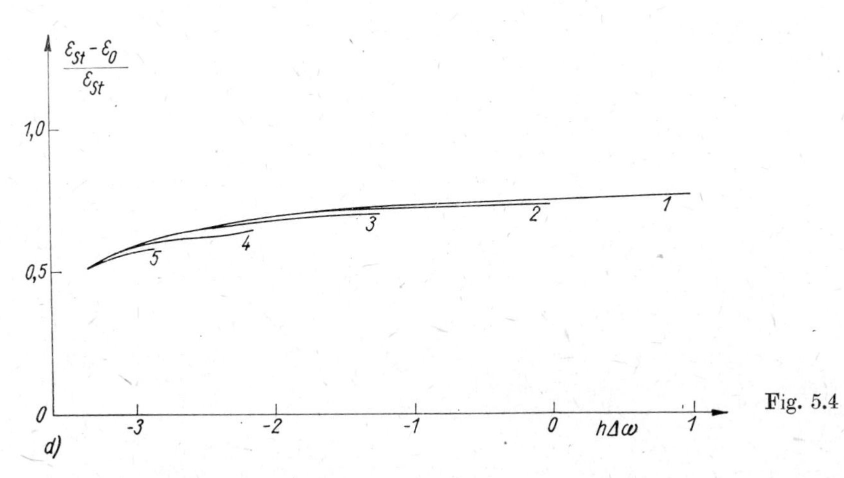
(εSt - ε0) / εSt
1,0
0,5
0
-3
-2
-1
0
1
hΔω
1
2
3
4
5
d)

Fig. 5.4

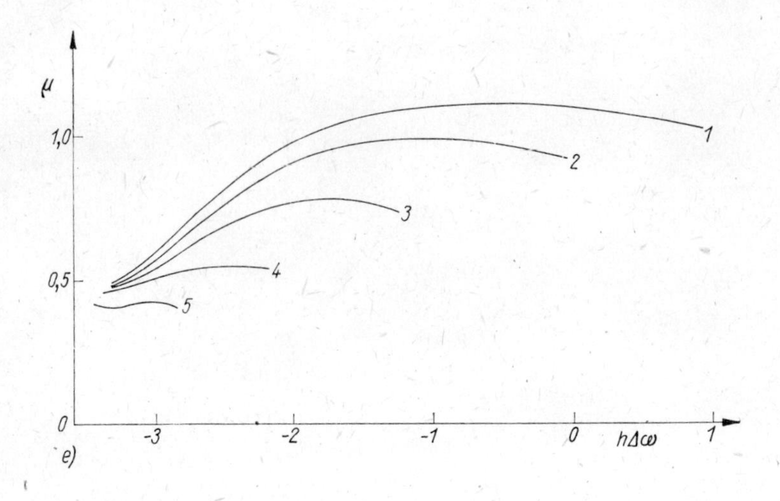

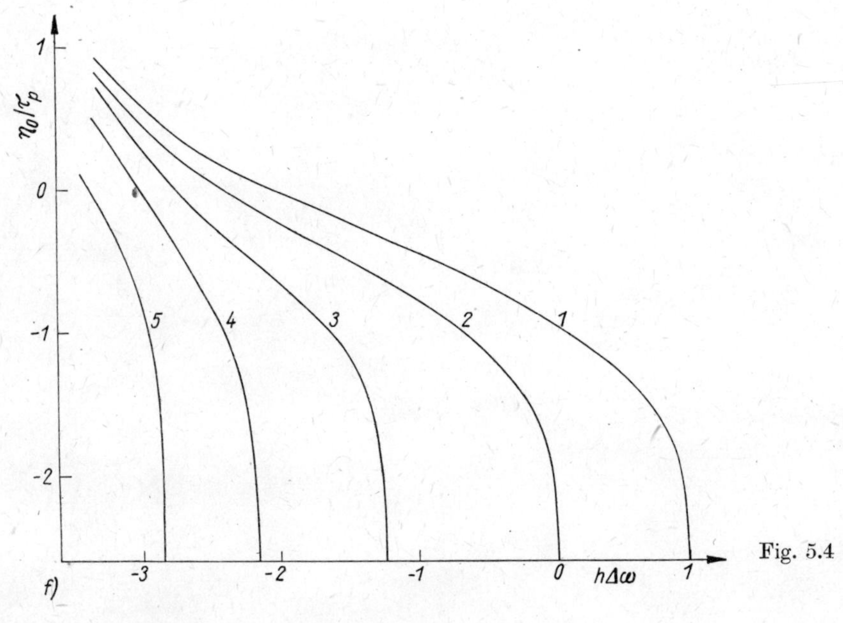

Fig. 5.4

5.2.2 *Discussion of the Solutions for the Steady-State Regime*

5.2.2.1 *Laser Threshold*

When the active medium is excited with a train of short pulses, the laser threshold differs from that discussed in chapter 2. We can characterize the laser threshold by $G(\eta)\ R = 1$ and a vanishing dye laser intensity $I_L \to 0(\mathscr{E}_{St} \to 0, \mathscr{E}_0 \to 0, I_0 \to 0)$. In this case the amplification $G(\eta)$ reaches its maximum at the end of the pump pulse ($\eta \to \infty$). Therefore, we obtain for the maximum amplification

$$G_{max} = \exp \{\tilde{\mathscr{E}}_P^{tot}\} \{1 + [\exp (\tilde{\mathscr{E}}_P^{tot}) - 1] \exp (-\sigma_{14} N L^a)\}^{-1}. \tag{5.21}$$

Thus, the laser threshold is exceeded, if the condition

$$\sigma_{14}NL^{a} \geqq \ln \{[1 - \exp(-\tilde{\mathscr{E}}_{P}^{tot})] [R - \exp(-\tilde{\mathscr{E}}_{P}^{tot})]^{-1}\} \tag{5.22}$$

is satisfied. For example, for $\tilde{\mathscr{E}}_{P}^{tot} = 1$, $R = 0.82$ we obtain $\sigma_{14}NL^{a} > 0.31$.

5.2.2.2 *Stable Pulse Regime (Modelocking Region)*

A set of real solutions of the equation (5.18) exists only in a limited region of the time mismatch h or the corresponding resonator detuning δL (compare Fig. 5.3 and 5.4). This region corresponds to a stable single pulse regime or the region of good modelocking. Within this variation range of the resonator length of the dye laser, the laser pulse and the pump pulse circulate with the same period. This is possible because of the pulse delay by the frequency selective element and the forward displacement of the pulse by the amplifier. The shift of the pulse peak caused by the amplifier occurs because the trailing edge of the pulse is cut off due to gain depletion. Thus, inside the stability range the synchronization condition for the round trip times of both lasers $u = u_P$ adjusts itself automatically. In this way the system is able to regulate by itself a steady spacing of laser pulse and pump pulse in each round trip. For small amplification coefficients ($\tilde{\mathscr{E}}_{P}^{tot} \leqq 2$) the delay effect dominates ($h < 0$ or $\delta L < 0$), whereas with greater amplification the delay effect is compensated, and the stability range shifts itself into the range of positive time mismatch ($h > 0$) (compare Fig. 5.4).

The boundaries of the stability range depend directly on the time spacing between the maximum of the dye laser pulse and the pump pulse (compare Fig. 5.3f) and 5.4f)). With decreasing resonator lengths the delay η_0 between pump and laser pulse becomes smaller, whereby the part of the pump pulse that is responsible for the amplification is reduced, and thus a smaller gain coefficient results. At a certain laser resonator length the laser threshold is not exceeded and therefore the generation of modelocked pulses is no longer possible. Accordingly, the left boundary of the stability ranges (the smallest possible laser resonator lengths) is characterized in Fig. 5.3 and 5.4 by a vanishing laser intensity. The right boundary of the range is produced because at a certain largest possible laser resonator length the total pump pulse energy is used for the amplification and thus the maximum possible displacement of the laser pulse is achieved by the amplifier. A further extension of the laser resonator length can then no longer be compensated by the forward shift of the pulse in the amplifier and leads to a distortion of the synchronization. At this point η_0 approaches $-\infty$.

To illustrate the results obtained for the stability range, we take as an example a very small linewidth of the frequency selective element of $\Delta\omega = 5 \times 10^{11}\,\text{s}^{-1}$, a reflectivity of $R = 93$ percent and a relative pump energy $\tilde{\mathscr{E}}_{P}^{tot} = 1$. From figure 5.3, curve (1) we obtain for the width of the stability range for the resonator detuning $\Delta\delta L = 2.4$ mm. This result is in good agreement with the corresponding experiments [5.28].

5.2.2.3 *Pulse Parameters*

In Figs. 5.3 and 5.4, the normalized pulse parameters such as the reciprocal pulse duration ($1/\theta\Delta\omega$), the steady-state pulse energy ($\tilde{\mathscr{E}}_{St}$), the intensity ($\sigma_{32}I_0/\Delta\omega$), the asymmetry factor μ as a function of the time delay ($h\Delta\omega$) and the resonator detuning $\delta L = hc/2$, respectively, are shown for different reflectivities R (Fig. 5.3) and pump energies $\tilde{\mathscr{E}}_{P}^{tot}$ (Fig. 5.4).

As seen from the figures, with increasing pump energy or higher reflectivity, shorter and more intense pulses are generated. At low reflectivities or small pump energies the reciprocal pulse length and the intensity exhibit a monotonic behavior; their maximum values are reached at the end of the stability range at the greatest possible laser resonator length. At higher reflectivities or greater pump energies the reciprocal pulse lengths and the intensity possess a maximum within the modelocking region. This extreme value is of particular practical interest for the choice of laser parameters, since the pulse length has its minimum there and small changes of the resonator detuning have no significant influence on the modelocking process and laser parameters. If we consider as an example a spectral width $\Delta\omega = 10^{14}\,\mathrm{s}^{-1}$ of the optical filter, a reflectivity of $R = 93\%$ and a relative pump energy of $\tilde{\mathscr{E}}_{\mathrm{P}}^{\mathrm{tot}} = 1$, we obtain from Fig. 5.3a (curve (1)) a pulse duration of $\tau_{\mathrm{L}} \approx 0.12$ ps. If we vary the bandwidth of the frequency selective element and adjust the resonator detuning to obtain the shortest pulses ($h\Delta\omega = \mathrm{const}$), then according to Fig. 5.3 the following relation between the pulse duration $\tau_{\mathrm{L}} \approx \theta\sqrt{2}$ and the bandwidth $\Delta\omega$ can be found:

$$\tau_{\mathrm{L}} \propto \frac{1}{\Delta\omega}. \tag{5.23}$$

The pulse energy $\tilde{\mathscr{E}}_{\mathrm{St}}$ increases monotonically with increasing laser resonator length and reaches its maximum at the end of the stability range (compare Fig. 5.3c), 5.4)). A higher reflectivity results in a larger stability range as well as in higher pulse energies. In comparison, a direct influence of the pump energy on the energy of the laser pulse is small; there is, though, a strong indirect influence because with increasing pump energy the stability range is enlarged considerably, whereby the energy of the laser pulse also is enhanced at greater resonator lengths. If we take for example the parameter values $R = 71$ percent, $\tilde{\mathscr{E}}_{\mathrm{P}}^{\mathrm{tot}} = 1$, $\tau_{\mathrm{P}}\Delta\omega = 1000$, then we obtain a normalized pulse energy of $\tilde{\mathscr{E}}_{\mathrm{St}} = 2$. The value $\sigma_{21} \approx 10^{-16}\,\mathrm{cm}^2$ and a photon energy of 10^{-19} Ws correspond to an energy per unit area of $2 \cdot 10^{-3}\,\mathrm{Ws\,cm}^{-2}$ or to a pulse energy of $2 \cdot 10^{-9}$ Ws, if a beam area of $(10\,\mu\mathrm{m})^2$ is assumed.

The shape of the pulse is roughly described by the asymmetry factor μ (compare Figs. 5.3e), 5.4e)). As a global measure for the asymmetry of the pulse the value $(\mathscr{E}_{\mathrm{St}} - \mathscr{E}_0)/\mathscr{E}_{\mathrm{St}}$ can also be considered (Figs. 5.3d), 5.4d)). Due to the gain depletion the leading edge of the pulse is amplified more than the trailing edge. For this reason the leading edge is steepened, which is evident in Figs. 5.3e) and 5.4e) by the positive values of the asymmetry factor μ. For small pump energies and low reflectivities the pulse is nearly symmetrical ($\mu \approx 0$), while with increasing pump energy and reflectivity the pulse asymmetry also increases. With constant pump energy and varying pump pulse duration the parameters of the laser pulse are hardly influenced.

5.2.3 *Evolution of Ultrashort Pulses from Noise and Formation of Satellite Pulses*

Let us now investigate the evolution of pulses from noise and the properties of laser radiation outside the stability range, where multiple pulses may appear or nonsteady-state regimes are possible. According to equation (5.16) the recursion formula

$$I_{\mathrm{L}}^{K}(\eta + h) = R\left\{\frac{\Delta\omega^2}{4}\int_{-\infty}^{\eta}(\eta - \eta')\exp\left[-\frac{\Delta\omega}{2}(\eta - \eta')\right]\sqrt{G^{K-1}(\eta')\,I_{\mathrm{L}}^{K-1}(\eta')}\,\mathrm{d}\eta'\right\}^2 \tag{5.24}$$

with $G^{K-1}(\eta)$ from (5.12), holds for the pulse shape at the K-th round trip. In [5.12], beginning with a certain noise intensity distribution, this equation has been solved directly with the help of a computer. The results proved to be independent of the kind of simulation of the noise intensity that was described for the one by a stochastic Gaussian process and for the other by a constant intensity distribution. In Fig. 5.5 the pulse evolution is plotted for two different sets of the resonator parameters. The steady-state regime is achieved after about 200 to 300 round trips. At higher reflectivity or higher pump powers the evolution process occurs faster. If we compare the results obtained in the steady-state range with those obtained using the approximate ansatz (5.19), then within the region of stable modelocking a good agreement is found. The deviations in various examples that were investigated are about 10 percent and in every case less than 20 percent.

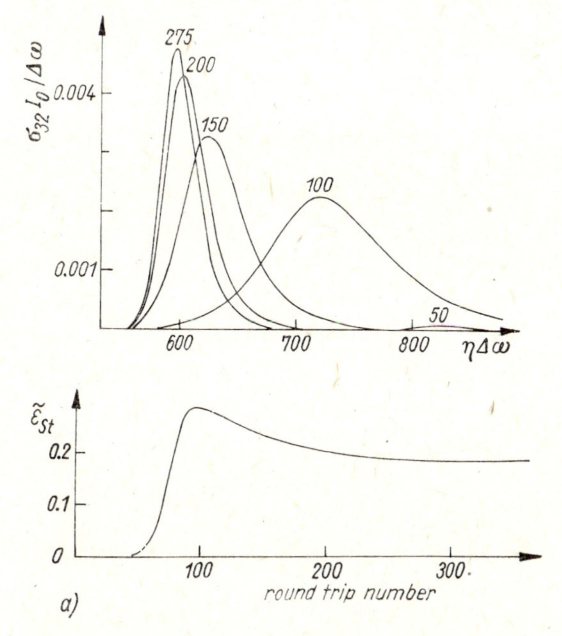

Fig. 5.5. Formation of the pulse shape and pulse energy of a synchronously pumped dye laser for two different sets of parameters

The numbers at the curves designate the resonator round trip numbers. The following were chosen as parameters:

a) $\sigma_{14}NL^{\mathrm{a}} = 10$, $\tilde{\mathscr{E}}_{\mathrm{P}}^{\mathrm{tot}} = 1$, $\tau_{\mathrm{P}}\Delta\omega = 1000$, $R = 0.5$; $h\Delta\omega = -3.2$;

b) $\sigma_{14}NL^{\mathrm{a}} = 10$, $\tilde{\mathscr{E}}_{\mathrm{P}}^{\mathrm{tot}} = 2.5$, $\tau_{\mathrm{P}}\Delta\omega = 1000$, $R = 0.83$; $h\Delta\omega = 0.4$.

(from Herrmann and Motschmann [5.12])

It is of particular interest to consider the solution of equation (5.24) in various ranges of the resonator detuning, since in contrast to the treatment in the last two sections, the description of multiple pulses is also possible as well as the investigation of the radiation field outside the stability range. In Fig. 5.6 the results of the numerical solution of (5.24) for fixed laser parameters R, $\tilde{\mathscr{E}}_{\mathrm{P}}^{\mathrm{tot}}$, $\Delta\omega$, τ_{P} and $\sigma_{14}NL^{\mathrm{a}}$ are shown with respect to

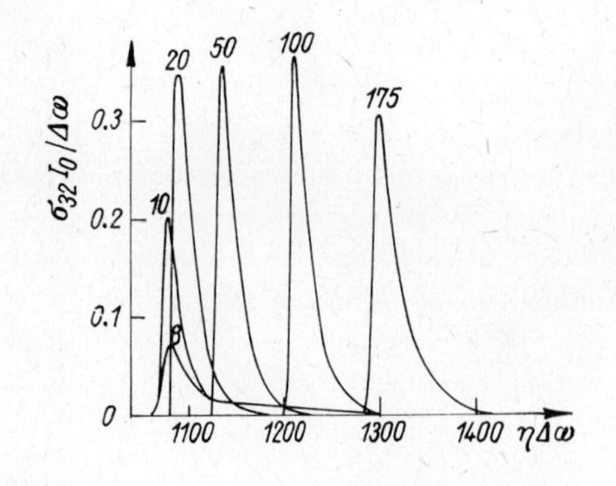

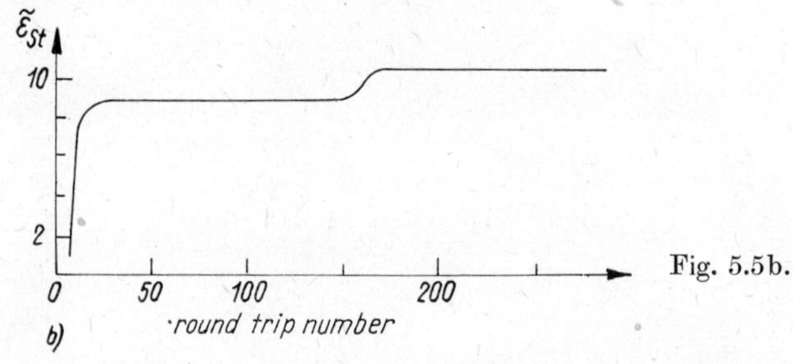

Fig. 5.5b.

the resonator detuning $\delta L \propto h$. If we choose the resonator detuning $\delta L = hc/2$ near the left boundary of the stability range, double pulses appear in the steady-state regime, even within a small range of the stable pulse region. In Fig. 5.6 this can be seen in the curves (B) and (C). Additionally, in Fig. 5.7 the stability range calculated according to section 5.2.2 is plotted once again, whereby the position of the chosen examples with respect to the stability range becomes evident. The range in which double pulses appear in the steady-state regime can be characterized in Fig. 5.7 by the condition $\eta_0 \geqq 0$, that is, the dye laser pulse passes through the amplifier before the maximum of the pump

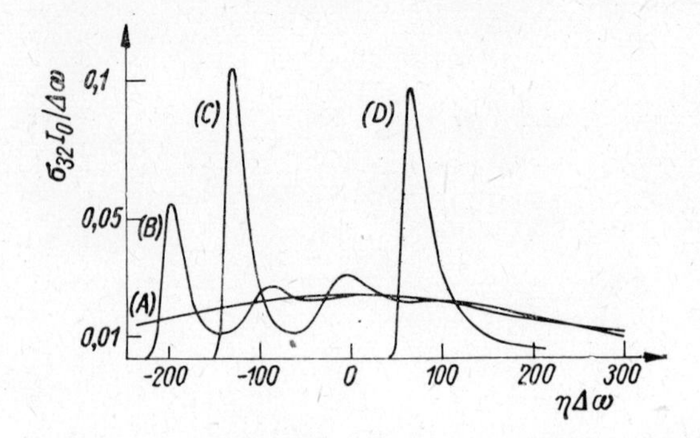

Fig. 5.6. Steady-state pulse shape for various values of the resonator detuning
The ordinate values of curve (B) must be multiplied by the factor 2. The following were chosen as parameters: $\sigma_{14}NL_a = 10$, $\tilde{\mathscr{E}}_P^{tot} = 2.5$, $\tau_P\Delta\omega = 300$, $R = 0.83$, (A): $h\Delta\omega = -4.8$, (B): $h\Delta\omega = -3.0$, (C): $h\Delta\omega = -2.4$, (D): $h\Delta\omega = 0$.
(from Herrmann and Motschmann [5.12])

pulse. From Figs. 5.3f and 5.4f. we see that a range of this kind only exists for high reflectivity and high pump energies. For other laser parameters, which have no domain $\eta_0 > 0$ within the stability range (compare e.g. curves (5) and (6) in Fig. 5.3f), no double pulse regimes should be expected. The appearance of secondary pulses can be explained in the following way: If the dye laser pulse is generated before the maximum of the pump pulse, the subsequent part of the pump pulse enlarges the gain again beyond the threshold, whereby a second or even more pulses are generated. In accordance with these results secondary pulses were observed after a shortening of the laser resonator length.

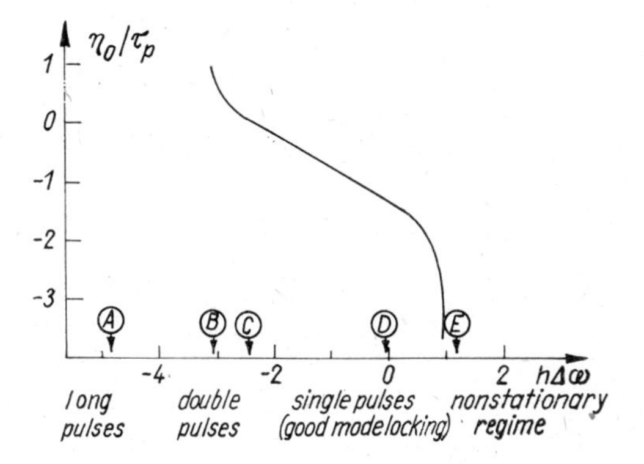

Fig. 5.7. Delay between the pump pulse and laser pulse as a function of the resonator detuning for steady state solutions of the single pulse regime
The points (A), (B), (C) and (D) refer to the curves in Fig. 5.6, and (E) refers to the example discussed which lies outside the right boundary of the stability range.

If we shorten the dye laser resonator length beyond the boundary of the stability ranges, then according to the theory presented in section 5.2.1 no steady-state pulse regime is to be expected. The computer calculation shows however that a steady-state regime of long pulses exists in which the dye laser pulse is about as long as the pump pulse (example (A) in Fig. 5.6a). The generation of long pulses occurs because, due to the shorter length of the dye laser, only the trailing part of the laser pulse is amplified and greatly extended, since the pulse is not cut off by a gain depletion at the trailing edge. If we increase the resonator length of the dye laser beyond the right boundary of the stability range (example (E) in Fig. 5.7), then no steady-state regime is produced. According to the computer calculation after about 150 round trips a stable pulse shape builds up in which, however, the delay between pump and laser pulse increases with each round trip. With a resonator detuning that lies outside the right boundary of the stability range the amplification is too small to compensate for the delay of the dye laser pulse (which results from the greater laser resonator length). Therefore, the pulses spread further and further apart. Due to the influence of the relaxation and to the possibility that new pulses develop from noise near the initial pulse, the pulse does not retain its stable shape in the further evolution, that is, in this range of resonator detuning no steady-state pulse generation is possible.

5.2.4 The Formation of Phase Modulated Pulses

Until now we have investigated the solution of the integral equation (5.16) in connection with (5.17) for the case $\Delta_{32} = 0$. We now consider the formation of phase modulated pulses which are generated at quasi-resonant interaction between the laser pulse and the amplifier ($\Delta_{32} \neq 0$). We follow here an analysis by Schubert, Stamm and Wilhelmi [5.32] who, beginning with the results of the work from [5.11] and [5.12], solved the equations (5.16) and (5.17) for $\Delta_{32} \neq 0$ using a computer. First of all, we consider qualitatively the formation of a chirp by calculating the change of the phase during a resonator round trip. After passing through the active medium the pulse possesses a time dependent phase given by

$$\varphi_L(1, \eta) = \varphi_L(0, \eta) - \frac{\Delta_{32}}{2} \ln [G(\eta)/R] \tag{5.25}$$

in accordance with equation (5.11b) even in the case of a constant phase $\varphi_L(0, \eta) = \text{const}$ at the input. A time-varying phase means a change in the frequency of the pulse, which is proportional

$$\frac{d\varphi_L(1, \eta)}{d\eta} = \frac{d\varphi_L(0, \eta)}{d\eta} - \frac{\Delta_{32}}{2} \frac{d}{d\eta} \ln [G(\eta)/R] \tag{5.26}$$

and represents a quantitative measure for the developing phase modulation. At each resonator round trip, the contributions regarding the phase add up until a further increase of the phase modulation is limited due to the dispersive action of the filter, so that a steady frequency variation develops. According to the relation (5.26), if $\Delta_{32} > 0$, a frequency decrease, a so-called down-chirp, occurs at the leading edge of the pulse $\left(\frac{\partial G(\eta)}{\partial \eta} > 0\right)$, while at the trailing edge $\left(\frac{\partial G(\eta)}{\partial \eta} < 0\right)$ a frequency increase (i.e. an up-chirp) results. If $\Delta_{32} < 0$, conversely an up-chirp occurs at the leading edge and a down-chirp at the trailing edge.

The quantitative evaluation of equations (5.16) and (5.17) verifies these conclusions. In Fig. 5.8, the photon flux density and the change in the frequency are plotted for various laser parameters. In the case of the formation of single pulses, the value $\frac{\partial \varphi}{\partial \eta}$ increases continuously from negative to positive values along with the pulse (Fig. 5.8a and b). At the formation of multiple pulses, however, a more complicated frequency variation results according to Fig. 5.8c. The results of these calculations thus provide a broadening of the spectra of dye laser pulses due to the developing phase modulation, which agrees well with the pulse-duration-bandwidth-products measured by Kuhl et al. [5.9], Ryon et al. [5.33] and Heritage and Jain [5.8]; the measured values of these products were greater than those for bandwidth limited pulses. A frequency detuning $\Delta_{32} \neq 0$ leads to a narrowing of the range of stable single pulse generation. With increasing Δ_{32}, the ratio between the maximum intensity of the main pulse and that of possible satellite pulses decreases, which corresponds to a reduction of the stability range. In addition, variations in the pulse shape and phase increase, so that under certain conditions a steady regime in the strict sense does not exist, which is indicated by irregular changes with respect to the intensity of the maximum pulse as well as the satellite pulses.

In connection with the treatment of passive modelocking in 6.2 and 6.3, we will

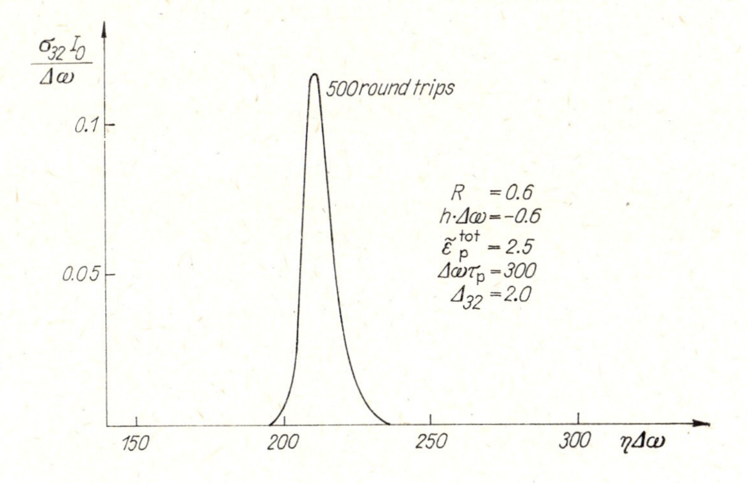

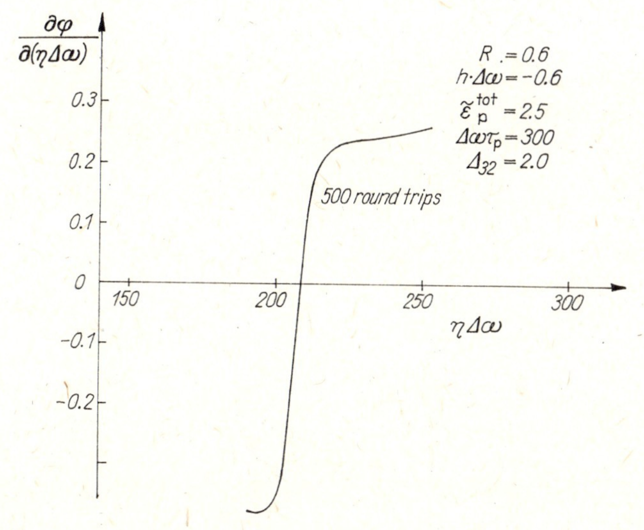

a

Fig. 5.8. Shape and chirp of pulses from a synchronously pumped dye laser operating under off-resonance conditions ($\Delta_{32} \neq 0$) for two different sets of parameters
(from Schubert, Stamm and Wilhelmi [5.32])

discuss in detail possibilities of chirp compensation using dispersive linear optical elements.

In the conclusion of this section, we will note that spontaneous fluorescence noticeably influences the pulse generation not only beyond the right side of the stability range. As investigations in [5.31] and [5.32] showed, spontaneous fluorescence also affects the pulse generation process within the stability range, if the resonator length of the dye laser is longer than that of the pump laser. Since in this case the laser pulses arrive late compared to the maximum of the pump pulse, a small fluorescence signal receives the complete unsaturated gain at the leading edge of the laser pulse, and therefore the maximum of the laser pulse shifts forward.

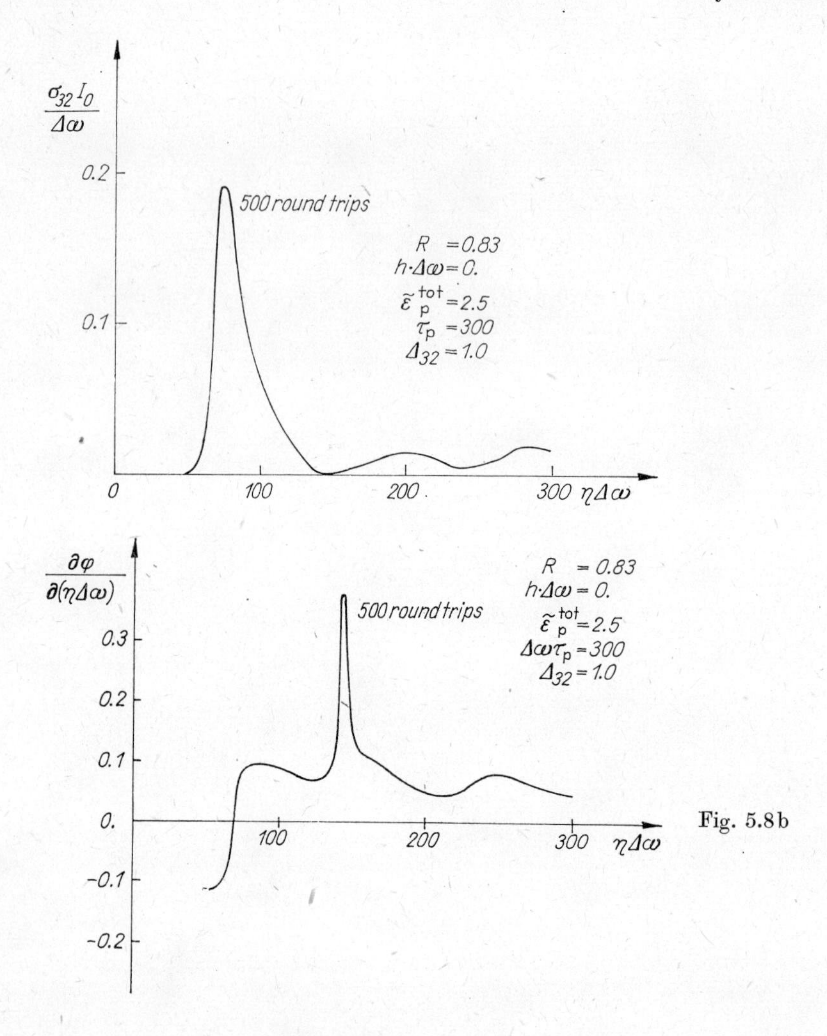

Fig. 5.8b

5.3 Experimental Set-ups and Results

5.3.1 *Basic Design of Synchronously Pumped Lasers*

A typical basic arrangement of a synchronously pumped dye laser is depicted schematically in Fig. 5.9. As pump source, an argon or krypton ion laser which is actively modelocked by means of an acoustooptical modulator is usually used. It is essential here that the frequency and phase of the HF generator, which supplies the electrical signal for the modulator, are highly stable. The relative fluctuation of the frequency should be less than 10^{-7}. Using suitable laser mirrors and dispersive components (e.g. a prism) a specific laser transition of the pump laser is selected. Various wavelengths of the near ultraviolet and visible spectral range, which are well-suited for the pumping of dyes, can be generated. In place of the noble gas ion laser an optically cw-pumped, actively modelocked solid state laser, for example, a Nd:YAG can be employed, whereby the fundamental wave ($\lambda_1 = 1.06$ μm) as well as higher harmonics ($\lambda_2 = 0.53$ μm, λ_3

$= 0.353\ \mu m \ldots$) of the laser radiation find application. Advantages in the application of solid state lasers may develop, because the fluctuations of the pulse parameters at high frequencies (above 100 kHz) are smaller than in noble gas ion lasers due to the long relaxation times of the active materials (table 1.2). A comparison between both laser types was conducted, for example, in [5.34].

With the noble gas ion laser as pump source a continuous, stable train of short pulses is generated having durations of 50 to 300 ps and a mean power of 0.1 to 10 W. These pulses excite the dye molecules serving as the active material in the dye laser. In the first experiments the dye rhodamin 6G was used predominantly as active material but meanwhile many other dyes have also been used. The dye laser resonator is commonly based on an arrangement in which the beam path is folded by mirrors. Mirror M_1 serves

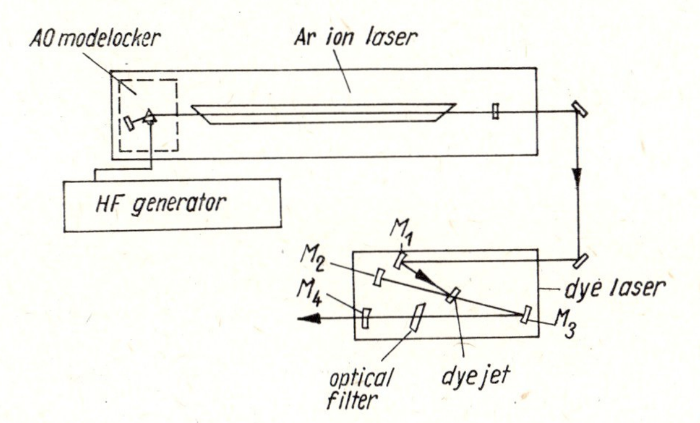

Fig. 5.9. Scheme of a synchronously pumped dye laser

to couple the pump pulse train into the dye laser resonator. Pump and laser radiation pass at a small angle through the dye. The laser radiation is deflected by the folding mirror M_3 and is partially reflected and partially coupled out at the output mirror M_4. The mirrors in the laser resonator are arranged such that the dye laser radiation in the resonator passes back and forth between mirrors M_2 and M_4, while the pump radiation, which is engaged at a small angle against the beam direction of the dye laser, leaves the resonator after passing through the dye. It is important that there is a very good overlapping of the waists of the pump radiation beam and the dye laser beam in the active material. Therefore, the smallest possible angle between the two beams and an arrangement with good compensation of the astigmatism is recommendable (compare 2.3.5).

In the cw operation it is necessary to hinder the molecules from accumulating in the triplet state. The relaxation time of the singlet-triplet transition ($S_1 \rightarrow T_1$) of typical laser dyes lies in the order of 10^{-6} s, whereas the transition from T_1 to S_0 is a considerably slower process. Therefore, after a longer time ($\approx$ 100 round trips) the molecules accumulate in the T_1-level, which affects the laser process. In order to reduce this effect, a triplet quencher that accelerates the transition $T_1 \rightarrow S_0$ can be added to the active medium. Even more effective is the method of fast dye exchange. For this purpose dye pumped through a cell, or a free flowing dye jet stream of good optical homogeneity is used as the active medium. With flow velocities of about 10 m/s and a laser waist diameter in the active material of 10 μm, the dye is exchanged within 10^{-6} s, which is sufficient under the conditions discussed above.

In the experimental set-up shown in Fig. 5.9 the frequency of the laser can be continuously changed using a tuning element, such as a Lyot filter, a Fabry-Perot etalon or a wedge interference filter. (The last of these is a dielectric multilayer system whose layer thickness changes linearly along an axis lying in the layer, whereby a translation of the filter in this direction results in a change of the laser wavelength.) Alternative devices of an additionally folded resonator can be used if a prism is used as a tuning element. With the use of various dyes it is possible to generate picosecond and subpicosecond pulses, whose wavelengths can be continuously tuned over a wide spectral range of about 420 nm—1000 nm in synchronously pumped lasers. Special attention must be given here to the relatively precise matching of the length of the dye laser resonator to the pulse repetition frequency of the pump laser, which also requires a high thermal

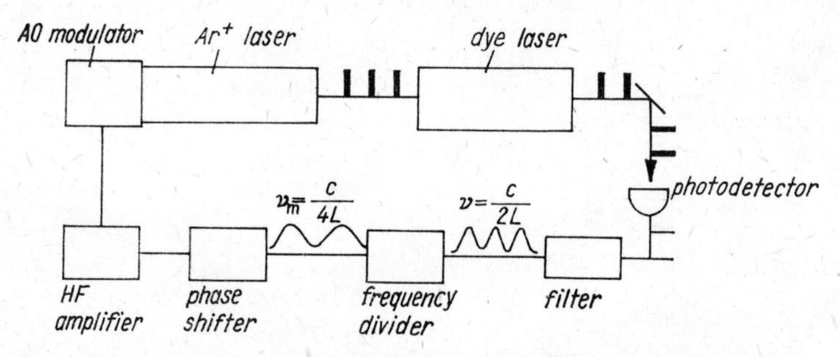

Fig. 5.10. Scheme of a synchronously pumped dye laser in which the modulator signal is obtained from the pulse repetition frequency of the dye laser pulse train

and mechanical stability of the laser system. It should be pointed out that the pulse repetition frequency of the pump laser is determined by the frequency of the active modulator and can deviate somewhat from the round trip frequency $c/(2L)$ of the corresponding "cold" resonator (that is, a resonator that is not in the laser operation). Therefore, the optimal length of the dye laser must be matched to the modulator frequency, whereby a relative accuracy of about 10^{-7} is required. If the modulator frequency and the length of the dye laser resonator cannot be adjusted by an automatic control, then both quantities must be held constant at about 10^{-7}. For this purpose generators of high stability and phase purity are necessary. The resonators are mounted on vibration damped tables and protected by glass tubes from the fluctuations of the surrounding air. Changes in the length of the resonator due to thermal expansion are largely compensated. The temperature of the optical components should be held as constant as possible, in order that the changes in the optical length do not exceed 0.1 μm. To adjust the resonator length, the output mirrors of the dye lasers for example can be mounted on a carriage with a differential micrometer screw, whereby a precise and reproducible control of the laser resonator length within several tenths of a micron is possible.

In order to achieve very short pulses greater effort must be made to improve the stabilization in synchronously pumped lasers. Therefore, it proves to be more useful not to adjust the modulator frequency and the dye laser length independently, but to obtain the modulator frequency from the pulse repetition frequency of the dye laser. A corresponding arrangement is shown in Fig. 5.10. A portion of the dye laser radiation is supplied to a photoreceiver. From the photoelectrical signal the mode spacing frequency $\delta\nu_{\mu}$ is filtered out. After halving the frequency, the electrical signal is amplified and drives the acoustooptical modulator. The phase shifter, with which we can set the most favorable position of the dye laser pulses in relation to the pump pulses for given experimental conditions is important. Here it is particularly advantageous to control the adjustment of the phase shifter which, for example, can be set at the minimum pulse width or maximum intensity.

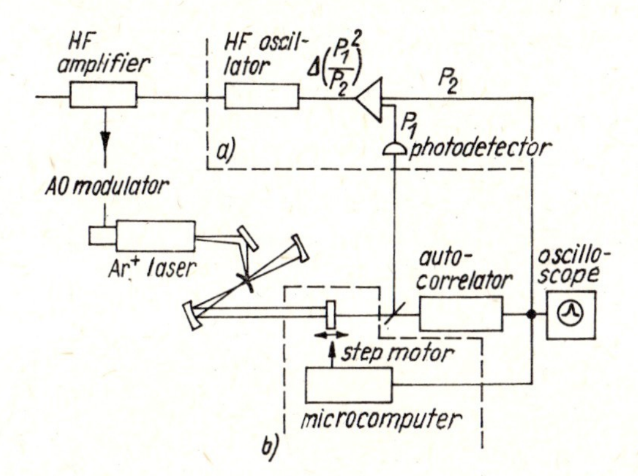

Fig. 5.11. Arrangement for optimizing a synchronously pumped dye laser using two control loops (from [5.30])

A device for generating very stable ultrashort light pulses was suggested by Rotman et al. [5.30]. A scheme of their arrangement is given in Fig. 5.11. Two automatic controls have been applied here. A fast control loop is based on the measuring of the mean power of the laser radiation P_1 and the radiation of the second harmonic P_2, which is generated in a KDP crystal (compare 3. and 8.). The peak power of the pulses of the second harmonic is proportional to the square of the peak power of the laser pulses. Therefore $P_2 \sim (P_1)^2/\tau_L$ holds, where τ_L is the duration of the laser pulses, from which the relation $(P_1)^2/P_2$ follows as a measure for the pulse duration. For comparison, the pulse duration is determined with an autocorrelator that is also based on the generation of the second harmonic (compare 3.). The experimental results show that a resonator detuning influences only little the mean power of the fundamental wave P_1, whereas P_2 and accordingly also τ_L change greatly (at a detuning of $\delta\nu_m = 1$ kHz, P_1 changes by less than 10 percent, P_2 by more than an order of magnitude; at a 10 Hz detuning, the change of the pulse duration is still detectable). The measured signal change $\delta[(P_1)^2/P_2]$ is utilized for the control of the oscillator frequency, in order to obtain pulses of minimal duration. (Here, the frequency of the modulator is adapted to the length of the dye laser.) Besides the fast response analogue control loop a slower control loop is applied with which, using the pulse width determined from the autocorrelation measurement, the tracking of the output mirror is calculated with a microcomputer. Using these automatic controls, pulses that are stable over a long time and that have a duration of 0.7 picoseconds can be generated. In addition the fluctuations of the pulse parameters in the low frequency range decrease greatly.

The synchronous pumping of dye lasers has gained many uses in the meantime, which is the result of its favorable pulse parameters and the availability of appropriate radiation sources as well as its easy handling. To be noted are the synchronously pumped dye lasers of the firms Spectra Physics [5.19] and Coherent (USA) [5.20, 5.21], which use powerful argon or krypton ion lasers as sources. On the basis of the argon ion laser ILA 120 of VEB Carl Zeiss Jena [5.22], D. Schubert and J. Schwarz constructed a synchronously pumped dye laser [5.24, 5.26], where the dye laser is now produced at the Center of Scientific Instruments of the Academy of Sciences of the GDR [5.25, 2.8].

Synchronously pumped lasers are also constructed in ring laser arrangements. Since in such arrangements the radiation in both circulation directions experiences the same amplification, a direction selection must be made by additional losses. This can, for example, be achieved with a Faraday rotator in connection with polarizers (see e.g. [5.21]). In a linear resonator one direction is generally favored, because the amplifier is located not in the center of the resonator, but near to one of the end mirrors. The pulse circulating in one propagation direction finds accordingly a greater residual amplification on the return trip, by which it obtains more favorable amplification conditions than the pulse circulating in the opposite direction. However, it should be noted that already counterpropagating pulses of relatively small energy can considerably disrupt the evolution of the main pulses by interacting in the active medium, for which reason additional pulse suppression measures have a favorable effect. For example these undesired pulses can be more strongly suppressed by adding a saturable absorber of small concentration to the active medium (compare 6.3.5).

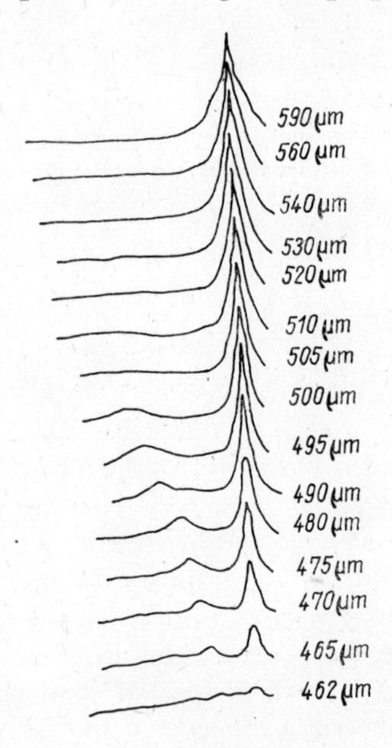

Fig. 5.12. Pulse autocorrelation measurements of synchronously pumped dye lasers with respect to the resonator detuning δL (from Ausschnitt, Jain and Heritage [5.18])

5.3.2 *Experimental Investigations of the Pulse Parameters of Synchronously Pumped Lasers*

The characteristic properties of synchronously pumped dye lasers, particularly the dependence of the pulse duration and the intensity on the resonator detuning, have been investigated experimentally by various authors (see e.g. [5.15 to 5.18]). As an example, a series of autocorrelation measurements conducted by Ausschnitt, Jain and Heritage [5.19] is presented in figure 5.12. These show the dependence of the pulse duration on the resonator detuning δL of a synchronously pumped rhodamine 6G laser (whose frequency was tuned using a birefringent plate). The numbers on the right side give the resonator length detuning $\delta L = hc/2$. As seen from the lower curves, for a resonator length of the laser that is too short a long pulse with a weak amplitude modulation is generated. If we increase the resonator length, a short main pulse develops out of this long pulse followed by a broader satellite pulse. With a further increase of the resonator length, the intensity of the main pulse is increased and its duration decreased. The satellite pulse, however, becomes weaker, broader and appears at a later moment. At a relative resonator detuning of 520 μm the satellite pulse is completely suppressed. A further increase of the resonator length leads to a further increase in the intensity and a broadening of the pulse. The pulse intensity has its maximum at a resonator detuning of 540 μm and decreases with a further increase of the resonator length. The pulse becomes broader and broader until at a resonator detuning of 590 μm the appearance of "shoulders" in the autocorrelation trace indicates the presence of substructures in the pulse. The total resonator length variation was 0.13 mm at an absolute resonator length of 1.8 m.

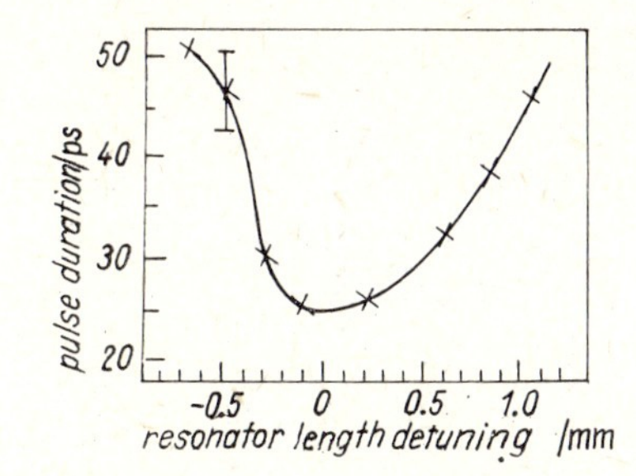

Fig. 5.13. Measured pulse durations as a function of the resonator detuning δL (from Kuhl, Lambrich and von der Linde [5.28])

These experimental results from [5.18] are in good agreement with the theory presented in section 5.2. At a resonator length of the laser that is too small, a regime of long pulses exists that changes into one of shorter double pulses and finally into a stable single pulse regime. At a resonator length that is too long a nonsteady-state regime results which is indicated by pulse substructures (compare section 5.2). The occurence of double pulses is to be expected here only at high reflectivity or greater pump pulse energies, according to the theoretical results in section 5.2, whereby there must be a range within the stability region in which the dye laser pulse is generated before the maximum of the pump pulse ($\eta_0 > 0$). Experimental investigations of the pulse length with respect to the resonator length detuning are also found in other papers. Fig. 5.13, for example, shows the results from Kuhl, Lambrich and von der Linde [5.28] who investigated the temporal structure of picosecond pulses from an oxazine laser pumped by an actively modelocked krypton ion laser. At a transmission coefficient of 2 percent for the outcoupling mirror and a spectral width of the optical filter of $\Delta\omega = 5 \cdot 10^{11}$ Hz, Fig. 5.13 yields for the shortest pulse duration $\tau_L \approx 25$ ps while the stability range comprises a resonator detuning of $\Delta(\delta L) \approx 1.8$ mm. If we compare Fig. 5.13 with Fig. 5.4a), curve (1), we can recognize a qualitatively consistent behavior of the experimental and theoretical curves with a minimum of the pulse length within the stability range. If we calculate the absolute value of the shortest pulse lengths at the given spectral halfwidth of the optical filter, we obtain $\tau_L \approx 28$ ps and a stability range of $\Delta(\delta L) \approx 2.4$ mm for $R = 93$ percent. A more accurate comparison of theory and experiment, however, would require a determination of the value of the pump parameter $\tilde{\mathscr{E}}_P^{tot}$. Experimental investigations on the dependence of the pulse width on the pump energy are plotted in Fig. 5.14 according to results from [5.9]. The decrease of the pulse duration with increasing pump energy corresponds to the theoretical results in Figs. 5.5a, b).

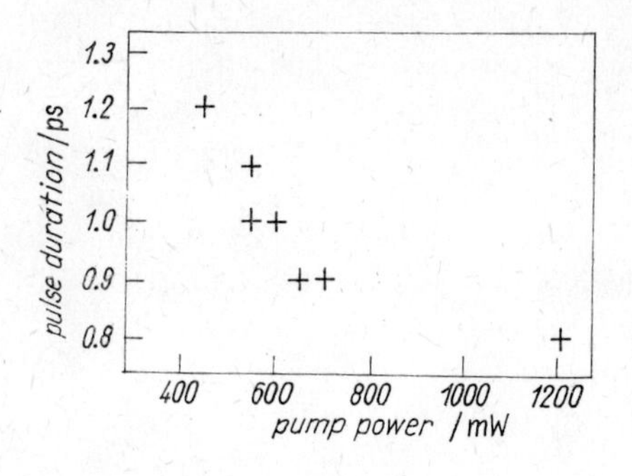

Fig. 5.14. Experimental investigation of the dependence of the pulse length on the pump power (from Kuhl, Klingenberg and von der Linde [5.29])

In Fig. 5.15 the relative position between the pump pulse and the dye laser pulse is plotted according to the experimental results from Horn et al. [5.26, 5.27] who investigated the cross correlation of the pump pulse and laser pulse that was obtained by generating the sum frequency $\omega_P + \omega_L$ in a nonlinear optical crystal. Since the pulse duration of the dye laser pulse is small compared to that of the pump pulse, the temporal profile of the pump pulse is determined by the cross correlation function. If we change the resonator detuning δL, then the resulting temporal displacement of the maximum is thus a measure for the change of the time delay between the pump pulse and

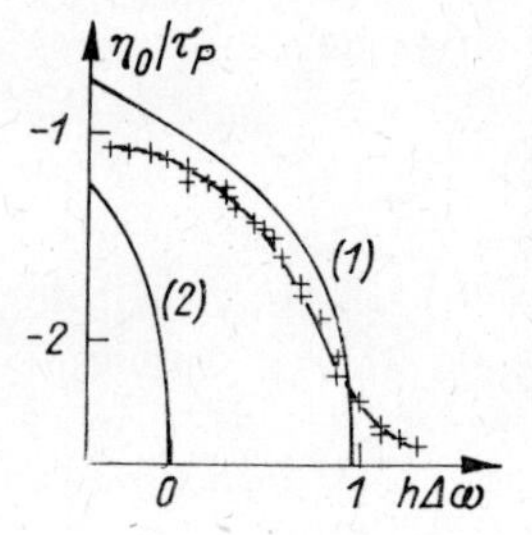

Fig. 5.15. Comparison of theoretical and experimental results for the change of the delay time between pump pulse and dye laser pulse with respect to the resonator length detuning δL (from Horn [5.27]; theoretical results from [5.11]).
Parameters: curve (1): $\tilde{\mathscr{E}}_P^{tot} = 2.5$; curve (2): $\tilde{\mathscr{E}}_P^{tot} = 2.0$; experiment: $\tilde{\mathscr{E}}_P^{tot} = 2.3$.

the dye laser pulse. In Fig. 5.15 the change in the time delay between the two pulses determined from the cross correlation function is plotted with respect to the resonator length detuning δL and compared with the theoretical results of section 5.2. Good agreement between theoretical and experimental values is found in the range $0 \leq 2\delta L\Delta\omega/c < 0.8$. With values greater than 0.8 it is assumed that satellite pulses simulate a time displacement that is too small.

5.3.3 *Cavity Dumping*

By further modifying a synchronously pumped laser, the pulse energy can be increased by more than an order of magnitude, and the time difference between the single pulses can be increased and variably adjusted, if instead of the output mirror M_4 in Fig. 5.9 a so-called cavity dumper is used. It may, for example, consist of an acoustooptical modulator (see Fig. 5.16) that deflects the pulse out of the resonator on the n-th round trip, while on the other round trips the radiation passes undeflected through the modulator and remains in the resonator. In this manner energy storage occurs. The dye laser resonator can be fitted with a highly reflecting end mirror which, at given pump powers, leads to high pulse powers. The Bragg cell in the cavity dumper must guarantee small losses in the "resting state" and a high diffraction efficiency ($\geqq 0.7$) after applying the driving signal. (In order to increase the diffraction efficiency, a double passage through the Bragg cell — as shown in Fig. 5.16 — is frequently used.) Furthermore, a rise time that is smaller than the pulse spacing $2L/c$ must be guaranteed. In the Bragg cell, which (in

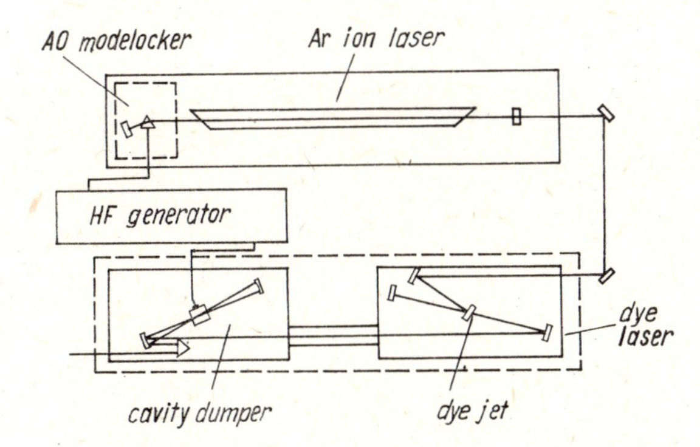

Fig. 5.16. Scheme of a synchronously pumped dye laser with cavity dumper

contrast to the acoustooptical modelocker) can only make use of the diffraction at travelling sound waves, the rise time of the diffraction efficiency is determined not only by the rise time of the HF power, but also by the transit time of the ultrasonic wave through the region of the beam waist in the center of the cell. At a waist diameter of 20 μm the transit time comes to about 5 ns. The HF power electronics must guarantee a rise time of the switching signal $\lessapprox 5$ ns, in order that the previously mentioned requirement for typical resonator lengths is satisfied. The HF power electronics must, moreover, ensure an accurately reproducible switching moment relative to the dye laser pulses, which is achieved by a close synchronization to the modulator signal.

In lasers having a cavity dumper a pulse power can be achieved which exceeds the usual values by about a factor of 10 and lies in the range of $10^3 \div 10^4$ W. The pulse repetition frequency can be chosen between 0 Hz and several MHz. Due to the process of cavity dumping, though, a pulse broadening — to several picoseconds, for the most part — occurs.

5.3.4 *Amplification*

Certain applications require even higher powers than those achieved by means of cavity dumping. Powers in the gigawatt range, for example, are needed to investigate nonlinear optical effects of higher order or to use these effects for efficient frequency conversion (compare 8.). Fig. 5.17 shows the scheme of a laser amplification system, which was constructed by Rotman and others, for amplifying the pulses of a synchronously pumped dye laser [5.30]. The amplification occurs in four consecutive dye cells that are pumped by the second harmonic ($\lambda = 0.53$ μm) of a Q-switched Nd : YAG laser. The dye laser is operated here without a cavity dumper. The selection of pulses having lower repetition rates occurs through the amplification process, which is trig-

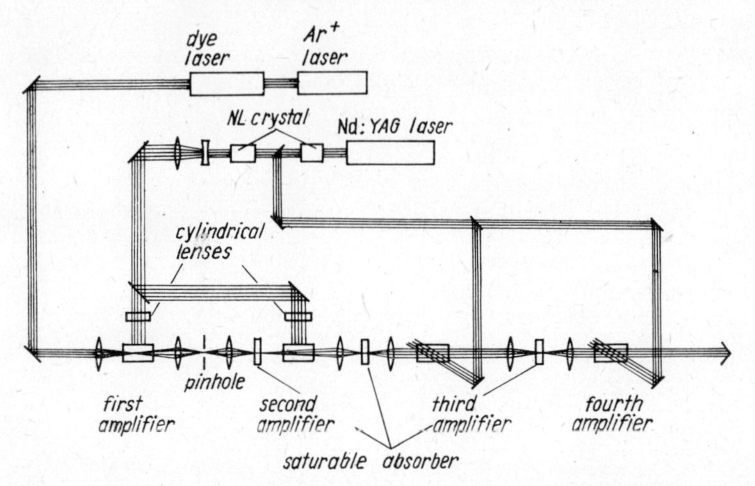

Fig. 5.17. Synchronously pumped dye laser with subsequent four-stage amplifier (from [5.30])

gered by the Nd : YAG laser with a repetition rate of about 10 Hz. The pulse duration of a Q-switched Nd : YAG laser is about 10 ns, whereby without special synchronization with the argon ion laser one or two pulses of the dye laser are amplified depending on the random phase position. Through an additional synchronisation of the two pump lasers a regime can be achieved where only one pulse is always amplified, which passes through the amplifier at the moment of maximum occupation number inversion. To pump the amplifier stages a relatively high pulse energy of the second harmonic of the Nd : YAG laser ($\gtrsim$ 50 mJ) is required. In order to obtain high conversion rates of the fundamental wave, two crystals are employed. The radiation of the harmonic, which is generated in the first crystal, is used to pump the final stages. The remaining 1.06 μm radiation passes through the second crystal and the 0.53 μm radiation generated here pumps two preamplifier stages. To achieve a good transverse structure of the amplified radiation, a spatial frequency filtering using a diaphragm is made after the first stage. The individual amplifier stages are decoupled by saturable absorbers. In this manner self-oscillations of the amplification system are avoided, which could otherwise occur due to the very high total amplification. Furthermore, the saturable absorbers favorably influence the shape of the dye laser pulses. The total amplification lies at several 10^6, where pulse powers in the gigawatt range are attained.

6. Passive Modelocking of Dye Lasers

6.1 Principle of Operation

Modelocking of a dye laser using saturable absorbers was first achieved by Schmidt and Schäfer [6.1]. They observed the formation of a train of short pulses in a rhodamine 6G laser pumped by a flash lamp, after inserting in the resonator a dye cell that acted as a saturable absorber. Bradley and O'Neill [6.2] reproduced these results and determined the duration of the pulses at about 5 ps by means of the two-photon fluorescence technique (see 3.). A possible arrangement of a passively modelocked dye laser is given in Fig. 6.1. The dye cell is pumped here using a flashlamp. All surfaces are inclined at the Brewster angle or at a wedge-angle to one another in order to avoid Fabry-Perot (subcavity)

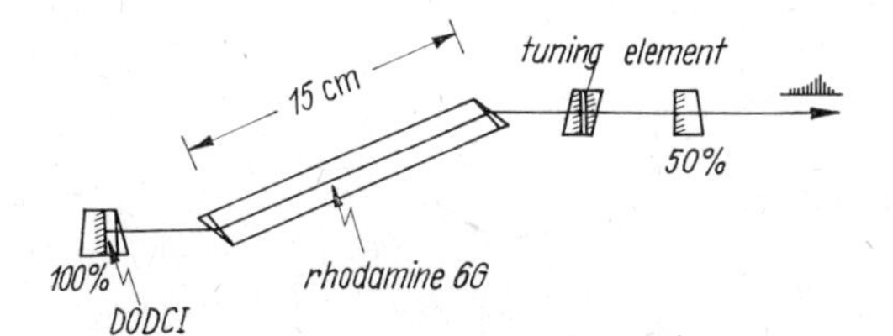

Fig. 6.1. Resonator configuration of a flashlamp pumped dye laser

resonances. The saturable absorber is placed as near as possible to the end mirror, in order that the reflected leading edge and the incoming trailing edge of the pulse can superimpose inside the absorber (compare 6.2.3). In this manner the absorber is saturated even at low intensities, and so the absorber position near the mirror is conducive to the process of modelocking. Using a frequency selective element the spectral width of the laser radiation can be narrowed and its maximum shifted, whereby a tuning of the frequency of the ps-pulses becomes possible. In this way bandwidth limited pulses can be generated which, for example, are tunable from 580 to 700 nm in rhodamine and cresyl violet dye lasers with the appropriate saturable absorbers.

The flashlamp pumped dye lasers operate in a quasi-continuous regime whose duration is limited by the pump duration of the flashlamp. A real continuous operation (cw-regime) was achieved for the first time by Ippen et al. [6.3] and by O'Neill [6.4], who continuously pumped a rhodamine 6G laser with an argon ion laser. With this laser a continuous pulse train of pulses having a duration of 1.5 ps was generated. In recent experiments the shortest pulses that could be obtained in such lasers reached into the femtosecond range [6.5–6.7, 6.30, 6.31].

As with active modelocking the mechanism of passive modelocking rests on the temporal modulation of the losses in the resonator. The main difference, however, exists in the fact that in passive modelocking the system itself determines the moment at which the losses reach a minimum. The pulse formation process in the dye laser can be characterized as follows. After the pump radiation has exceeded the laser threshold, the laser radiation is amplified from spontaneous noise in the resonator whereby, in the multimode operation considered here, the radiation field consists of a statistical superposition of many fluctuation peaks. Due to the large emission cross section σ_{32}^{a} of the laser dye, the radiation is amplified by stimulated emission after only 20 to 30 round trips to a value at which the saturation of the absorber becomes important. The absorber favors those fluctuations or groups of fluctuations that possess maximum energy, since these experience smaller losses due to the saturation of the absorption. Because the mean dura-

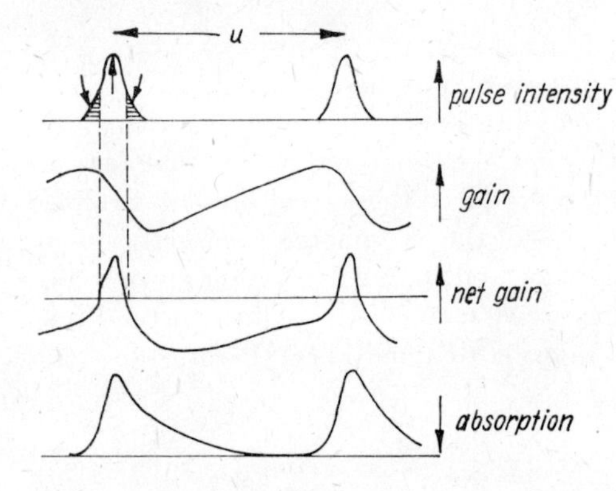

Fig. 6.2. Scheme of the pulse compression in a passively modelocked dye laser
The combined action of the saturation in the amplifier and absorber produces a net loss at both edges and a net gain at the center of the pulse.

tion of the fluctuations or of the pulse that ultimately builds up is considerably smaller than the fluorescence lifetime of the absorber (T_{21}^{b}) and the amplifier (T_{32}^{a}), the saturation of both media is not controlled by the intensity, but by the energy of the fluctuation pulses. The saturation of the absorber and the depletion of the amplification, respectively, begin to play a part when the energy per unit area reaches the order of $\mathscr{E}_{S} = \hbar\omega/(\sigma_{32}^{a})$. With an absorber of this kind the leading edge of the pulse and the fluctuation group are absorbed respectively, until the energy has reached a value at which the absorption is greatly diminished due to the saturation. For this reason, the trailing edge of the pulse remains unweakened and thus the pulse obtains a steeper leading edge (see Fig. 6.2). For the build-up of a short pulse, however, it is beneficial if the trailing edge of the pulse is suppressed as well. This occurs in passively modelocked lasers, because above a certain pulse energy the amplification is decreased due to the depletion of the occupation inversion and falls below the value of the threshold amplification, whereby the pulse is absorbed at its trailing edge. Due to the combined action of the absorber and amplifier, saturation conditions arise under which a group of fluctuations having the highest energy is preferred with respect to the amplification and to the continous shortening at the leading and trailing edges, so that a single ultrashort pulse can ultimately build up. It is necessary here that the absorption cross section (σ_{14}^{b}) of the absorber is greater than that of the amplifier (σ_{32}^{a}), in order that the absorber is saturated with less energy than the amplifier and that the pulse peak is subsequently amplified more than the edges. Furthermore, it is necessary that the occupation inversion in the amplifier is not completely built up again during a round trip of the pulse, in order that the losses exceed the gain at the leading edge. This determines an upper limit for the ratio of the resonator round trip time and the lifetime of the amplifier T^a_{31}. On the other hand, the resonator must not be too short, in order that the time between subsequent pulse passages is sufficient for the recovery of the occupation inversion by means of pumping. These requirements lead to conditions for the laser and resonator parameters, which define a so-called stability range within which the generation of an ultrashort pulse is possible. This process will be explained in more detail in the next section.

After the pulse has undergone a rapid amplification and the pulse shortening has reached a certain value, the bandwidth-limiting action of the optical filter (or, with careful removing of all passive bandwidth-limiting elements, the influence of the finite linewidth of the laser transition) begins to play an important part. As with other modelocked types of lasers in the cw-range, a steady-state pulse shape evolves in this manner, which does not change during further round trips of the pulse.

6.2 Theory

6.2.1 Basic Equations

Theoretical investigations of passively modelocked dye lasers were first conducted by New on the basis of rate equations [6.8, 6.9]. He showed that through the combined action of saturable absorption and gain depletion a fast pulse-shortening process may occur, if the laser parameters are chosen such that the pulse experiences an effective loss at both edges, though its maximum is amplified (this parameter range is also called the static pulse compression zone). Bandwidth-limiting frequency dependent effects were neglected in this analysis. Thus, no steady-state regime could be described and no conclusions about the theoretically achievable pulse duration, pulse shape, etc. could be drawn (in the rate equation approximation the pulse duration approaches zero as the number of round trips increases). A simple analytical description of a steady-state pulse regime was given by Haus, who considered the frequency-dependent action of an optical filter [6.10]. However, he used various approximations, such as small pulse energies (compared with the saturation energy of the amplifier and absorber) and small loss and gain per round trip, which greatly limited the validity of his solution. In this manner, the laser parameters are limited to a small range that in many cases do not contain typical experimental values.[1]) In our treatment let us follow a paper by Herrmann and Weidner, in which the modelocking process was investigated under more general conditions and no restrictions were made for the pulse energy, the gain factor and the losses [6.11].

Our analysis refers to a linear resonator configuration as shown in Fig. 6.3. The active medium is excited by a continuous pump source, and therefore we can use the relations derived in the last section for the description of the amplification process, where $I_P(0, \eta) = \text{const}$ is now assumed. In addition, we can simplify the equation if we assume that, during the short duration of the laser pulse, the change in the occupation inversion due to the pump radiation is negligibly small, and thus it is caused only by the

---

[1]) Further theoretical investigations are found in [6.22—6.25].

laser pulse itself. During the passage of the laser pulse, we can accordingly neglect I_P as opposed to I_L in equation (5.8). Further, we assume resonant light-matter interactions (that is $\omega_L \approx \omega_{32}^a \approx \omega_{31}^b$) resulting in a time-independent phase ($\varphi(\eta) = \text{const}$). The integration of (5.8) yields then

$$\gamma(\eta, z) = \exp\left\{\sigma_{32}^a \int_{-\infty}^{\eta} I_L(0, \eta')\, d\eta'\right\} + \dot{\gamma}(\eta = -\infty, z) - 1, \tag{6.1}$$

where $\gamma(\eta = -\infty, z)$ can be determined from the initial condition before the instant of the pulse passage ($\eta \to -\infty$)[1], at which an occupation of the upper laser level $N_3{}^a = N_3^a\,(\eta = -\infty, z)$ exists which is determined by the preceding pump process. Using (5.3) we can write $\gamma(\eta \to -\infty, z)$ as

$$\gamma(\eta \to -\infty, z) = \exp\left\{-\sigma_{32}^a \int_0^z N_3{}^a(\eta \to -\infty, z')\, dz'\right\} \equiv \big(V(z)\big)^{-1}. \tag{6.2}$$

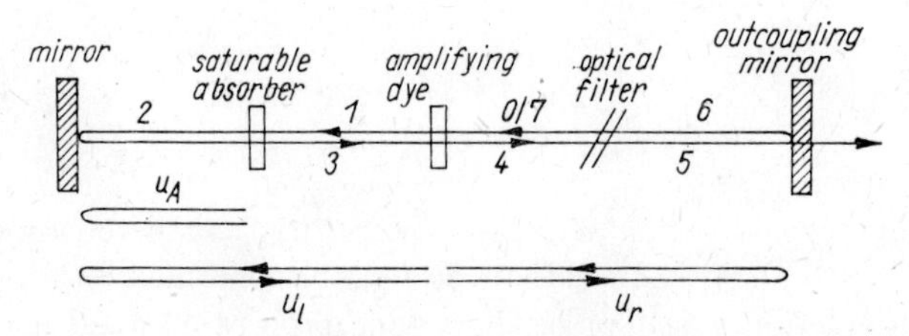

Fig. 6.3. Scheme of a modelocked dye laser
u_A, u_l, u_r are the transit times of the pulse along the plotted path. The numbers 0 to 7 indicate positions within the resonator.

$V(z)$ describes here the gain of the pulse at the leading edge, whereby we have to consider that due to the condition $T_{31}^a \simeq u$ the occupation inversion depends on the preceding pulse passage and on the position of the amplifier within the resonator. For the change of the pulse shape at the passage of the pulse through the active element from position (0) in Fig. 6.3 to position (1), we obtain from (5.11a), (5.12), (6.1) and (6.2)

$$I_L(1, \eta) = \frac{V_r I_L(0, \eta)\, e^{\tilde{\mathscr{E}}(0,\eta)}}{K(\eta)}, \tag{6.3}$$

where

$$K(\eta) = 1 + V_r\big[\exp\big(\tilde{\mathscr{E}}(0, \eta) - 1\big)\big] \tag{6.4}$$

is introduced as an abbreviation and $V_r(z)$ signifies the gain for the leading edge of the pulse if the pulse is going through the amplifier from the right. The pulse energy normalized to the saturation energy of the amplifier $\tilde{\mathscr{E}}(1, \eta) = \sigma_{32}^a \int_{-\infty}^{\eta} I_L(1, \eta')\, d\eta'$ is given by

$$\tilde{\mathscr{E}}(1, \eta) = \ln K(\eta) \tag{6.5}$$

[1]) The time η is only defined in the interval $-\frac{u}{2} \leqq \eta \leqq \frac{u}{2}$, consequently the notation $\eta \to -\infty$ means $\eta \ll -\tau_L$.

at the output of the amplifier. The change of the pulse at the passage through the saturable absorber can be described similarly. The saturable absorber is considered here as a three-level system (Fig. 6.4), in which level 3 represents the excited vibrational, level of the excited electron level whose occupation is negligibly small due to the very fast relaxation time T_{32}^{b}. Because of $\tau_L \ll T_{21}^{b}$, we can furthermore neglect the relaxation to the ground level. The appropriate rate equations are then

$$\frac{\partial I_L}{\partial z} = -\sigma_{13}^{b} N_1^{b} I_L, \tag{6.6}$$

$$\frac{\partial N_1^{b}}{\partial \eta} = -\sigma_{13}^{b} N_1^{b} I_L, \quad N_1^{b} + N_2^{b} = N^{b}. \tag{6.7}$$

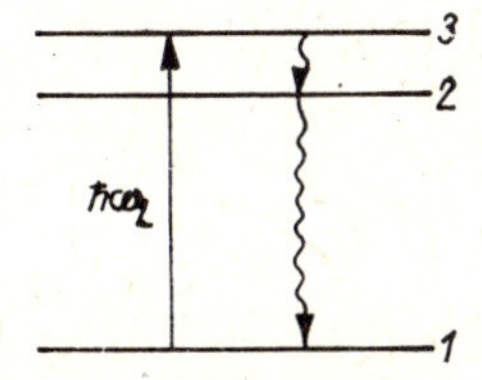

Fig. 6.4. Energy levels of the saturable absorber
Level 3 is a very short-lived excited vibrational state of the S_1 state (level 2), whose occupation number density N_3 is negligibly small $(T_{32}^{b} \ll T_{21}^{b}, \tau_L)$.

The solution to the equations can be found in a similar manner to that of the amplifier; accordingly, for the intensity at position (2) we obtain

$$I_L(2, \eta) = \frac{B_r I_L(1, \eta) \exp\left(m\tilde{\mathscr{E}}(1, \eta)\right)}{1 + B_r\left[\exp\left(m\tilde{\mathscr{E}}(1, \eta)\right) - 1\right]},$$
$$B(z) = \exp\left\{-\sigma_{13}^{b} \int_0^z N_1^{b}(\eta \to -\infty, z')\, dz'\right\}. \tag{6.8}$$

$B(z)$ is here the transmission of the absorber at the leading edge of the pulse: $B_r(z)$ is the transmission when the pulse coming from the right side passes through the absorber. The factor $m = q\sigma_{13}^{b}/\sigma_{32}^{a}$ contains the ratio of the absorption cross sections $\sigma_{13}^{b}/\sigma_{32}^{a}$ and the ratio of the beam cross sections $q = q^{a}/q^{b}$ in the amplifying and absorbing medium. With the use of a telescope the value of m can be increased by a greater value of q. As in (6.4) the energy at position (2) is given by

$$\tilde{\mathscr{E}}(2, \eta) = \frac{1}{m} \ln\left\{1 + B_r\left[\exp\left(m\tilde{\mathscr{E}}(1, \eta)\right) - 1\right]\right\}. \tag{6.9}$$

We obtain similar equations for describing the return trip of the pulse through the amplifier and the absorber, in which V_l and B_l replace V_r and B_r. There are relations among the various transmission coefficients that we need in the following. We must take into consideration that the relaxation time of the amplifier T_{31}^{a} lies in the order of the resonator round trip times u_r, u_l and $u = u_r + u_l$, whereas that of the absorber is small compared to u_l and u_r, though larger or of the same order as u_A (the absorber is placed near the end mirror). We designate the occupation number density of the upper laser level before the passage of the pulse coming from the left or right with $N_{3r,i}^{a}$ and $N_{3l,i}^{a}$, respectively, and after the passage of the pulse with $N_{3r,f}^{a}$ and $N_{3l,f}^{a}$, respectively. We denote the equilibrium occupation number caused by the constant pumping with N_{3e}^{a}.

Due to the relaxation of the molecules, we have

$$N^{\mathrm{a}}_{31,\mathrm{i}} = N^{\mathrm{a}}_{3\mathrm{e}} - [N^{\mathrm{a}}_{3\mathrm{e}} - N^{\mathrm{a}}_{3\mathrm{r,f}}] \exp\left(-\frac{u_{\mathrm{l}}}{T^{\mathrm{a}}_{31}}\right) \tag{6.10}$$

for the change of the occupation number during the round trip u_{l} (see Fig. 6.3) of the pulse coming from the right after its passage through the active medium up to its arrival from the left. Using $\ln V_{\mathrm{l}} = \sigma^{\mathrm{a}}_{32} \int\limits_0^{L^{\mathrm{a}}} N^{\mathrm{a}}_{31,\mathrm{i}}(z')\,\mathrm{d}z'$, we obtain for the gain at the leading edge of the pulse coming from the left the relation

$$\ln V_{\mathrm{l}} = \ln V_0 - \left[\ln V_0 - \sigma^{\mathrm{a}}_{32} \int\limits_0^{L^{\mathrm{a}}} N^{\mathrm{a}}_{3\mathrm{r,f}}(z')\,\mathrm{d}z'\right] \exp\left(-\frac{u_{\mathrm{l}}}{T^{\mathrm{a}}_{31}}\right), \tag{6.11}$$

where $V_0 = \exp(\sigma^{\mathrm{a}}_{32} N^{\mathrm{a}}_{3\mathrm{e}} L^{\mathrm{a}})$ represents the small signal amplification. From (5.2a) and (6.3), we can derive the relation

$$\sigma^{\mathrm{a}}_{32} \int\limits_0^{L^{\mathrm{a}}} N^{\mathrm{a}}_{3\mathrm{r,f}}(z')\,\mathrm{d}z' = \ln \frac{I_{\mathrm{L}}(1, \eta = \infty)}{I_{\mathrm{L}}(0, \eta = \infty)} = \ln \frac{V_{\mathrm{r}} \mathrm{e}^{\tilde{\mathscr{E}}_{\mathrm{st}}}}{1 + V_{\mathrm{r}}(\mathrm{e}^{\tilde{\mathscr{E}}_{\mathrm{st}}} - 1)}$$

where $\tilde{\mathscr{E}}_{\mathrm{st}} = \tilde{\mathscr{E}}(0, \infty)$. Using this relation and (6.11) we obtain

$$V_{\mathrm{l}} = V_0 \left[\frac{V_{\mathrm{r}}\, \mathrm{e}^{\tilde{\mathscr{E}}_{\mathrm{st}}}}{V_0[1 + V_{\mathrm{r}}(\mathrm{e}^{\tilde{\mathscr{E}}_{\mathrm{st}}} - 1)]}\right]^{\exp\left(-\frac{u_{\mathrm{l}}}{T^{\mathrm{a}}_{32}}\right)}. \tag{6.12}$$

Similarly, we can calculate the change in the occupation inversion of the active medium during the round trip time u_{r} of the pulse propagating from right to left, whereby the change of the pulse shape and energy according to (6.8) and (6.9), which is caused by the saturable absorber, has to be taken into account. Hence, we have

$$V_{\mathrm{r}} = V_0 \left\{\frac{V_{\mathrm{l}}\left(1 - B_0 B_{\mathrm{l}} + B_0 B_{\mathrm{l}}[1 + V_{\mathrm{r}}(\mathrm{e}^{\tilde{\mathscr{E}}_{\mathrm{st}}} - 1)]^m\right)^{\frac{1}{m}}}{V_0\left[1 - V_{\mathrm{l}} + V_{\mathrm{l}}\left(1 - B_0 B_{\mathrm{l}} + B_0 B_{\mathrm{l}}[1 - V_{\mathrm{r}}(\mathrm{e}^{\tilde{\mathscr{E}}_{\mathrm{st}}} - 1)]^m\right)^{\frac{1}{m}}\right]}\right\}^{\exp\left(-\frac{u_{\mathrm{r}}}{T^{\mathrm{a}}_{31}}\right)}. \tag{6.13}$$

For the transmission of the absorber we proceed in the same way. If we still consider $u \gg T^{\mathrm{b}}_{21}$, then we have

$$B_{\mathrm{r}} = B_0 \tag{6.14}$$

$$B_{\mathrm{l}} = B_0 \left\{\frac{[1 + V_{\mathrm{r}}(\mathrm{e}^{\tilde{\mathscr{E}}_{\mathrm{st}}} - 1)]^m}{1 - B_0 + B_0[1 + V_{\mathrm{r}}(\mathrm{e}^{\tilde{\mathscr{E}}_{\mathrm{st}}} - 1)]^m}\right\}^{\exp\left(-\frac{u_{\mathrm{A}}}{T^{\mathrm{b}}_{21}}\right)}, \tag{6.15}$$

where $B_0 = \exp(-N^{\mathrm{b}} L^{\mathrm{b}} \sigma^{\mathrm{b}}_{31})$ is the small signal transmission of the absorber. By inserting the expressions (6.3), (6.5), (6.8) and (6.9) in one another, the change that the pulse undergoes in the forward and the return passage through the amplifier and the saturable absorber can be determined. Hence,

$$I_{\mathrm{L}}(4, \eta) = G(\eta) \frac{1}{R} I_{\mathrm{L}}(0, \eta) \tag{6.16}$$

where

$$G(\eta) = \frac{V_r V_l B_0 B_l R e \tilde{\mathscr{E}}(\eta)\, K^{m-1}(\eta)\, \{1 - B_0 B_l (1 - K^m(\eta))\}^{\frac{1-m}{m}}}{1 - V_l + V_l \{1 - B_0 B_l (1 - K^m{}_{(\eta)})\}^{\frac{1}{m}}}. \quad (6.17)$$

The influence of the frequency-selective elements can be described in a way similar to that in section 5.2. We assume again that the spectral width of the pulse is small compared to the halfwidth $\Delta\omega$ of the frequency-selective element. After a double passage through this element the pulse intensity is given by

$$I_L(7, \eta) = R \left\{ I_L(4, \eta) - \frac{4}{\Delta\omega} \frac{d}{d\eta} I_L(4, \eta) + \frac{4}{\Delta\omega^2} \left[3 \frac{d^2 I_L(4, \eta)}{d\eta^2} - \frac{1}{2 I_L(4, \eta)} \left(\frac{d I_L(4, \eta)}{d\eta} \right)^2 \right] \right\}. \quad (6.18)$$

The steady-state pulse shape that evolves in the cw pumped dye laser can be described by the assumption that the pulse reproduces itself after each round trip. Hence,

$$I_L(7, \eta) = I_L(0, \eta + h) \approx \left[1 + h \frac{\partial}{\partial\eta} + \frac{h^2}{2} \frac{\partial^2}{\partial\eta^2} \right] I_L(0, \eta); \quad (6.19)$$

h is here a possible time shift of the maximum which results from the action of the amplifier, the saturable absorber and the frequency-selective element. Combining (6.16), (6.18) and (6.19), we obtain a nonlinear integro-differential equation for the steady-state pulse shape $I_L = I_L(0, \eta)$ which is identical with (5.18) where, however, the gain factor $G(\eta)$ is given by (6.17). To solve this equation the ansatz (5.19) can be used again, since at the pulse edges $G(\eta)$ is only weakly dependent on time. As in section 5.2. this results in six transcendental equations for the laser parameters. In addition, we have (6.13) as a seventh equation, whereby the seven unknown quantities $\tilde{\mathscr{E}}_{St}$, $\tilde{\mathscr{E}}_0$, I_0, μ, θ, h and V_r can be determined as functions of the given laser parameters V_0, B_0, m, u_r/T_{31}^a and u_l/T_{31}^a. The other quantities B_r, B_l and V_l are explicit functions of V_r, V_0, B_0 and $\tilde{\mathscr{E}}_{St}$ and can be substituted in (6.17) by the expressions (6.12), (6.14) and (6.15).

6.2.2 *Solutions for the Steady-State Regime*

The laser parameters calculated from the numerical solution of the transcendental equations are presented in figure 6.6. The amplifier was assumed to be located in the center of the resonator ($u_r \approx u_l$) and the saturable absorber near one of the end mirrors ($u_A = 0$). These are optimal positions of both elements.

6.2.2.1 *The Stable Single-Pulse Regime*

Real solutions of the transcendental equations for the pulse parameters, which correspond to a stable single-pulse regime, exist only within a limited range of the laser parameters V_0, B_0, R, u/T_{31}^a and m, whereby non-physical solutions, such as $\tilde{\mathscr{E}}_0 > \tilde{\mathscr{E}}_{St}$, are ruled out. In figure 6.5a, the parameter range of a stable single pulse regime (mode-locking region) is represented by the two solid lines in the V_0,B_0-plane at constant m, R, u/T_{31}^a. For small absorption losses ($1 - B_0 \ll 1$) and small gain coefficients ($V_0 - 1 \ll 1$)

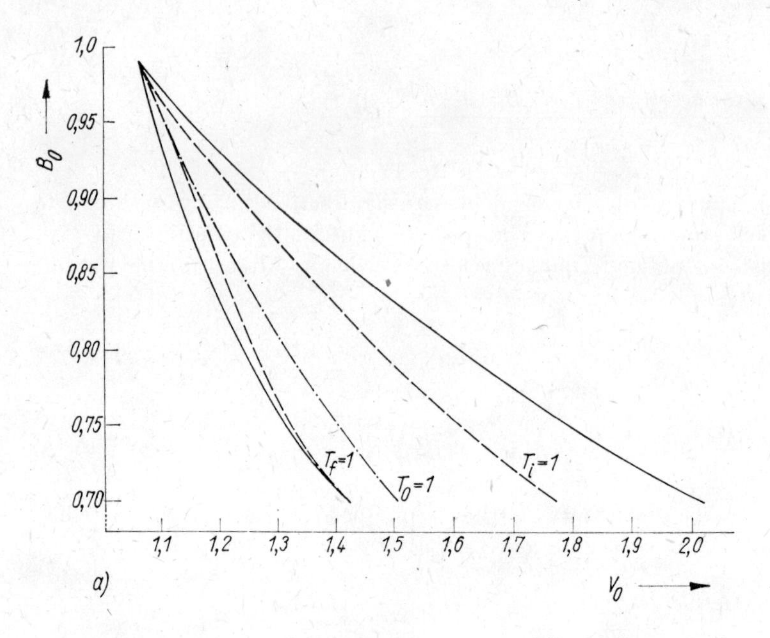

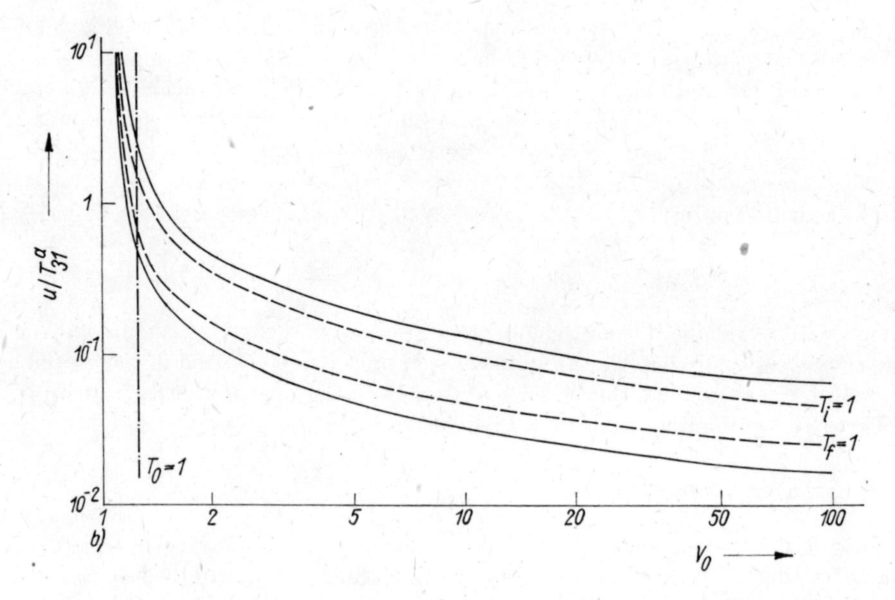

Fig. 6.5. Stability range for single pulses with respect to the B_0, V_0-plane at $u = T^a_{31}$ (a) and the V_0, u/T^a_{31}-plane at $B_0 = 0.85$ (b)

Parameters: $R = 0.9$, $m = 7$, $u_A = 0$, $u_r = u_l$.

(from Herrmann and Weidner [6.11])

the parameter range for a stable single-pulse regime is small. Greater absorption losses lead to a larger stability range, whereby, of course, the minimum pump power needed to exceed the laser threshold increases. With the chosen laser parameters the lower boundary of the modelocking region lies below the laser threshold $T_0 = V_0{}^2B_0{}^2R = 1$, which is represented by the dash-dotted line in Fig. 6.5. This occurs because, due to the condition $T_{31}^{\mathrm{a}} \simeq u$, the occupation inversion is not completely depleted by the preceding pulse passage. In order for the system to lase, the pump power must first be above the threshold. However, while the laser oscillates, the amplification can be decreased even to somewhat below the threshold without leaving the region of modelocking. It should be noted that the region of modelocking is not identical with the region of the static pulse compression zone, which is characterized by a negative gain at both pulse edges. The boundaries of this range are defined by the curves $\mathsf{T}_{\mathrm{i}} = G(\tilde{\mathscr{E}} = 0) = 1$ and $\mathsf{T}_{\mathrm{f}} = G(\tilde{\mathscr{E}} = \tilde{\mathscr{E}}_{\mathrm{St}}) = 1$. They are represented in Fig. 6.5 by dotted lines and lie within the range of the stable single-pulse regime. This means that the requirement of a negative net gain at both pulse edges that is frequently assumed as a criterion for modelocking, as introduced by New [6.8], is not a necessary condition for the appearance of modelocked pulses. A steady-state ultrashort pulse can also develop if it experiences a gain at one of the edges and a loss at the other. Whether, and under which conditions, a single short pulse develops in the resonator follows, in the theory presented here, from the requirement that equation (5.18) with the gain factor from (6.17) has a real self-consistent solution. This condition can be considered a necessary criterion for the occurrence of modelocking. On the other hand the figure shows that the region of static pulse compression provides at least an approximate description of the modelocking region.

In Fig. 6.5b) the range of the stable single-pulse regime is depicted in the u/T_{31}^{a}, V_0-plane with constant m, R and B_0. As we see, the width of the stability range depends very sensitively on the ratio of the resonator round trip time and the relaxation time u/T_{31}^{a}. The requirement that the ratio u/T_{31}^{a} is not too large means that the occupation inversion of the amplifier does not completely recover after the preceding pulse passage, so that the pulse experiences a net loss at the leading edge or, at least, (corresponding to the remarks above regarding the necessary criterion for modelocking) not too great a gain. Otherwise intensity fluctuations appearing in front of the pulse leading edge are amplified and lead to the generation of multiple pulses. On the other hand, the ratio u/T_{31}^{a} must not be too small, because otherwise there is insufficient time for the pump radiation to build up the occupation inversion after the gain has been depleted by the amplification at the preceding pulse passage. For this reason a very large amplification must be chosen, in order that the region of modelocking is reached. This can, however, also lead to the appearance of undesired satellite pulses which are not taken into account in the analysis presented here. As is evident from figure 6.5b) we can estimate an optimal value $u/T_{31}^{\mathrm{a}} \approx 0.2 \cdots 0.5$ at the parameters chosen. In general, under typical experimental conditions, the condition

$$0.1 < u/T_{31}^{\mathrm{a}} < 10 \tag{6.20}$$

must be satisfied as a prerequisite to modelocking. A further prerequisite can be derived from the condition that according to Fig. 6.5a) the small signal absorber transmission B_0 must be smaller than a certain maximum value $B_0{}^{\max}$. At the absorber transmission $B_0{}^{\max}$, the boundaries of the stability range and the three curves $\mathsf{T}_{\mathrm{i}} = 1$, $\mathsf{T}_{\mathrm{f}} = 1$ and $\mathsf{T}_0 = 1$ coincide and end at one point. This point is furthermore characterized by $\tilde{\mathscr{E}}_{\mathrm{St}} = 0$

and $dT_i/d\tilde{\mathscr{E}}_{St}|_{\tilde{\mathscr{E}}_{st}=0} = dT_f/d\tilde{\mathscr{E}}_{St}|_{\tilde{\mathscr{E}}_{st}=0} = 0$. From this condition we obtain

$$B_0^{\max} = \sqrt{1 + \frac{1}{4R}\left(\frac{1-R}{m-1}\right)^2} - \frac{1-R}{2\sqrt{R}\,(m-1)}. \tag{6.21}$$

With given B_0 (6.21) can be rewritten as a lower limit for the value m, namely

$$m_{\min} = 1 + \frac{(1-R)\,B_0}{\sqrt{R}\,(1-B_0^2)}. \tag{6.22}$$

(6.22) represents a requirement on the ratio between the absorption cross-sections of the absorber and amplifier and means that the absorber must be saturated before the amplifier. At the leading edge of the pulse the absorption is not yet saturated, whereby the leading portion of the pulse is attenuated. The trailing edge of the pulse is suppressed through the depletion of the amplification. The pulse peak, however, can only be favored, if the absorber is already saturated in the time range over which the amplifier is still unsaturated.

6.2.2.2 Pulse Parameters

Let us now discuss the dependence of the pulse parameters on the amplifier, absorber and resonator parameters. In Fig. 6.6a the normalized reciprocal pulse length $W = 2/(\Delta\omega\theta)$ is depicted as a function of the small signal amplification V_0 for different values of the small signal transmission B_0 and for constant values $m = 7$, $R = 0.9$. The end points of the curves indicate the end of the stable single-pulse regime. The boundaries of the static pulse compression zone $T_i = 1$ and $T_f = 1$ are represented as dotted lines and lie — as already mentioned — within the stability range. At small values of V_0 the pulse length becomes smaller with increasing pump intensity (i.e. increasing V_0). After the pulse duration has achieved a minimum value, a further increase of the pump intensity leads to greater pulse durations. Smaller absorber transmissions B_0 lead to the generation of shorter and more intense pulses. For smaller B_0, the position of the minimum of the pulse length tends toward the left boundary of the modelocking region. The pulse length is again inversely proportional to the bandwidth of the frequency-selective element. The pulse energy increases monotonically with the small signal amplification V_0 (see Fig. 6.6b). At smaller absorber losses the intensity also increases monotonically with increasing pump intensity, whereas at higher absorber losses the intensity possesses its maximum value within the stability range (see Fig. 6.6c). The asymmetry factor μ is shown in Fig. 6.6d. For values of $B_0 \gtrapprox 0.8$, and at pump powers at which the pulse durations attain their smallest values, the pulses are nearly symmetric ($\mu \approx 0$). If we reduce the pump power below the value that corresponds to the shortest pulse durations, then steeper trailing edges of the pulses occur $\left(\mu < 0,\ \tilde{\mathscr{E}}_0 > \frac{1}{2}\tilde{\mathscr{E}}_{St}\right)$, whereas a greater small signal amplification leads to steeper leading edges $\left(\mu > 0, \tilde{\mathscr{E}}_0 < \frac{1}{2}\tilde{\mathscr{E}}_{St}\right)$. This behavior is due to higher pump powers that lead to a higher gain at the leading edges of the pulses.

As an example we calculate the pulse parameters for the typical values of the rhodamine 6G laser assuming a bandwidth of the frequency selective element $\Delta\omega = 5 \cdot 10^{13}\,s^{-1}$,

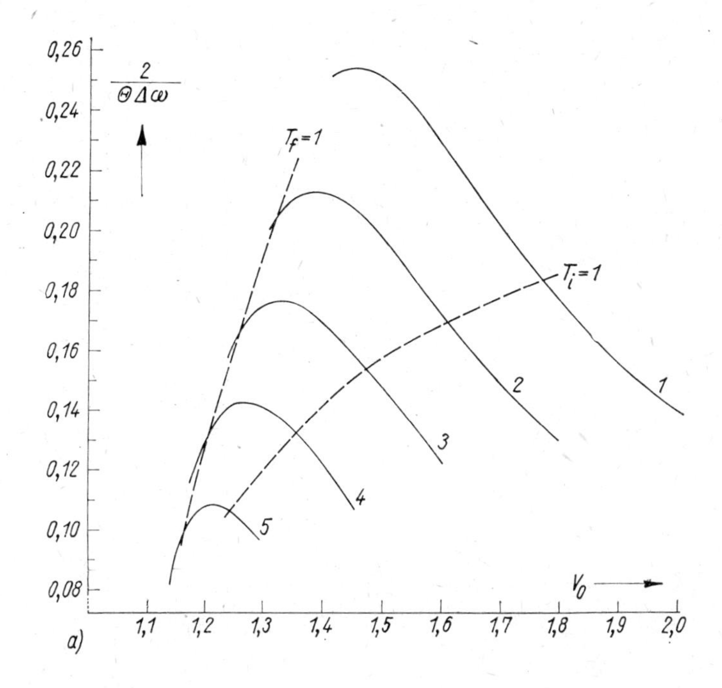
0,26
0,24
0,22
0,20
0,18
0,16
0,14
0,12
0,10
0,08
$\frac{2}{\Theta \Delta\omega}$
$T_f=1$
$T_i=1$
1
2
3
4
5
V_0
1,1 1,2 1,3 1,4 1,5 1,6 1,7 1,8 1,9 2,0
a)

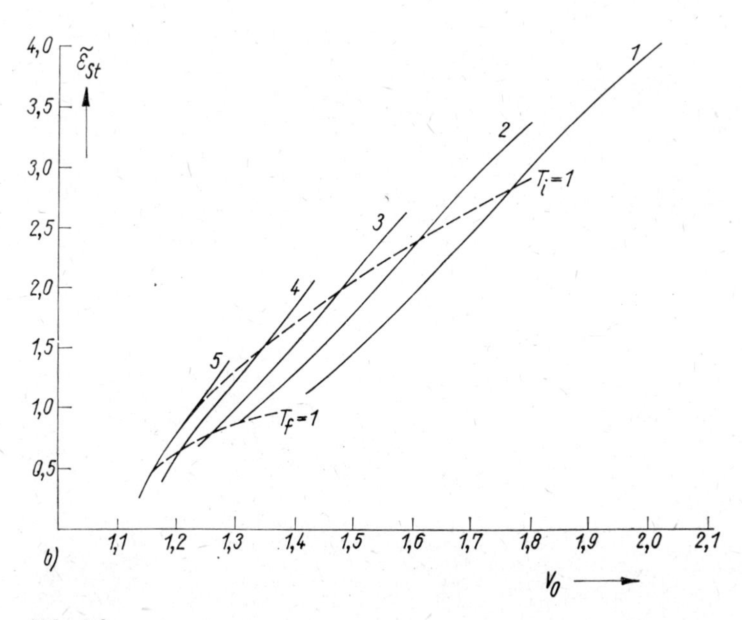
4,0
3,5
3,0
2,5
2,0
1,5
1,0
0,5
$\tilde{\varepsilon}_{St}$
1
2
3
4
5
$T_i=1$
$T_f=1$
1,1 1,2 1,3 1,4 1,5 1,6 1,7 1,8 1,9 2,0 2,1
b)
V_0

Abb. 6.6.

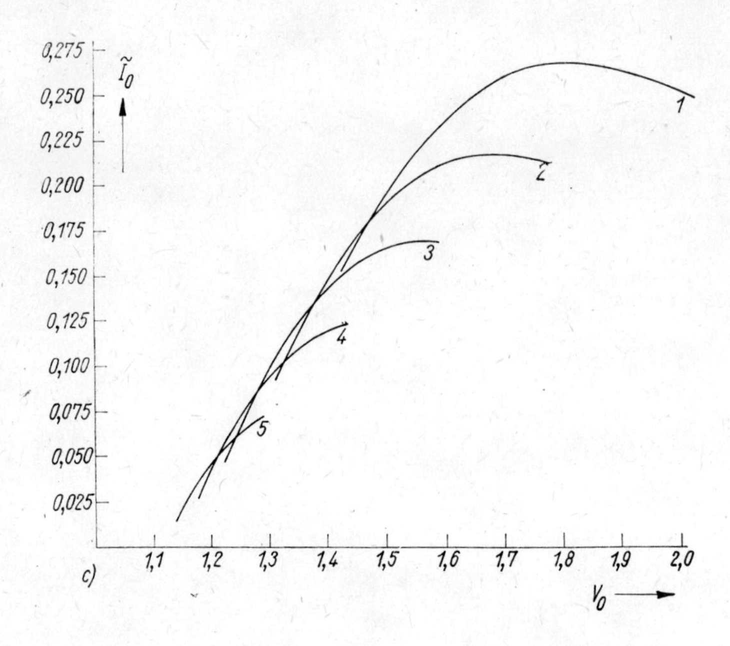
0,275
0,250
0,225
0,200
0,175
0,150
0,125
0,100
0,075
0,050
0,025
$\tilde{I}_0$
1
2
3
4
5
1,1 1,2 1,3 1,4 1,5 1,6 1,7 1,8 1,9 2,0
c)
V_0

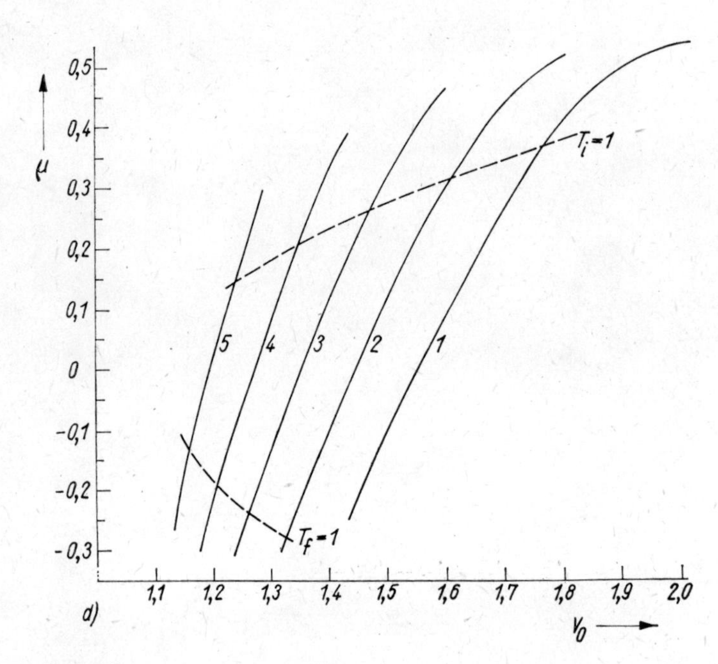
0,5
0,4
0,3
0,2
0,1
0
-0,1
-0,2
-0,3
μ
$T_i = 1$
$T_f = 1$
5 4 3 2 1
1,1 1,2 1,3 1,4 1,5 1,6 1,7 1,8 1,9 2,0
d)
V_0

Abb. 6.6.

a cross section of the amplifier $\sigma_{32}^{a} = 3 \cdot 10^{-16}$ cm², a beam diameter of 15 µm within the amplifier and a laser wavelength $\lambda = 600$ nm. The other laser parameters were chosen as follows:

$V_0 = 1.3$, $B_0 = 0.85$, $R = 0.9$ $m = 7$, $u_A = 0$ and $u_l = u_r = 0.5T_{31}^{a}$. For the pulse length we have $\tau_L = \sqrt{2}\,\theta = 0.4$ ps. The pulse energy and the peak power outside the resonator are 2.4×10^{-10} J and 470 W. The pulse is nearly symmetric ($\mu = 0.079$).

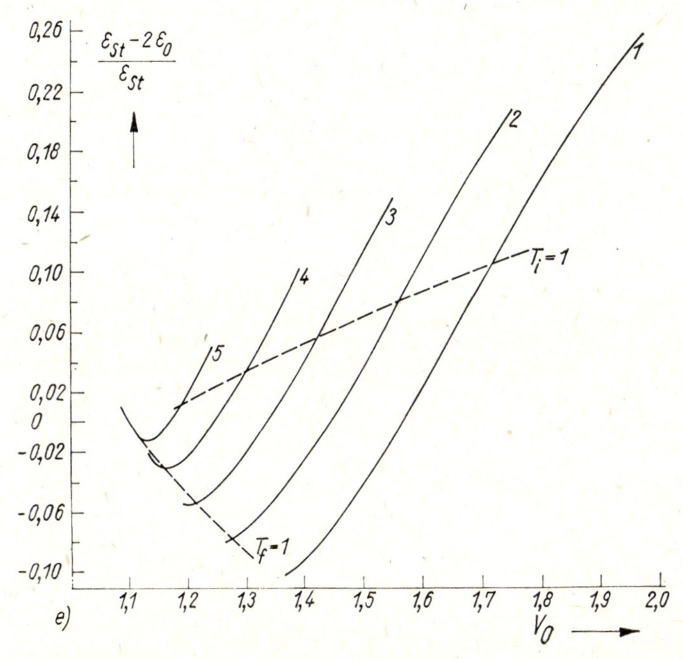

Fig. 6.6. a) to e) show characteristic pulse parameters with respect to the small signal amplification V_0 and the absorber transmission B_0

Parameters: curve 1: $B_0 = 0.7$, curve 2: $B_0 = 0{,}75$, curve 3: $B_0 = 0.8$, curve 4: $B_0 = 0.85$, curve 5: $B_0 = 0.9$. The other laser parameters are: $R = 0.9$, $m = 7$, $u_A = 0$, $u_l = u_r = 0.5T_{32}^{a}$. (from Herrmann and Weidner [6.11])

6.2.3 *Influence of Coherent Interaction between Counterpropagating Pulses in Passive Modelocking*

6.2.3.1 *Basic Equations*

In the previous investigations we have disregarded particularities that could develop due to the positioning of the absorber very near to the highly reflective mirror. Experiments have shown that a resonator configuration with a thin absorber cell in close contact with the highly reflective mirror improves the stability of the pulse generation and leads to pulse shortening (see e.g. [6.12]). This favorable influence of a short contacted absorber rests on the fact that the absorber is already saturated at lower pulse intensities or pulse energies due to the superposition of the forward and back propagating pulse.

The coherent superposition of two pulses occurs very favorably in experiments in a ring resonator in which two counterpropagating pulses precisely overlap in a thin absorber (colliding pulse modelocking (CPM)) [6.6], [6.7], [6.32], [6.36]–[6.38]. In this manner pulses of 50 fs duration and, most recently, even of 27 fs duration [6.44] were attained directly from a laser (compare 6.3.4). The precise superposition of the two counterpropagating pulses in the absorber is controlled by the system itself, since this regime possesses the optimal generation conditions provided that both pulses receive the same amplification in the active medium. This is achieved by separating the amplifier and absorber by a quarter of the round trip length. In this section, let us derive the basic equations for the description of the coherent superposition of two counterpropagating pulses in a laser including both the case of a contacted absorber in a linear resonator and of a ring laser in the colliding pulse modelocking regime (CPM) ([6.13], [6.29]). The optical components in the case of the linear resonator should be arranged according to figure 6.3 where $u_A = 0$ and with the optimal position of the amplifier in the center of the resonator ($u_r = u_l$). In the case of the CPM ring laser the return route does not lead through the resonator components again, but directly to position 2 via a triangular-arrangement. The distance from absorber to amplifier should also come to $\frac{u}{4}$. In this case, both directions are equally favored, and therefore the two pulses that collide in the absorber are equally amplified in the gain medium. In the following, we will denote both cases as well as other similar ones as CPM-regimes, whereby the following treatment is based on the papers of Herrmann, Weidner, Wilhelmi [6.13], as well as those of Kühlke, Rudolph, Wilhelmi [6.29]. The laser field strength for two oppositely travelling pulses can be expressed as the superposition of one pulse propagating to the left $(A_l(z, t))$ and one to the right $(A_r(z, t))$:

$$E(z, t) = \frac{1}{2} [A_l(z, t) e^{ik_L z} + A_r(z, t) e^{-ik_L z}] e^{i\omega_L t} + \text{C.C.} \tag{6.23}$$

If we insert (6.23) in equation (6.7) for the occupation number density $N_1{}^b$ of the absorber ground level, then fast oscillating terms, e.g. proportional to exp $(2ik_L z)$, exp $(4ik_L z), \ldots$ result. For the first coefficients q, p of the spatial Fourier expansion $N_1{}^b = q + p\, e^{2ik_L z} + p^*\, e^{-2ik_L z}$ we obtain the equations

$$\frac{\partial q}{\partial t} = -\beta[q(|A_l|^2 + |A_r|^2) + pA_l^*A_r + p^*A_lA_r^*], \tag{6.24}$$

$$\frac{\partial p}{\partial t} = -\beta[p(|A_l|^2 + |A_r|^2) + qA_lA_r^*], \tag{6.25}$$

and from the wave equation (compare 1.3) we find

$$-\frac{\partial A_l}{\partial z} + \frac{1}{v}\frac{\partial A_l}{\partial t} = -\frac{1}{2}\sigma_{13}^b(qA_l + pA_r)\,\mathscr{L}_b \tag{6.26}$$

and

$$\frac{\partial A_r}{\partial z} + \frac{1}{v}\frac{\partial A_r}{\partial t} = -\frac{1}{2}\,\sigma_{13}^{b}(qA_r + p^*A_l)\,\mathcal{L}_b \tag{6.27}$$

where $\beta = \tau_{13}^{b}\,|\mu_{13}^{b}|^2/2\hbar^2[1 + 4(\omega_{31}^{b} - \omega_L)^2/\Delta\omega^2]$, and $\mathcal{L}_b = \mathcal{L}_b' + i\mathcal{L}_b'' = [1 + 2i(\omega_{31}^{b} - \omega_L)/\Delta\omega_b]^{-1}$, $\sigma_{13}^{b} = \dfrac{\tau_{13}^{b}\,|\mu_{13}^{b}|^2\,\omega_L\sqrt{\mu_0}}{\hbar\sqrt{\varepsilon_0}}$ were introduced. For the sake of simplification, let us solve equations (6.24) to (6.27) solely for the limiting case of small absorption and amplification as well as small pulse energies (normalized to the saturation energy of amplifier and absorber, i.e. we assume $\tilde{\mathcal{E}}_{St} \ll 1$, $m\tilde{\mathcal{E}}_{St} \ll 1$, $B_0 = e^{-\varkappa_0} \approx 1 - \varkappa_0$, $V_0 = e^{a} \approx 1 + a$). A model of this kind with respect to the case of a non-contacted absorber or a unidirectional regime was first reported by Haus [6.10]. The range of validity of the approximations considered here restricts the laser parameters to a relatively small range, as can be seen from a comparison of the results from sections 6.2.1, 6.2.2 which are based on more general assumptions. For simplification we will nevertheless investigate in the following the effects of the coherent superposition in this approximation, whereby for comparison with experimental results these assumptions always have to be checked for their validity. The treatment of a single pulse in a laser with a contacted absorber differs here from that of the CPM ring laser solely because the small signal absorption $\varkappa_0 = \sigma_{13}^{b}N^{b}L^{b}$ is replaced by $\varkappa_0 = 2\sigma_{13}^{b}N^{b}L^{b}$ for the contacted absorber, since in this case the pulse passes through the absorber twice. For successive approximation we set in a first iteration step in (6.26), (6.27) $p = 0$ and replace q by the occupation of the absorber ground state before the arrival of the pulse. After two iteration steps we obtain

$$A_r(\eta, 3) = A_r(\eta, 2)\left\{1 - \frac{\varkappa}{2}\left(1 - m_1\tilde{\mathcal{E}}(\eta) + \frac{m_2}{2}\,\tilde{\mathcal{E}}^2(\eta)\right)\right\} \tag{6.28}$$

for the pulse amplitudes at the output of the absorber, where $\tilde{\mathcal{E}}(\eta) = \sigma_{32}^{a}\mathcal{L}_a' \int_{-\infty}^{\eta} I(\eta')\,d\eta'$ and $\varkappa = \varkappa_0\mathcal{L}_b$. For a short absorber ($L^{b} \ll \tau_L v$) in the CPM-regime $m_1 = 3m$ and $m_2 = 10m^2$ hold, where $m = \sigma_{31}^{b}\mathcal{L}_b'/\sigma_{32}^{a}\mathcal{L}_a'$. For a long absorber $L^{b} \gg \tau_L v$ in the CPM-regime, $m_1 = 2m$ and $m_2 = 4m^2$ have to be replaced, whereas in the unidirectional operation $m_1 = m$ and $m_2 = m^2$ have to be replaced. The long absorber in the CPM-regime is thus saturated, as if by a pulse of double energy. The short absorber is saturated even earlier in the CPM-regime due to the formation of standing waves, whereby in the approximation given here this corresponds to about a tripling of the cross section σ_{31}^{b} as opposed to the unidirectional operation ($m \to 3m$).

With similar assumptions concerning the action of the amplifier we obtain

$$A_r(\eta, 4) = \left[1 + \frac{a}{2}\left(1 - \tilde{\mathcal{E}}(\eta)\right)\right] A_r(\eta, 3) \tag{6.29}$$

where

$$a = a_0\mathcal{L}_a\left\{1 - \tilde{\mathcal{E}}_{St}\left[1 + \exp\left(\frac{u}{2T_{31}^{a}}\right)\right]\left[\exp\frac{u}{T_{31}^{a}} - 1\right]^{-1}\right\} \tag{6.30}$$

represents the gain for the leading edge and $a_0 = \sigma_{32}^{a}N_{32}^{a}L^{a}$ the small signal gain in the center of the gain profile.

As in chapter 5., let us assume that the radiation in the resonator is influenced by frequency-selective elements (a frequency filter). This frequency filter can be provided by an additional component (e.g. by a prism) for frequency tuning. It can, however, also represent approximately the effective bandwidth-limiting properties of the amplification. If the spectral bandwidth of the pulse is small compared to the spectral width of the filter, and the laser frequency ω_L is determined by the center frequency of that filter, then after passing through this element the pulse is given by

$$A_r(\eta, 5) = \left[1 - \frac{2}{\Delta\omega}\frac{d}{d\eta} + \frac{4}{\Delta\omega^2}\frac{d^2}{d\eta^2}\right] A_r(\eta, 4) \tag{6.31}$$

according to (5.13) and (5.14)

As we shall see later, for the generation of the shortest pulses possible it is beneficial under certain circumstances to insert in the resonator an additional optical element (e.g. a glass plate) with group velocity dispersion whose thickness must be optimized for chirp compensation. This element is also assumed to include all contributions to group velocity dispersion of the other resonator elements. The change of the field amplitude in a linear optical element due to the group velocity dispersion is described by equation (1.50′) (where $\bar{P}_1' = 0$). If we integrate this equation with respect to z, using the method of successive approximation and account additionally for the loss coefficient γ of the output mirror, we obtain

$$A_r(\eta, 6) = \left(1 - \frac{\gamma}{2}\right) A_r(\eta, 5) + \frac{ir}{2}\frac{d^2 A_r(\eta, 5)}{d\eta^2} \tag{6.32}$$

where $r = L_0 \dfrac{d^2k}{d\omega^2} = \dfrac{4\pi^2 c}{\omega_L^3} L_0 \dfrac{d^2n}{d\lambda^2}$ and L_0 is the effective length of the glass plate. This plate is assumed to be placed between the filter and the output mirror. Under the condition that in the steady-state regime the pulses reproduce after each round trip, we obtain from (6.28) to (6.32) a nonlinear integro-differential equation with complex coefficients for the amplitude $A_r(\eta, 2) = A_l(\eta, 2) = A(\eta)$, namely

$$A_r(\eta, 6) = A_r(\eta + h, 2) \approx \left[1 + h\frac{d}{d\eta} + \frac{h^2}{2}\frac{d^2}{d\eta^2}\right] A_r(\eta, 2). \tag{6.33}$$

The real and imaginary parts may be separated by

$$A(\eta) = \tilde{A}(\eta)\, e^{i\varphi(\eta)}$$

and we obtain the following system of equations for calculating the phase $\varphi(\eta)$ and the real amplitude $\tilde{A}(\eta)$

$$\left\{g'(\eta) + \frac{r}{2}\frac{d^2\varphi}{d\eta^2} - \left(\frac{4}{\Delta\omega^2} - \frac{h^2}{2}\right)\left(\frac{d\varphi}{d\eta}\right)^2 - \left(\frac{2}{\Delta\omega} + h - r\frac{d\varphi}{d\eta}\right)\frac{d}{d\eta} + \left(\frac{4}{\Delta\omega^2} - \frac{h^2}{2}\right)\frac{d^2}{d\eta^2}\right\}\tilde{A}(\eta) = 0, \tag{6.34a}$$

$$g''(\eta) + \frac{r}{2}\left(\frac{d\varphi}{d\eta}\right)^2 - \left(\frac{2}{\Delta\omega} + h\right)\frac{d\varphi}{d\eta} + \left(\frac{4}{\Delta\omega^2} - \frac{h^2}{2}\right)\left[\frac{d^2\varphi}{d\eta^2} + 2\frac{d\varphi}{d\eta}\frac{d}{d\eta}\ln\tilde{A}(\eta)\right]$$

$$-\frac{r}{2}\frac{1}{\tilde{A}}\frac{d^2\tilde{A}}{d\eta^2} = 0, \tag{6.34b}$$

where

$$g(\eta) = g'(\eta) + \mathrm{i}g''(\eta) = \frac{1}{2}(a - \varkappa - \gamma) + (m_1\varkappa - a)\,\tilde{\mathscr{E}}(\eta)/2 - \frac{\varkappa}{2}\,m_2\tilde{\mathscr{E}}^2(\eta)\,.$$

6.2.3.2 *Solution of the Basic Equations for the Case of Resonance*

Let us first consider the simplest case in which the center frequency of the laser pulse ω_L is equal to the mid-frequencies of absorber and amplifier ($\omega_\mathrm{L} = \omega_{32}^\mathrm{a} = \omega_{31}^\mathrm{b}$), and the dispersion of the group velocity is negligible ($r = 0$). In this case we obtain $g''(\eta) = 0$, and as a possible solution $\mathrm{d}\varphi/\mathrm{d}\eta = 0$ follows from (6.34b). Equation (6.34a) can then be solved using the ansatz

$$\tilde{A}(\eta) = \sqrt{\frac{\tilde{\mathscr{E}}_\mathrm{St}}{2\beta^\mathrm{a}\tau_\mathrm{L}'}}\,[\cosh(\eta/\tau_\mathrm{L}')]^{-1} \tag{6.35}$$

where, the unknown parameters $\tilde{\mathscr{E}}_\mathrm{St}$ and τ_L' can be determined by equation (6.34a) [6.13] as

$$\tilde{\mathscr{E}}_\mathrm{St} = \frac{4[\varkappa_0 m_1 - a_0(1 + 2\delta)]}{3\varkappa_0 m_2}\left[1 \pm \sqrt{1 + \frac{3\varkappa_0 m_2(a_0 - \varkappa_0 - \gamma)}{[\varkappa_0 m_1 - a_0(1 + 2\delta)]^2}}\right], \tag{6.36}$$

$$\tau_\mathrm{L}' = \frac{8}{\Delta\omega\tilde{\mathscr{E}}_\mathrm{St}}\cdot\frac{1}{\sqrt{\varkappa_0 m_2}}, \tag{6.37}$$

where $\delta = \left[1 + \exp\left(\frac{u}{2T_{32}^\mathrm{a}}\right)\right]\left[\exp\left(\frac{u}{T_{32}^\mathrm{a}}\right) - 1\right]^{-1}$ holds in the CPM-regime and $\delta = \left[1 + \exp\left(\frac{u}{T_{32}^\mathrm{a}}\right)\right]\left[\exp\left(\frac{u}{T_{32}^\mathrm{a}}\right) - 1\right]^{-1}$ in the unidirectional regime. As is obvious from (6.37), the pulses from the CPM-regime with a short absorber are shorter by a factor of $\sqrt{10} \approx 3$ at equal energy. In Fig. 6.7, for the cases of a short absorber ($m_1 = 3m$, $m_2 = 10m^2$), a long absorber ($m_1 = 2m$, $m_2 = 4m^2$) in the CPM-regime as well as for an absorber in the unidirectional regime ($m_1 = m$, $m_2 = m^2$), the static pulse shortening zone (which is here characterized by a negative effective gain at the leading edge of the pulse ($g_\mathrm{i} < 0$) and at its trailing edge ($g_\mathrm{f} < 0$)) and the laser threshold $g_0 = a_0 - \varkappa_0 - \gamma = 0$ are plotted. As can be seen, the CPM-laser with the short absorber possesses the most favorable stability range: the range is broader than that of the unidirectional regime and the left boundary of the stability range ($g_\mathrm{f} = 0$) is near the laser threshold ($g_0 = 0$). Therefore the laser needs to be pumped only slightly above the threshold to obtain a stable single pulse regime. In Fig. 6.8 the pulse energy and pulse duration are depicted for a set of parameters within the stability range. The absorber loss is chosen such that the left boundary of the stability range ($g_\mathrm{f} = 0$) is the same in all three cases, in order that the pulse parameters can be compared at the same amplification coefficients a_0 and the same pump energies. Due to the smaller absorption coefficient produced in this manner, the pulse energy is greater in the unidirectional operation, though the pulse duration is three times smaller in the CPM-regime with the short absorber. Thus, in agreement with the experimental results [6.12], the pulse duration as well as the stability range is more favorable in the case of CPM. If the distance between the absorber and the amplifier deviates from the optimal value $(uc)/4$, the counterpropagating pulses attain different energies $\mathcal{E}_{L1}$ and $\mathcal{E}_{L2}$, respectively, as a result of the differing amplification conditions. The ratio of the pulse durations here is given by

$$\frac{\tau_{L2}}{\tau_{L1}} = \sqrt{\frac{3 + 6\zeta + \zeta^2}{1 + 6\zeta + 3\zeta^2}}$$

where $\zeta = \mathcal{E}_{L1}/\mathcal{E}_{L2}$.

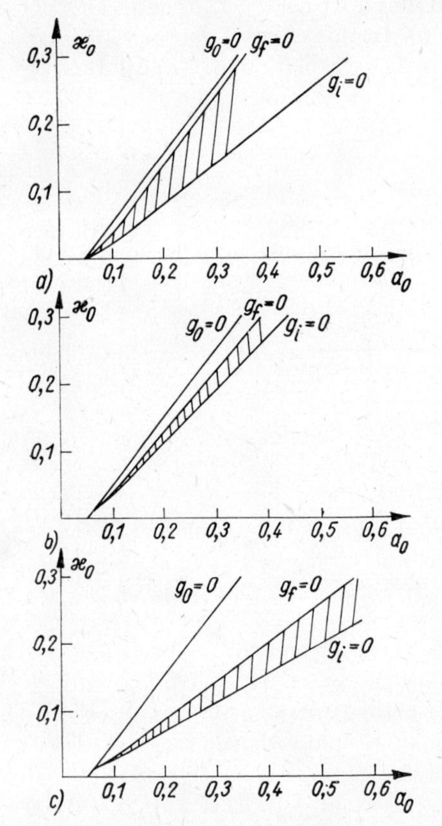

Fig. 6.7. Stability range of a dye laser with a) a short, b) a long absorber in the CPM-regime and c) an absorber in a unidirectional regime

As parameters $u/T_{31}^{a} = 0.8$, $u_r = u_l$, $m = 4$, $\gamma = 0.1$ were chosen. The hatched area indicates the static pulse compression zone which is limited by the curves $g_f = 0$ and $g_i = 0$.
(from Herrmann, Weidner, Wilhelmi [6.13])

Simultaneously, the stability range defined as the average of the stability ranges of the two oppositely travelling pulses becomes smaller. For increasing deviations of the distance between amplifier and absorber from the value $(uc)/4$, at given a_0 and $\varkappa_0$, laser regimes are achieved in which only unidirectional operation occurs. Accordingly the pulse duration is about 3 times greater than in the case of two counterpropagating pulses.

6.2.3.3 *Solution for the Case of Quasiresonant Conditions: Chirp Formation and Chirp Compensation*

Let us now investigate the influence of a detuning of the laser frequency from the absorber and amplifier mid-frequency
$(\omega_L \neq \omega_{32}^{a}, \omega_L \neq \omega_{31}^{b})$.

Equation (6.34) can be solved in this case using (6.35) for the amplitude $\tilde{A}(\eta)$ and

$$\frac{d\varphi(\eta)}{d\eta} = -\frac{c_1}{\tau_L'} - \frac{c_2}{\tau_L'} \tanh(\eta/\tau_L') \tag{6.38}$$

for the time dependent phase. The term $c_2(\tau_L')^{-1} \tanh(\eta/\tau_L')$ provides the instantaneous frequency shift. $c_2 > 0$ thus indicates a "down" and $c_2 < 0$ an "up" chirp. The quan-

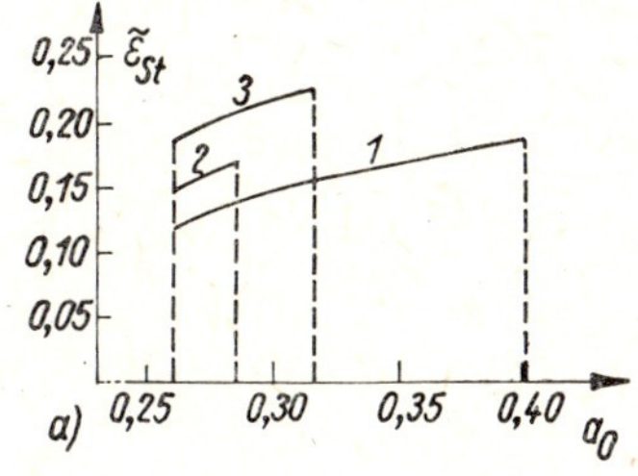

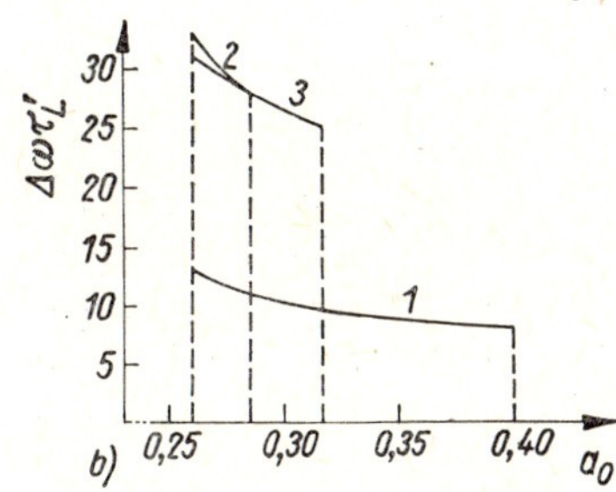

Fig. 6.8. a) Pulse energy $\tilde{\mathscr{E}}_{St}$ and b) pulse duration ($\tau_L = 1.76\tau_P'$) with respect to the amplification a_0 for a short absorber (curve 1), a long absorber in the CPM regime (curve 2) and an absorber in an unidirectional regime (curve 3).
The absorption coefficients are: $\varkappa_0 = 0.2$ (curve 1), $\varkappa_0 = 0.176$ (curve 2) and $\varkappa_0 = 0.12$ (curve 3).
(from Herrmann, Weidner, and Wilhelmi [6.13])

tity $\dfrac{c_1}{\tau_L'}$ describes a deviation of the instantaneous frequency at the moment $\eta = 0$ from the frequency $\omega_L = k_L v$. If we insert (6.35) and (6.38) in (6.34), we obtain independent equations for the unknowns τ_L', $\tilde{\mathscr{E}}_{St}$, h, c_1, c_2, ω_L [6.33] after comparing coefficients in front of linearly independent hyperbolic functions:

$$\frac{1}{2}(a' - \varkappa' - \gamma) + \frac{\tilde{\mathscr{E}}_{St}}{4}(m_1\varkappa' - a') - \frac{1}{8}\tilde{\mathscr{E}}_{St}^2 m_2 \varkappa' - \frac{rc_2}{\tau_L'^2}$$
$$+ \frac{1}{\tau_L'^2}\left(\frac{4}{\Delta\omega^2} - \frac{h^2}{2}\right)(1 - c_1^2 - c_2^2) = 0, \tag{6.39a}$$

$$\frac{\tilde{\mathscr{E}}_{St}}{4}(m_1\varkappa' - a') - \frac{1}{8}\tilde{\mathscr{E}}_{St}^2 m_2\varkappa' - \frac{c_1 r}{\tau_L'^2} + \frac{1}{\tau_L'}\left(\frac{2}{\Delta\omega} + h\right) - \frac{2c_1c_2}{\tau_L'^2}\left(\frac{4}{\Delta\omega^2} - \frac{h^2}{2}\right) = 0, \tag{6.39b}$$

$$\frac{1}{16}\tilde{\mathscr{E}}_{St} m_2\varkappa' + \frac{3}{2}\frac{rc_2}{\tau_L'^2} + \frac{c_2^2 - 2}{\tau_L'^2}\left(\frac{4}{\Delta\omega^2} - \frac{h^2}{2}\right) = 0, \tag{6.39c}$$

$$\frac{1}{2}(a'' - \varkappa'') + \frac{\tilde{\mathscr{E}}_{St}}{4}(m_1\varkappa'' - a'') - \frac{1}{8}\tilde{\mathscr{E}}_{St}^2 m_2\varkappa'' + \frac{c_1}{\tau_L'}\left(\frac{2}{\Delta\omega} + h\right)$$
$$+ \frac{r}{2\tau_L'^2}(1 + c_1^2 + c_2^2) + \frac{2c_2}{\tau_L'^2}\left(\frac{4}{\Delta\omega^2} - \frac{h^2}{2}\right) = 0, \tag{6.39d}$$

$$\frac{\tilde{\mathscr{E}}_{\mathrm{St}}}{4}(m_1\varkappa'' - a'') - \frac{1}{10}\tilde{\mathscr{E}}^2_{\mathrm{St}} m_2 \varkappa'' - \frac{c_1 c_2 r}{\tau_{\mathrm{L}}'^2} + \frac{c^2}{\tau_{\mathrm{L}}'}\left(\frac{2}{\Delta\omega} + h\right) + \frac{2c_1}{\tau_{\mathrm{L}}'^2}\left(\frac{4}{\Delta\omega^2} - \frac{h^2}{2}\right) = 0\,, \tag{6.39e}$$

$$\frac{1}{16}\tilde{\mathscr{E}}^2_{\mathrm{St}} m_2 \varkappa'' - \frac{r}{\tau_{\mathrm{L}}'^2}\left(1 - \frac{c_2{}^2}{2}\right) - \frac{3c_2}{\tau_{\mathrm{L}}'^2}\left(\frac{4}{\Delta\omega^2} - \frac{h^2}{2}\right) = 0 \tag{6.39f}$$

$(a = a' + \mathrm{i}a'',\quad \varkappa = \varkappa' + \mathrm{i}\varkappa'')$.

For the present, we shall neglect the influence of linear optical elements ($r = 0$). From (6.39c) and (6.39f) we then immediately obtain for the chirp parameter

$$c_2 = -\frac{3\varkappa'}{2\varkappa''} \pm \sqrt{\left(\frac{3\varkappa'}{2\varkappa''}\right)^2 + 2} = \frac{-1.5}{\tau^{\mathrm{b}}_{31}(\omega^{\mathrm{b}}_{31} - \omega_{\mathrm{L}})} \pm \sqrt{2 + \left(\frac{1.5}{\tau^{\mathrm{b}}_{31}(\omega^{\mathrm{b}}_{31} - \omega_{\mathrm{L}})}\right)^2} \tag{6.40}$$

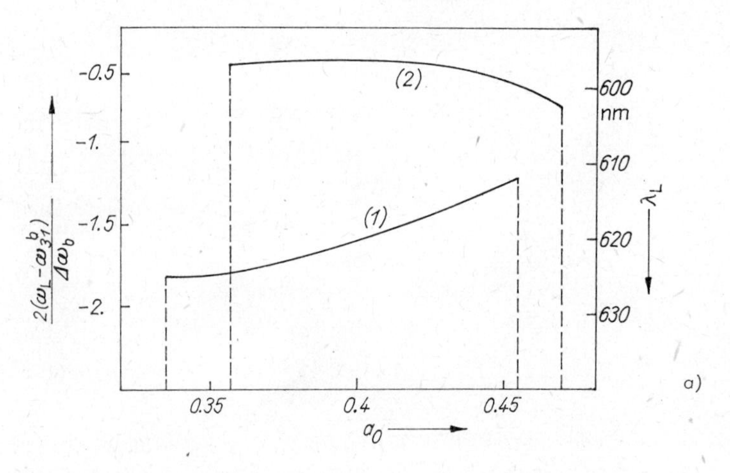

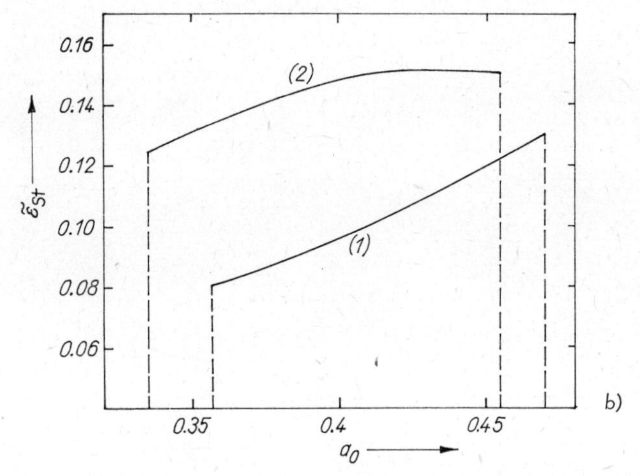

Fig. 6.9. Laser frequency ω_{L} (a) and pulse enery $\tilde{\mathscr{E}}_{\mathrm{St}}$ (b) as functions of the gain coefficient $a_0(\Delta\omega_{\mathrm{b}} = 2/\tau^{\mathrm{b}}_{31})$
Parameters: $m_0 = 6$, $\varkappa_0 = 0.4$, $\gamma = 0.015$, $u/T^{\mathrm{a}}_{32} = 0.8$, $(\omega^{\mathrm{b}}_{31} - \omega^{\mathrm{a}}_{32})\ \tau^{\mathrm{b}}_{21} = -0.174$, $\tau^{\mathrm{b}}_{31}/\tau^{\mathrm{a}}_{32} = 1.56$
The dashed curve indicates the boundaries of the stability range. (from Kühlke and Rudolph [6.33])

where the $(\pm)$ sign holds for $\omega_L \gtrless \omega_{31}^b$, since for $\omega_L \to \omega_{32}^b$ the parameter c_2 must approach zero. As expected, the sign of the chirp depends on whether the laser frequency lies above or below the resonance frequency of the absorber. In Fig. 6.9, for certain para-

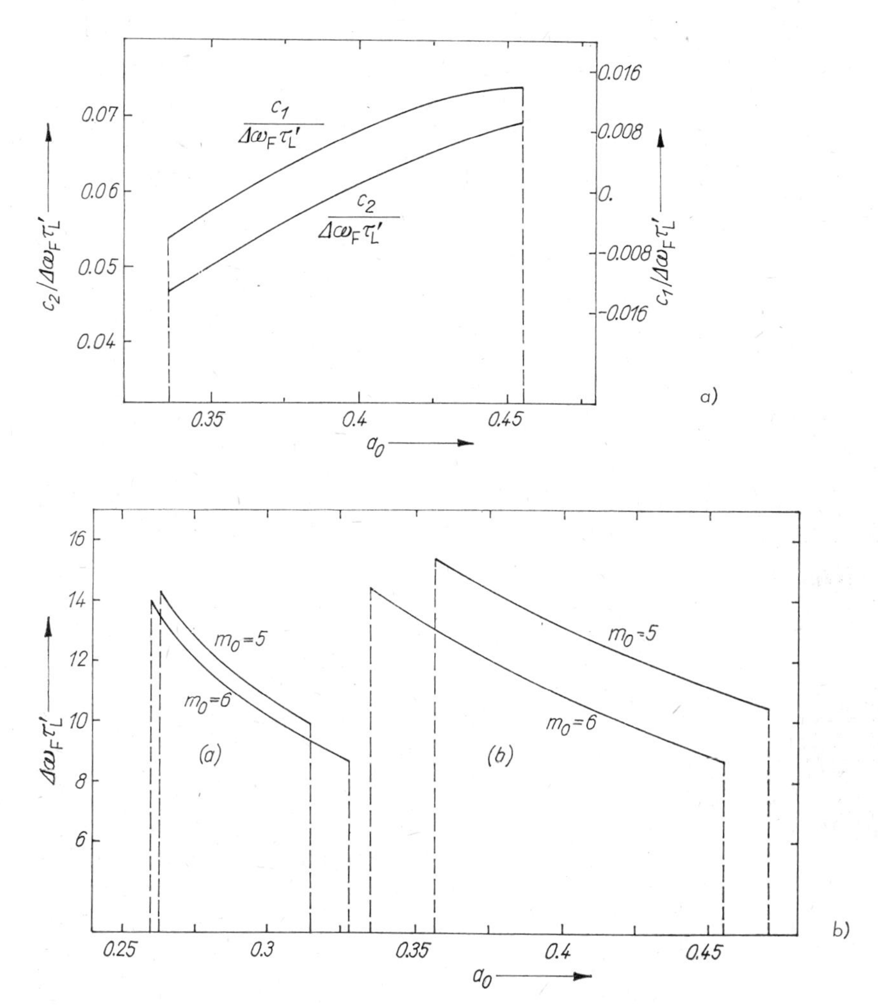

Fig. 6.10. Frequency shift and chirp (a) and pulse duration (b) as functions of the gain coefficient In Fig. 6.10b) curve (a) corresponds to $u/T_{32}^a = 2.3$ and curve (b) to $u/T_{32}^a = 0.8$. The other parameters are the same as in Fig. 6.9.
(from Kühlke and Rudolph [6.33])

meters the solutions of (6.39) are shown, where two different solutions result for a fixed set of parameters. In Fig. 6.10 the frequency shift, the chirp parameter and the pulse duration are depicted for the solution with the greater energy. For a numerical evaluation $\omega_{32}^a \approx 3.24 \times 10^{15}\,\mathrm{s}^{-1}$, $\omega_{31}^b \approx 3.22 \times 10^{15}\,\mathrm{s}^{-1}$, $\Delta\omega^a = 2/\tau_{32}^a \approx 3.6 \times 10^{14}\,\mathrm{s}^{-1}$, and $\Delta\omega^b = 2/\tau_{31}^b \approx 2.3 \times 10^{14}\,\mathrm{s}^{-1}$ can be estimated from the fluorescence profile of Rh6G and from the absorption profile of DODCI. Using these values a modelocking region of 612–624 nm follows from figure (6.10) for the solution with the greater energy. Like the computed concentration dependence of the laser wavelength this result also agrees with experimental findings [6.37]. With a filter bandwidth of $\Delta\omega \approx 0.31 \times 10^{15}\,\mathrm{s}^{-1}$ pulse widths $\tau_L = 1.76\tau_L'$ of 50 to 90 fs result, which corresponds with the observed values ([6.34]–[6.37]). Assuming $c_2/\tau_L' \approx (1.7 \cdots 2.2) \times 10^{13}\,\mathrm{s}^{-1}$ (Fig. 6.10a) a wavelength sweep within the pulse width of $(4.9 \cdots 6.2)$ nm can be estimated.

In order to generate shorter pulses it is necessary to compensate the approximately constant part of the down chirp using a linear optical element with a positive group velocity dispersion (e.g. glass, where $\frac{d^2n}{d\lambda^2} > 0$), through which an effective pulse compression is achieved. Thus, the laser parameters must be chosen such that the chirp parameter c_2 in (6.38) vanishes. From equation (6.39f) it can be seen that this case is achieved precisely when the dispersion parameter $r = L_0 \left.\frac{\mathrm{d}^2k}{\mathrm{d}\omega^2}\right|_{\omega_L}$, which describes the dispersion of the additional linear optical element, obeys the relation

$$r = \frac{5}{8}\,\tau_L'^2 \tilde{\mathscr{E}}_{St}^2 m^2 \varkappa_0 \mathscr{L}_b'' . \tag{6.41}$$

In conclusion, papers by Kobayashi and Yoshizawa [6.40] and by Ippen and Stix [6.39] should be mentioned in which starting with the basic equations described here for $\omega_L = \omega_a = \omega_b$ they carried out computer simulations, where some of Haus' approximations could be dropped. The results of both papers agree with those presented here with respect to the functional dependences and the pulse parameters estimated.

6.2.3.4 *Influence of the Transverse Relaxation Time of Amplifier and Absorber*

In order to obtain the shortest pulses the bandwidth of the additional optical elements in the resonator must be kept as large as possible, so that eventually the resulting amplification profile itself acts as the bandwidth limitation. To a rough approximation, the bandwidth of the "frequency-selective filter" can be substituted by the bandwidth of the effective amplification profile. In detail, however, the action of a linear optical filter distinguishes itself from the effective bandwidth limitation of the amplification profile whose behavior is determined by saturable, hence nonlinear, optical elements. Rudolph and Wilhelmi investigated this case [6.35], in which they did not neglect $\frac{\partial \varrho_{12}}{\partial t}$ in the density matrix equation for the off-diagonal elements ϱ_{12} throughout (compare e.g. equation (1.60)), but calculated two correction terms resulting from $\frac{\partial \varrho_{12}}{\partial t}$ by means of successive approximation. As a result, they obtained similar equations to (6.39) that have additional terms taking into account the bandwidth-limiting properties of the absorption and amplification profile. For the case in which the pulse chirp was compensated in the pulse center pulse solutions with $\frac{\mathrm{d}^2\varphi}{\mathrm{d}\eta^2} = \frac{\mathrm{d}\varphi}{\mathrm{d}\eta} = 0$ were sought. For the value of the dispersion parameter r of the linear optical element at which the chirp

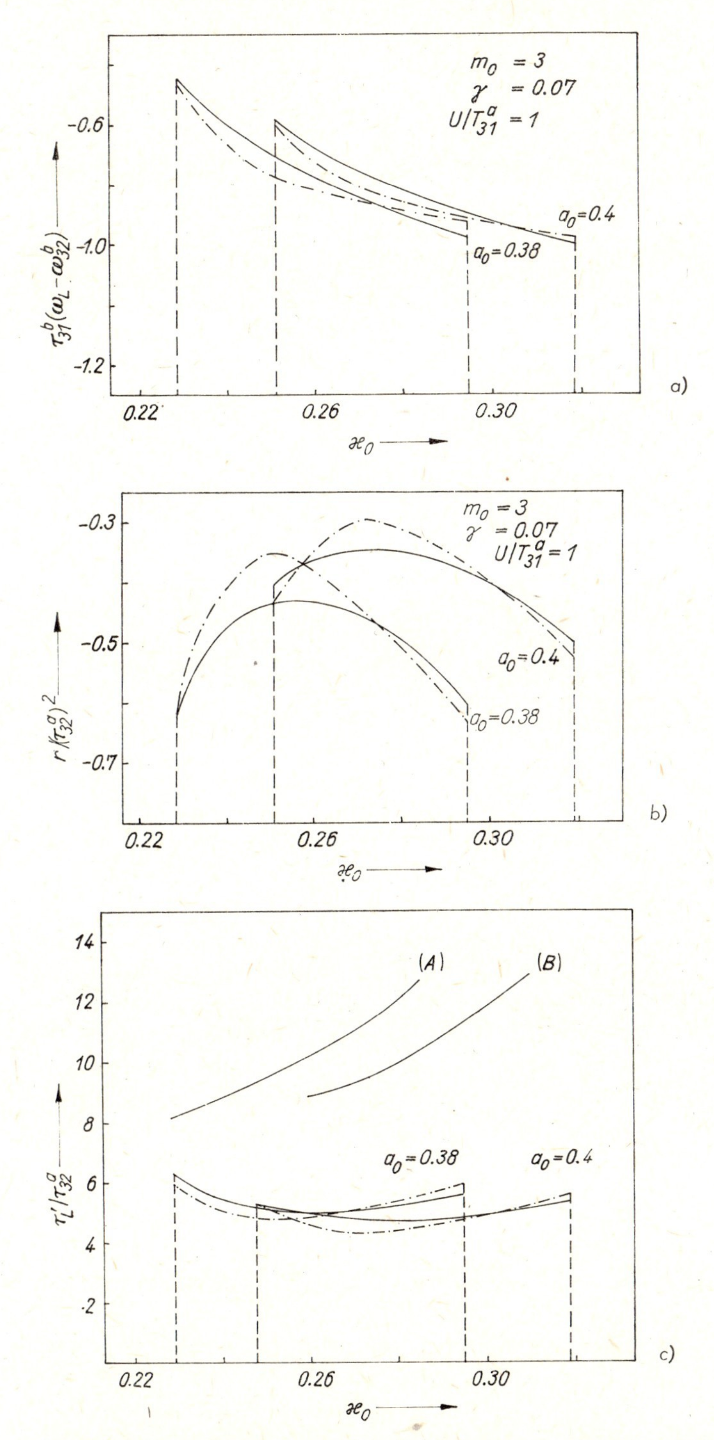

Fig. 6.11. Laser frequency ω_L (a), dispersion parameter r for chirp compensation (b) and pulse duration (c) as function of the absorption coefficient

Curves (A) and (B) correspond to a regime without phase modulation and chirp compensation. (from Rudolph and Wilhelmi [6.35])

is just compensated, the following relation can be derived:

$$r = \tau_L'^2 \left[\frac{5}{8} \tilde{\mathscr{E}}_{St}^2 \varkappa_0 \mathscr{L}_b'' m^2 - \frac{\varkappa_0}{2\tau_L'} \tilde{\mathscr{E}}_{St} \cdot 3m\tau_{31}^b \left(\gamma_b'' + \frac{1}{4} \gamma_b' \frac{\mathscr{L}_b''}{\mathscr{L}_b'} \right) \right.$$
$$\left. - a_0 \tau_{32}^a \left(\gamma_a'' + \frac{1}{4} \gamma_a' \frac{\mathscr{L}_a''}{\mathscr{L}_a'} \right) \right] + \varkappa_0 (\tau_{31}^b)^2 \, \delta_b'' - a_0 (\tau_{32}^a)^2 \, \delta_a'', \tag{6.42}$$

where

$$\gamma'_{a,b} = \mathrm{Re}\,(\mathscr{L}_{a,b})^2,\ \gamma''_{a,b} = \mathrm{Im}\,(\mathscr{L}_{a,b})^2,\ \delta'_{a,b} = \mathrm{Re}\,(\mathscr{L}_{a,b})^3,\ \delta''_{a,b} = \mathrm{Im}\,(\mathscr{L}_{a,b})^3.$$

The first term corresponds to the relation (6.41), and the remaining terms give the values that are produced by the finite transverse relaxation times τ_{21}^a and τ_{21}^b. In Fig. 6.11 some results are shown. The laser frequency ω_L decreases with increasing small signal absorption, which corresponds also to Fig. 6.11a and the experimental findings. The dispersion parameter r calculated as a function of $\varkappa_0$ and ω_L possesses an extreme value at certain values of $\varkappa_0$ and ω_L according to Fig. 6.11b. This result, which cannot be explained using the rate equation approximation, corresponds to the experimental results. The pulse durations are plotted in Fig. 6.11c, where for comparison the curves (A) and (B) were also plotted for a regime with resonant interaction ($\omega_L = \omega_{31}^b = \omega_{32}^a$), in which no chirp develops that must be compensated. It becomes evident that the quasi-resonant interaction in combination with the chirp formation and compensation leads to a significant pulse shortening.

6.3 Experimental Set-ups and Results

6.3.1 Basic Design of Flashlamp Pumped Dye Lasers

Modelocked dye lasers were first pumped by flashlamps. In Fig. 6.1 the resonator scheme for such a laser is shown. The laser is pumped here using a xenon flashlamp with a double-elliptical pumping reflector arrangement. The pump duration is about 1 μs and the pump energy about 100 J. The saturable absorber is contained in a cell that is in optical contact with the 100 percent mirror. As was already discussed, this position is favorable because it enables the forward and back propagating pulse to superpose coherently within the absorber, whereby the absorber is more easily saturated. For the laser dye rhodamine 6G the dye DODCI is an appropriate saturable absorber. The laser frequency can be tuned with a Fabry-Perot etalon. Since the amplification and absorption in the dye media change with a change of the laser wavelength, the concentration of the saturable absorber must be adjusted for each wavelength to keep the laser operating above threshold and in the stability range.

Due to the finite duration of the pumping by the flashlamp, which lies in the microsecond range, there also results a finite pulse train having about the same duration, and whose envelope varies with the detuning of the wavelength. At shorter wavelengths the pulse evolution occurs more rapidly than at longer wavelengths of the tuning range, which is caused by the different cross sections. The tuning range can be extended to both shorter and longer wavelengths by exchanging the laser dyes and the corresponding absorber dyes. In table 6.1, several combinations of laser dyes and corresponding absorbers with the respective possible frequency ranges are shown.

Table 6.1. Laser dyes and saturable absorbers (cw pumped) (according to [6.12])

Laser dye	Saturable absorber	Tuning range/nm
Rhodamine 6G	DODCI	598–615
	DQOCI	580–613
Rhodamine B	DODCI	610–630
	DQOCI	600–620
	Cresyl-violet	610–620
Natrium-fluorescein	Rhodamine 6G	564

6.3.2 *Basic Design of CW Dye Lasers*

With the development of cw dye lasers, it also became possible to generate a continuous train of picosecond and subpicosecond pulses through passive modelocking. A possible resonator arrangement is shown in Fig. 6.12 [6.14]. As pump source, a cw argon ion laser is used whose radiation passing the quartz prism is engaged in the resonator and focussed into the free-flowing laser dye jet by the spherical mirror. The saturable absorber is here in contact with the reflecting 100 percent mirror and flows through a cell in a thin layer

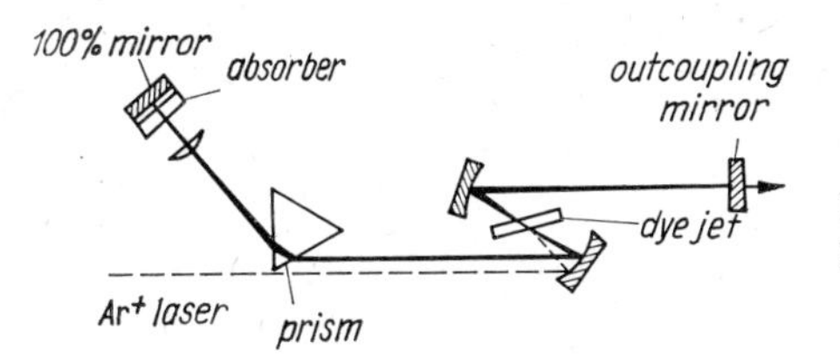

Fig. 6.12. CW pumped passively modelocked dye laser configuration

of variable thickness from 200 μm to 500 μm. The surfaces of the dye cell are appropriately inclined in order to avoid Fabry-Perot subcavity resonances. The dye laser beam is focussed by a lens in the absorber cell, whereby the intensity in the absorber is increased about fourfold. The laser dye flows in a jet through the center of the resonator. Using rhodamine 6G as laser dye and DODCI as saturable absorber in this configuration, pulses of about 1 ps duration result. The thickness of the absorber cell is about 0.5 mm and output mirror transmissions from 1 to 6 percent are employed. With a further shortening of the absorber cell length to 200 μm, the pulse length could be decreased to 0.3 ps [6.14].

An alternative arrangement for generating subpicosecond pulses differs from the one described in that the saturable absorber and the amplifier are combined in one solvent, which flows in a free-flowing dye jet through the center of the resonator [6.15]. For example, rhodamine 6G as amplifier and DODCI as saturable absorber can be mixed in a solvent of ethylene glycol. The use of this mixture simplifies the adjustment and the operation of the modelocked laser, since only one free-flowing jet and a few optical components are necessary. On the other hand, a disadvantage of this arrangement is that a telescope cannot be used to increase the intensity in the absorber independently of the amplifier. Also, the absorber is not in contact with the mirror. Because the ratio between the absorption cross sections of DODCI and rhodamine 6G is sufficiently large, the system can operate in a stable modelocking regime. Since the laser operation is now determined only by the pump intensity and the dye concentration, the adjustment becomes easier.

6.3.3 Experimental Results

The first experiments with passively modelocked dye lasers were carried out to investigate the conditions of modelocking and to explain the physical mechanisms of the pulse evolution as well as to develop methods for recording ps pulses and determining their duration. By means of two photon fluorescence and the generation of the second harmonic the duration of the pulses was measured (compare 3.). In the first experiments pulses of 1 to 5 ps duration were observed. Although the application of these methods (which rests on correlation measuring techniques) has not lost its importance for the measuring of pulse widths and represents the only possible method in the subpicosecond

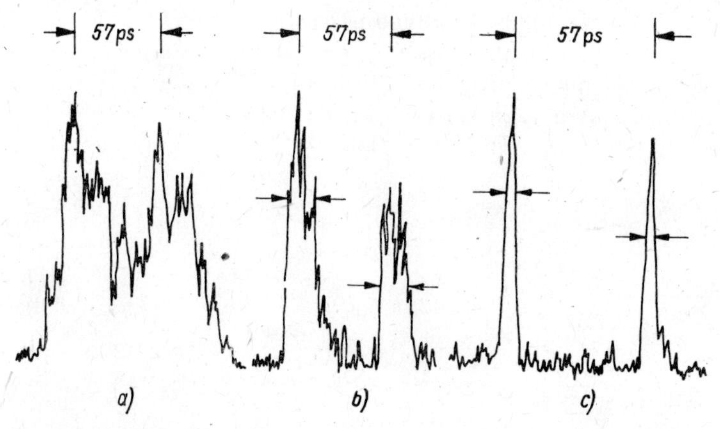

Fig. 6.13. Microdensitometer trace of streak camera records showing the evolution of an ultrashort pulse in a rhodamine 6G laser modelocked by DODCI
a) Noise fluctuation at the start of the laser oscillation
b) Envelope of the selected fluctuation group of 17 ps duration after 30 round trips
c) Single pulse after 35 round trips
Each record shows the intensity profile twice; the second profile was generated in an optical delay line from the original profile to provide a time calibration.
(from Adrain [6.18] and [16], respectively)

range, the development of direct time-resolved measuring methods — such as the ultrafast streak camera — for the investigation of the temporal structure of modelocked laser radiation has opened up new possibilities. In particular, the application of the streak camera has made it possible to study more accurately the temporal build up of modelocking from radiation noise [6.16 to 6.18]. Such investigations were of particular interest in explaining the mechanism of the formation of modelocked dye laser pulses, which differs considerably from that in passively modelocked solid state lasers. The length of the pulses generated in passively modelocked dye lasers proves to be several hundred times shorter than the energy relaxation time of the absorber dye. This result, which appeared to be paradoxic from the concept of passively modelocked solid state lasers, was explained by means of the time-resolved investigations mentioned above in connection with a suitable theoretical model [6.9]. Due to the combined action of saturable absorption and amplification depletion a rapid pulse shortening process occurs, which is not limited by the relaxation time of the absorber. Fig. 6.13 shows one example of such time-resolved measurements of the history of evolution of a single ultrashort pulse from noise in a rhodamine 6G laser (according [6.18]). In this case the laser is pumped by a flashlamp, whereby within the pump process the steady-state regime is achieved so that the conclusions drawn here are also valid for the cw pumped dye laser. The dye DODCI was used as the saturable absorber in these investigations. The wavelength was tuned to 605 nm. According to the microdensitometer trace of streak camera records a fluctuation noise burst of about 100 ps duration exists in the resonator after only a few

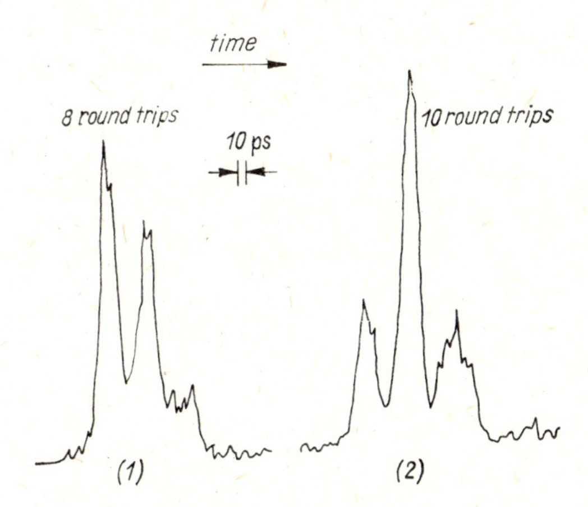

Fig. 6.14. Microdensitometer trace of two simultaneously recorded pulse profiles
The number of round trips is counted from the start of the pulse train.
(from Adrain [6.18])

round trips (about 20 ns) once the laser oscillation has begun. After about 25 round trips (120 ns) the noise burst is compressed considerably. In the further development, a rapid pulse compression process begins at the edges so that after about 35 round trips a single 2 ps pulse evolves. In these investigations two pulses initially separated by 2 round trips are imaged in one picture by means of an optical delay line (Fig. 6.14). This picture shows clearly the selection of a single fluctuation after two round trips. In particular, the important role of the amplification depletion is also demonstrated here, because the pulse is attenuated not only at the leading but also at the trailing edge. Further experimental investigations dealt, on the one hand, with improving and simplifying the experimental techniques for generating picosecond dye laser pulses as well as with the development of new laser dyes and the corresponding absorbers for extending the wavelength range (in particular the continuously tunable range) and, on the other hand, with the generation of even shorter pulses. It became possible to produce pulses with durations in the time range below 100 femtoseconds.

Ippen and Shank [6.19] attained pulse widths of about 0.5 ps using an experimental arrangement that contained two free-flowing dye jets as absorber and amplifier. In this laser the saturable absorber consisted of a mixture of two absorber dyes. DODCI alone makes possible the generation of short pulses, however a stable pulse operation exists only near the laser threshold. Accordingly a careful adjustment and stabilizing of the laser parameters is necessary. Adding malachite green to the DODCI solvent also permits pulse operation further above the laser threshold. However, malachite green alone cannot generate a stable pulse operation. The pulses generated using this arrangement can be characterized by an asymmetric pulse shape and exponentially decreasing edges. This supports the solution ansatz for the pulse shape which was derived in section 6.2 solely from the structure of the basic equations. On the other hand, the pulses measured in [6.19] were not bandwidth-limited but turned out to be phase (frequency)-modulated (chirp — compare 2.5, 6.3.4 and 8.3). By compressing the pulses in dispersive devices a pulse shortening to 0.3 ps could be achieved. In [6.14] the direct generation of bandwidth-limited pulses of 0.3 ps duration was reported. The laser resonator in this operation consisted of a free-flowing rhodamine 6G jet as amplifier and an absorber (DODCI) in contact with the mirror, which flowed through a thin cell. This resonator configuration has already been described in section 6.3.2 and is shown schematically in Fig. 6.9.

Diels et al. [6.5, 6.20] were able to achieve a further shortening of the pulses to 0.17 ps and 0.13 ps, respectively. They used a laser configuration in which — as described in section 6.3.2. — absorber and amplifier (rhodamine 6G, DODCI and malachite green) were mixed and contained in a single jet. In this way it was possible to reduce the number of optical components in the resonator from seven in [6.19] and eight in [6.14] to four (two focussing mirrors and a flat output mirror). The shorter pulses can develop, because now the bandwidth limitation of dispersive elements is avoided or greatly reduced, and thus the bandwidth of the laser transition can be more completely used. In this laser, though, the frequency of the laser is no longer tunable.

Even shorter pulses with a length of 70 fs were attained by Mourou and Sizer [6.30] using a synchronously pumped laser with double modelocking, and by Fork, Green and Shank [6.6] in a regime of counterpropagating pulses in a ring resonator (CPM), which we will describe in the following sections. Dietel, Fontaine and Diels [6.37] attained pulses as short as 53 fs by means of intracavity chirp compensation in a CPM laser. Finally, recently Valdmanis and Fork [6.44] succeeded in generating pulses of about 30 fs by similar means. Through subsequent pulse compression outside the resonator, pulses as short as 30 fs were attained by Shank, Fork and Yen [6.31], as short as 16 fs by Fujimoto, Weiner and Ippen [8.41a], 12 fs by Halbout and Grischkowsky [8.41b] and 8 fs by Shank et al. [8.41c] (see 8.3.2).

Detailed systematic investigations of the dependence of pulse parameters on the various laser parameters do not exist in the literature, so that a comparison of the theoretical predictions of section 6.2 with experimental results can only be qualitatively made. In several papers (e.g. in [6.14, 6.20]) the shortest pulses were reported to be found near the pump threshold, whereas with greater pump intensities the pulses became broader. This corresponds to the dependence of the pulse duration on the pump intensity as depicted in Fig. 6.6a for greater absorber losses (curves (1) and (2)), and contradicts the simplified theory by Haus [6.10] in which a decrease in the pulse duration was generally predicted with increasing amplification.

Combining theoretical and experimental results gives the following conditions for attaining the shortest possible pulses (which, however, are in part mutually exclusive):

— The spectral bandwidth of the laser transition must be large and be as fully utilized, as possible, i.e., selective effects that are produced by prisms, etalons or dye cells, should be reduced.

- The greatest possible effective value of the ratio between the cross sections of the absorber and amplifier $\left(m = \sigma_{13}^{\mathrm{b}} q^{\mathrm{a}} / (\sigma_{32}^{\mathrm{a}} q^{\mathrm{b}})\right)$ must be reached; this value can be controlled by a telescope. A similar effect can be achieved by a regime of counterpropagating pulses, as it builds up with a short contacted absorber or in a laser configuration in which two pulses collide in the absorber (which is described in 6.3.4).
- The laser is to be operated with the highest possible absorber losses, for which higher pump powers are required. The pump power must be adjusted slightly above the threshold in order to attain the shortest pulses.

However, it should be noted that at high pump powers secondary pulses not taken into consideration in the solution ansatz (5.19) in section 6.2 may develop.

As in the case of synchronous pumping the peak intensity of the pulse can be raised considerably if a cavity dumper is used for the outcoupling, whereby peak intensities of several kilowatts are reached at a pulse repetition frequency of 10^3 Hz. An amplification of the pulses is also possible, whereby powers in the Gigawatt range can be attained [6.12] (compare 5.3).

6.3.4 *Counterpropagating Pulses*

The optimal saturation behavior of a short absorber saturated by colliding pulses has already been pointed out several times. This effect can be very effectively utilized, if in the resonator two counterpropagating pulses overlap precisely in an absorber whose optical thickness is smaller or in the order of the pulse length.

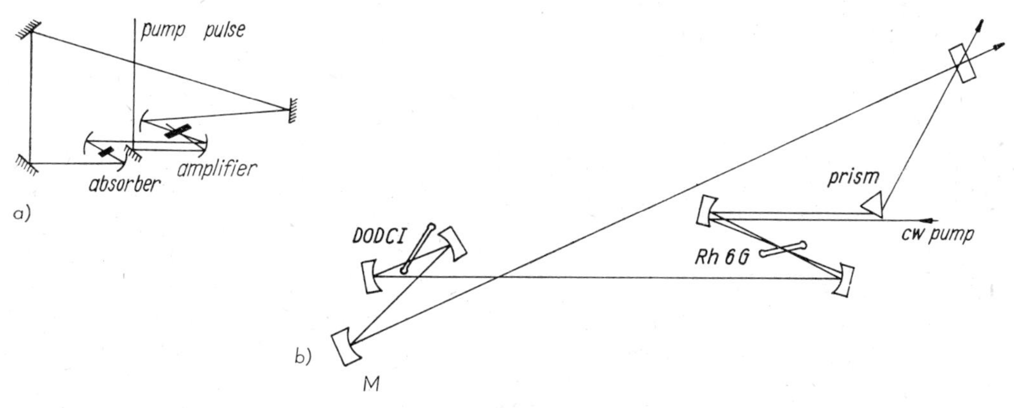

Fig. 6.15. Modelocked ring laser configuration for a regime of counterpropagating pulses
a) Without chirp compensation (from [6.6])
b) With chirp compensation by means of a prism (from [6.37])

A regime of this kind was achieved by Fork, Greene and Shank using a ring laser configuration, as shown schematically in Fig. 6.15a [6.6, 6.31]. Rhodamine 6G and DODCI dissolved in ethylene glycol were used as amplifier and absorber, respectively. To produce a free-flowing absorber jet of about 10 μm thickness, a specially formed nozzle was used. The amplifier was excited by a cw argon ion laser at powers of 3 to 7 W at 514.5 nm; the small signal loss of the absorber came to 20 percent and the outcoupling losses to 3 percent. Using this arrangement laser pulses as short as 65 to 90 fs and with a spectral width of (5 ± 1) nm were generated.

Pulses of the same duration were obtained by Dietel [6.7] who used a similar ring laser configuration, which was pumped by an argon ion laser ILA 120 at a power of about 1 W. In this investigation the pulse parameters attained with the ring laser were compared with those obtained with a Fabry-Perot laser under similar conditions regarding the pump laser, absorber, amplifier and the outcoupling losses. It was found that the pulses from the ring laser were about four times shorter.

The method considered here has several advantages compared with the method of the contacted absorber. To begin with, the complicated technical problems of optically contacting the absorber and the mirror are avoided. Secondly, the superposition of two pulses in the absorber is more favorable. Their exact timing is automatically regulated by the system, since two counterpropagating pulses that precisely overlap in the ab-

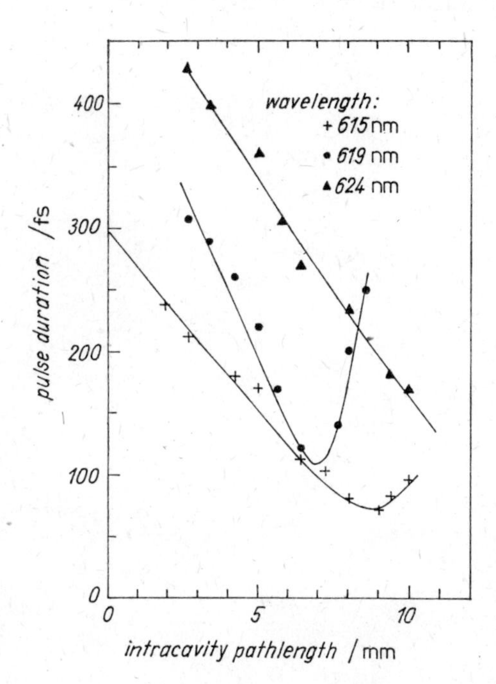

Fig. 6.16. Pulse duration τ_L versus the intracavity glass path for $\lambda = 615$ nm (SQ1), $\lambda = 619$ nm (Flint) and $\lambda = 624$ nm (SQ1) (from Dietel, Rudolph, Wilhelmi, Diels and Fontaine [6.36])

sorber maintain the optimal generation conditions. The theory of a CPM laser of this kind has been presented in detail in section 6.2.3. As calculated there, a chirp of the pulse develops in the case of a detuning of the laser frequency from the center of the absorption cross section or emission line. This was measured [6.21] and compensated outside the resonator by Dietel, Döpel, Kühlke, Rudolph and Wilhelmi. More favorable conditions for the generation of even shorter pulses result according to the calculations if the chirp is compensated within the resonator by a dispersive optical element. As in [6.21], Dietel et al. used a resonator arrangement in which, contrary to Fig. 6.15a, dispersive elements were not avoided, but were employed for chirp compensation according to Fig. 6.15b. The effect of the angular dispersion of the prism is reduced here by the mirror M or partially compensated by a second prism or by applying a four-prism sequency [6.44] whereby altogether a relatively weak bandwidth limitation of the passive resonator is attained. Using resonator configurations of this kind, Dietel, Diels and Fontaine [6.37] and Valdmanis and Fork [6.44] were able to generate pulses as short as 53 fs and 27 fs, respectively. The wavelength of the laser was determined mainly by the concentration of the absorber dye.

The pulse duration can be continuously controlled by moving the prism, whereby the intracavity glass path through which the pulses pass is changed. As shown in Fig. 6.16, at all wavelengths a minimum pulse duration results at a certain glass thickness at which the chirp is just compensated. The optical path length at the minimum pulse length varies here at various absorption concentrations and has itself a minimum at a

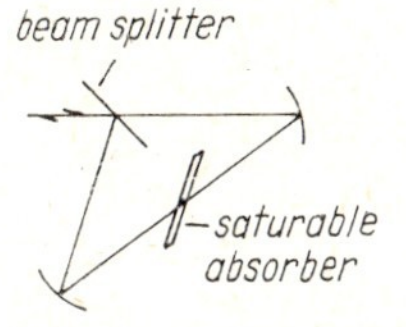

Fig. 6.17. Antiresonant ring with saturable absorber (from [6.41], [6.42])

certain concentration or at a corresponding wavelength of the laser. The shortest pulses below 100 fs were attained here in a wavelength range of 605 to 630 nm. The essential experimental findings could be verified by the theory presented in 6.2.7.

Moreover, a regime of colliding pulses can be achieved in a linear resonator, by replacing an end mirror with an antiresonant ring (see Fig. 6.17) having the saturable absorber at its center. An arrangement of this kind was first applied by Siegman et al. [6.41], [6.42] in a Nd:YAG laser in order to build up a CPM regime. Diels et al. [6.43] applied this principle to the modelocked dye laser, whereby in comparison to the ring arrangements a more compact design and a simpler adjusting can be achieved.

A regime of counterpropagating pulses can also be achieved using a linear resonator configuration. In this case the resonator length must be precisely an integral multiple n ($n = 2, 3, 4, \ldots$) of the distance from the absorber to the end mirror, whereby n pulses oscillate in the resonator [6.6].

As already mentioned pulses of minimal duration — until now to 8 fs — can be achieved through subsequent compression (see 8.3.2).

6.3.5 *Hybrid Modelocking*

The shortest pulses could be generated through the passive modelocking of dye lasers. On the other hand the method has several disadvantages, such as the very critical adjusting of the pump and resonator parameters for the generation of a stable regime and the restricted tuning range due to the saturable absorber. Synchronous pumping on the other hand has the advantage of a greater frequency tunability, and it requires no critical adjusting of the pump intensity. However, the generated pulses are not usually very short, and a very precise matching of the resonator length of the dye laser to the pump pulse separation is necessary. In order to combine the advantages of both lasers, a so-called hybrid modelocking regime was developed [6.26 to 6.28]. In this regime, synchronous pumping and moreover, by means of a saturable absorber, an additional passive modelocking were simultaneously employed. Using a single jet in which absorber and amplifier were mixed and which was pumped by an acoustoopticaly modelocked argon ion laser, pulses with a duration of 0.3 ps and tunable over a wavelength range of 574 to 611 nm were generated in [6.28]. The matching of the resonator length is here less critical than with pure synchronous pumping. As already mentioned in the treatment of synchronous pumping, the hybrid modelocking suppresses satellite pulses that pass through the gain medium during the passage of the pump pulse and simultanously with the main pulse but from the opposite direction. The suppression of the satellite pulses takes place on the return trip, because the main pulse and the satellites pass through the absorber at different moments. By means of hybrid modelocking, where the fast absorber DQOCI was added to the rhodamine 6G jet, Mourou and Sizer [6.30] obtained pulses of 70 fs duration. The system was excited by a frequency-doubled modelocked Nd:YAG laser.

6.3.6 *Amplification*

As with synchronously pumped dye lasers (compare 5.3.4) the weak light pulses from passively modelocked dye lasers can be amplified, whereby powers of some Gigawatt can be attained [6.12]. Pump pulses from frequency-doubled *Q*-switched Nd:YAG lasers [6.45], eximer lasers [6.46] and copper vapor lasers [6.47] have successfully been applied, where the latter are distinguished by their high repetition rate of about 10 kHz. By compensating the chirp resulting from various resonant and nonresonant optical components in the amplifier chain, high-power pulses as short as 70 fs have been obtained. Such pulses can be further compressed by nonlinear optical chirp generation in glass fibers and subsequent chirp compensation in dispersive media (compare 8.3).

In particular XeCl and ArF excimer lasers ($\lambda_{P1} = 308$ nm and $\lambda_{P2} = 193$ nm) provide the advantage that respectively the second and third harmonic of amplified dye laser pulses at $\lambda_{L1} = 616$ nm and $\lambda_{L2} = 579$ nm can be further amplified in the active medium of the pump laser (compare 8.2.1). Thus powerful femtosecond pulses can be generated in the UV, too. This fact is of importance for pumping other lasers, exciting nonlinear optical phenomena and inducing ultrafast chemical processes.

7. Passive Modelocking of Solid State Lasers

7.1 Principle of Operation

The first experiments in generating ultrashort pulses by passively modelocking a solid state laser were conducted by Macker and Collins [7.1] in 1965 using a ruby laser and by De Maria et al. [7.2] in 1966 using a neodymium glass laser. Whereas the pulse duration in solid state lasers is generally greater than that in passively modelocked dye lasers (about 15 to 20 ps in ruby lasers and 2 to 10 ps in neodymium glass lasers), the peak intensity of the pulses in solid state lasers is considerably larger and lies, for example, at several hundred MW/cm² in ruby lasers, and at several GW/cm² in Nd glass lasers.

A qualitative understanding of the principle of modelocking in the solid state laser was achieved by the so-called fluctuation model formulated by Letokhov [7.3, 7.4]. The fundamental idea of this model is that, due to the nonlinear action of the absorber, from the large number of intensity fluctuations that are present at the beginning of the amplification process only the most intensive fluctuation is selected and amplified, whereas the rest is suppressed. Similár conclusions were made independently by Fleck [7.5] who simulated the amplification process in the solid state laser using computer calculations. The model was further developed in a series of papers [7.6 to 7.13, 7.40 to 7.44]; the theoretic description of the modelocking process in the solid state laser in section 7.2 is based mostly on [7.12, 7.13].

According to the fluctuation model the pulse evolution process can be divided into three phases (see Fig. 7.1): The linear phase of the pulse evolution process (region I) begins if, at a certain time after the pùmp process begins, the laser threshold is exceeded and the gain exceeds the losses. The pump energy is chosen here such that the laser operates only slightly above the laser threshold. Nevertheless, due to the large linewidth of the laser transition a very large number of longitudinal modes begin to oscillate, which are amplified independently from each other. The superposition of the different modes with stochastic phase relations can be described by a stochastic Gaussian process. In the first instance, the total number of fluctuations is approximately the same as the number of resonator modes from which, however, only a small number significantly exceeds the mean intensity level. Due to the greater amplification of the modes in the center of the lasing bandwidth and because the modes at the wings of the gain profile consequently do not experience a sufficiently large amplification the spectrum of the radiation narrows during the linear phase (natural mode selection). In the time domain this means a smoothing and broadening of the amplitude fluctuations. Neodymium glass possesses, for example a bandwidth of $\Delta\nu_{21} = 7.5 \times 10^{12}$ Hz; accordingly, the duration of one fluctuation spike at the beginning of the linear phase is $\tau_{FL} \simeq 1/\Delta\nu_{21} \simeq 10^{-13}$ s. At a resonator length of 1 m and an effective gain of several percent the linear phase comprises about 2000 round trips. Due to the mode selection, the fluctuation duration increases to about 10^{-11} s during this time.

The nonlinear phase (region II) begins when the largest fluctuation peak of the radiation field reaches the intensity at which the nonlinearity of the absorber or the amplifier becomes effective. The saturation of the absorption in very fast recovering absorbers leads to a preference for the largest fluctuation peak in contrast to the others, because it experiences fewer losses than the remaining fluctuation maxima of lower intensity. The selection of the maximum peak is additionally favored because due to a small decrease in the amplification resulting from the depletion of the occupation in-

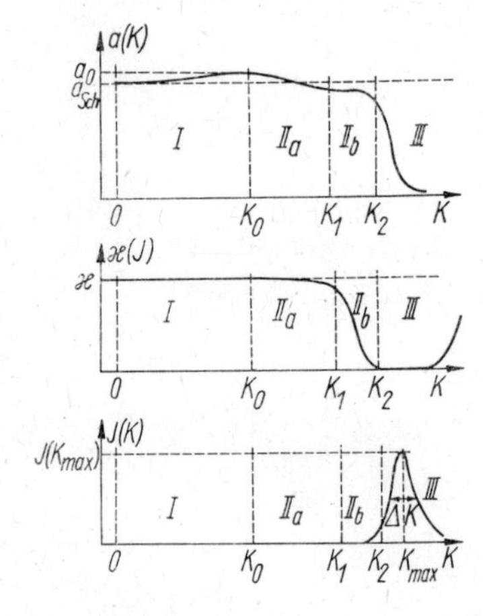

Fig. 7.1. Scheme of the temporal evolution of the gain $a(K)$, the absorber losses $\varkappa(J)$ and the peak intensity of the pulses $J(K)$. (K is the number of round trips of the pulse in the resonator.)

version in the amplifier, the lower intensity fluctuations fall with greater probability below the threshold, whereby the probability that secondary pulses will develop diminishes considerably. This leads simultaneously to the requirement that the pump energy must exceed a so-called second threshold. Otherwise, due to the reduction of the gain in the nonlinear phase, the largest pulse can also fall into the range of negative net gain. During the nonlinear phase the intensity of the maximum pulse increases rapidly within 100 round trips and exceeds the saturation intensity of the absorber, while the lower intensity fluctuations are suppressed. Simultaneously, a shortening of the pulses results due to the greater absorption at the pulse edges as opposed to that at the center by the nonlinear absorber. Though, this shortening is less than the pulse broadening during the linear phase.

If the nonlinear absorber is saturated, the nonlinear phase is finished and the pulse shaping process is, in principle, concluded. The saturation phase of the amplifier then follows (region III). In this stage the occupation inversion in the amplifier is completely depleted, and the process of generation is broken off. In passively modelocked solid state lasers, depending on the specifies of the generation process a steady state regime is not achieved, but rather a train of pulses spaced by the resonator round trip time and of varying parameters is emitted (see Fig. 7.6). The properties of the pulse train such as its mean duration and the intensity of the maximum result from the processes proceeding in region III. Due to the large intensity of the pulses, intensity-dependent refractive index effects such as self-phase modulation can play a role in this amplification phase, which may lead to a spectral broadening, a positive frequency sweep or a breaking up of the pulse into stochastically distributed multiple components. Although these effects can influence considerably the properties of the pulses, they appear only after the modelocking process has ended and can be avoided under certain circumstances by limiting the peak intensity.

7.2 Theory

7.2.1 Basic Equations

An important distinction between the passively modelocked solid state lasers and the previously described types of lasers is that in the solid state laser a continuous steady state regime is not achieved, and other proportions exist among the various time parameters of the system. The lifetime of the excited laser level in the solid state medium T_{31}^{a} (which is of the order of microseconds) is very long compared to the resonator round trip time u (which is of the order of nanoseconds),

$$T_{31}^{\mathrm{a}} \gg u \gg \tau_{\mathrm{L}}, \tag{7.1}$$

whereas the saturable absorber must relax very rapidly into the ground level; the pulse duration τ_{L} is greater than or of the same order as the relaxation time T_{21}^{b}:

$$\tau_{\mathrm{L}} \gtrapprox T_{21}^{\mathrm{b}}. \tag{7.2}$$

Simplifying, we assume that the relative intensity change of the laser radiation is small per round trip:

$$|I_{\mathrm{L}}^{(K+1)}(\eta) - I_{\mathrm{L}}^{K}(\eta)| \ll I_{\mathrm{L}}^{K}(\eta). \tag{7.3}$$

K is here the number of resonator round trips. The variable $\eta = t - z/v$ is restricted to the time interval of one resonator round trip u ($0 \leqq \eta \leqq u$). Now, we consider the change of the radiation in passing through the amplifier and absorber and after reflection at the mirrors, whereby we choose the arrangement of the resonator components similar to Fig. 6.3. A separate frequency-selective element is not considered here. Instead, we will take into account the finite bandwidth of the laser transition. For the description of the laser processes in a four-level solid state laser we can proceed from equations (4.1) to (4.3), if the homogeneous broadening is predominant (Nd:YAG laser). (In inhomogeneously broadened systems many of the conclusions drawn retain their validity.) As in chapter 1, for the investigation of pulse evolution from noise it is necessary as performed in chapter 1, to add a stochastic quantity to equation (4.2), which describes the fluctuation of the medium. Because of relation (7.1) we can assume, as in section 4.2, that the change in the occupation numbers during a round trip is small ($N_3{}^K \approx \overline{N}_3{}^K$). Accounting for the stochastic fluctuation $F(\eta)$ equation (4.5) then becomes

$$\frac{\partial A_{\mathrm{L}}{}^K}{\partial z} = \frac{\sigma_{32}^{\mathrm{a}} \overline{N}_3{}^K}{2} \left[A_{\mathrm{L}}{}^K - \tau_{32}^{\mathrm{a}} \frac{\partial A_{\mathrm{L}}{}^K}{\partial \eta} + (\tau_{32}^{\mathrm{a}})^2 \frac{\partial^2 A_{\mathrm{L}}{}^K}{\partial \eta^2}\right] + F(\eta) \frac{C}{2L^{\mathrm{a}}}, \tag{7.4}$$

where $C = -2\mathrm{i}\sqrt{\mu_0/\varepsilon_0}\,\omega_L \mu_{32}^a \tau_{32}^a N^a L^a$. Assuming (7.3) we can use the approach of successive approximation to solve (7.4). The field strength and occupation of the upper laser levels at position (1) are related to the corresponding values at position (0) in the following manner:

$$A_L{}^K(1,\eta) = A_L{}^K(0,\eta) + \frac{a^K(0)}{4}\left[A_L{}^K(0,\eta) - \tau_{32}^a \frac{\partial A_L{}^K(0,\eta)}{\partial \eta} + (\tau_{32}^a)^2 \frac{\partial^2 A_L{}^K(0,\eta)}{\partial \eta^2}\right] + \frac{C}{2} F(\eta), \tag{7.5}$$

$$\overline{N}_3{}^K(1) = \overline{N}_3{}^K(0) - \sigma_{32}^a u \bar{I}_L{}^K(0)\,\overline{N}_3{}^K(0) + N^a W_P - \left(W_P + \frac{1}{T_{31}^a}\right)\overline{N}_3{}^K(0), \tag{7.6}$$

where the natural logarithm of the amplification per round trip $a^K(0) = 2\sigma_{32}^a L^a \overline{N}_3{}^K(0)$, the pump parameter $W_P = \sigma_{14}^a \bar{I}_P$ and the intensity averaged over one resonator round trip $\bar{I}_L{}^K = \dfrac{1}{u\hbar\omega_L}\sqrt{\dfrac{\varepsilon_0}{\mu_0}}\dfrac{1}{2}\displaystyle\int_0^u |A_L{}^K(\eta)|^2\,\mathrm{d}\eta$ were introduced; for the sake of simplification we will refer in the following to a^K as "amplification". (The arrangement of the individual resonator elements is chosen similarly to that in Fig. 6.3). Because of condition (7.2) we can describe the change in the field strength due to the saturable absorber by

$$\frac{\partial A_L(\eta)}{\partial z} = -\frac{\sigma_{13}^b}{2} A_L(\eta)\, N_1{}^b, \tag{7.7}$$

$$N_1{}^b = \frac{1}{1 + I_L(\eta)/I_S{}^b} \tag{7.8}$$

where $I_S{}^b = (\sigma_{13}^b T_{21}^b)^{-1}$ is the saturation intensity of the absorber. Equation (7.7) is again solved by successive approximation: Hence,

$$A_L{}^K(2,\eta) = A_L{}^K(1,\eta) - \frac{\varkappa_0 A_L{}^K(1,\eta)}{4\left(1 + I_L{}^K(1,\eta)/I_S{}^b\right)} \tag{7.9}$$

where $\varkappa_0 = 2\sigma_{13}^b L^b N^b$. (The quantity $\varkappa_0$ represents the natural logarithm of the small signal transmission at a double passage.)

The change in the field strength during the return trip through the absorber and amplifier is calculated similarly and can be determined easily in the approximation considered here by substituting $2L^a$ for L^a and $2L^b$ for L^b. Finally, there are the losses at the mirror to consider:

$$A_L{}^K(6,\eta) = \left(1 - \frac{\gamma}{2}\right) A_L{}^K(5,\eta). \tag{7.10}$$

The linear loss per round trip is determined by the constant γ. By combining equations (7.5), (7.9) and (7.10) we obtain with the requirement $A_L{}^K(6,\eta) = A_L{}^{K+1}(0,\eta)$ a recursion formula for the field strength amplitude. Under assumption (7.3) we can transform this relation into a differential equation, namely $A_L{}^{K+1}(\eta) - A_L{}^K(\eta) \approx \partial A_L(K,\eta)/\partial K$. Altogether, then for the description of the passively modelocked solid state laser we obtain the following system of equations for the field strength amplitude $A_L(K,\eta)$ and the normalized amplification quantity $a(K) = 2\sigma_{32}^{a} L^{a} \overline{N}_3(K)$ at the K-th round trip:

$$\frac{\partial A_L(K,\eta)}{\partial K} = \frac{1}{2}\, a(K) \left[A_L(K,\eta) - \tau_{32}^{a} \frac{\partial A_L(K,\eta)}{\partial \eta} + (\tau_{32}^{a})^2 \frac{\partial^2 A_L(K,\eta)}{\partial \eta^2} \right]$$

$$- \frac{\gamma}{2} A_L(K,\eta) - \frac{\varkappa_0 A_L(K,\eta)}{2\left(1 + I_L(K,\eta)/I_S{}^{b}\right)} + CF(K,\eta), \tag{7.11}$$

$$\frac{\mathrm{d}a(K)}{\mathrm{d}K} = P - \sigma_{32}^{a} u a(K)\, \bar{I}_L(K). \tag{7.12}$$

$P = pu$ signifies the pump rate in units of the reciprocal round trip time u. According to (7.6), it is given by

$$P = u \left[W_P(K)\, a_{\max} - \left(W_P(K) + \frac{1}{T_{31}^{a}} \right) a(K) \right] \tag{7.13}$$

for the four-level laser, whereas we obtain

$$P = u W_P(K) \left(a_{\max} - a(K) \right) \tag{7.14}$$

in the case of the three-level laser (ruby laser). As abbreviation $a_{\max} = 2\sigma_{32}^{a} L^{a} N^{a}$ was introduced here, and in (7.14) the relation $W_P \gg 1/T_{31}^{a}$ was used.

Equation (7.11) describes the change in the complex field strength amplitude with respect to the round trip number K. The terms proportional to $a(K)$ describe the amplification of the radiation by the active medium, whereby the time derivatives in the brackets account for the finite spectral bandwidth of the laser transition. The term proportional to γ describes the linear losses at the mirrors, whereas the term proportional to $\varkappa_0$ represents the losses due to the saturable absorber, which decrease considerably at larger intensities of the radiation field. Finally the quantity $F(K, \eta)$ characterizes the fluctuations of the radiation noise from which the laser radiation develops. According to (7.12), the pump rate P, which describes the build-up of the occupation inversion due to the pump radiation, and the saturation of the active medium determine the temporal change in the amplification $a(K)$. It can be seen from (7.12), particularly, that the depletion of the amplification of the active medium is due to the total energy, of the radiation field summed up from the start of the laser oscillations.

After the laser threshold is reached, the pump rate P can be assumed to be nearly constant during the pulse evolution, because the total pulse formation takes place over a short time in comparison to the pump pulse duration.

For a more accurate calculation of the pump rate P at the laser threshold, however, the temporal change of $W_P(K)$ and the time interval up till reaching the laser threshold are important. Using (7.12) and (7.14), for the temporal change of the amplification $a(K)$ below the threshold, we obtain for example, in the case of the three-level laser:

$$a(K) = a_{\max} \left\{ 1 - 2 \exp\left[-u \int_{-\infty}^{K} W_P(K')\, \mathrm{d}K' \right] \right\}. \tag{7.15}$$

Now, from (7.15) the threshold value of W_P is calculated, which is sufficient to reach the laser threshold $a_{\mathrm{Th}} = \varkappa_0 + \gamma$ at the end of the pump process $K \to \infty$. Assuming a

Gaussian shaped pump pulse of pulse duration

$$W_P(K) = W_0 \exp\left[-4\ln 2\left(\frac{\tau_P}{u}\right)^2 K^2\right],$$

then (7.15) leads to

$$(W_0)_{Th} = \sqrt{\frac{\ln 2}{\pi}}\,\frac{2}{\tau_P}\ln\frac{2}{1-\frac{a_{Th}}{a_{max}}}. \tag{7.16}$$

The value of W_0 must now be chosen somewhat larger than $(W_0)_{Th}$, in order that a positive amplification is achieved at a finite round trip number K: $W_0 = (W_0)_{Th} + \delta W_0$. Using this W_0 the round trip number K_{Th}, at which the laser threshold is exceeded, can be calculated. With (7.16) we obtain from (7.14) the relation

$$\varphi\left[2\sqrt{2\ln 2}\,\frac{u}{\tau_P}\,K_{Th}\right] = 1 - 2\,\frac{\delta W_0}{(W_0)_{Th}}, \tag{7.17}$$

where $\varphi(x)$ is the error function.

For the pump rate at the moment the laser threshold is exceeded, we then find

$$P(K_{Th}) = u[(W_0)_{Th} + \delta W_0]\,e^{-4\ln 2\left(\frac{u}{\tau_P}\right)^2 K_{Th}^2}\,[a_{max} - a_{Th}]. \tag{7.18}$$

Choosing as an example a threshold exaggeration $\delta W_0/(W_0)_{Th} = 0.03$, we obtain a pump rate of $P = 3.6\times 10^{-5}$, at $u = 10^{-8}$ s, $\tau_P = 500$ μs and $a_{Th} = 0.5$, for $\delta W_0/(W_0)_{Th} = 0.01$, we find $P = 1.4\times 10^{-5}$. The calculation of the pump rate for the four-level laser is contucted similarly, whereby however the influence of the relaxation during the pump process must be considered. Using the relation $W_P \ll 1/T_{31}^a$, (7.15) can be written as

$$a(K) = a_{max}\exp\left(-\frac{u}{T_{31}^a}K\right) u \int_{-\infty}^{K} dK'\,W_P(K')\exp\left(\frac{u}{T_{31}^a}K'\right). \tag{7.19}$$

For $\delta W_0/(W_0)_{Th} = 0.03$ a pump rate of $P = 4.9\times 10^{-6}$ can be estimated, and for $\delta W_0/(W_0)_{Th} = 0.01$ $P = 3.7\times 10^{-6}$.

7.2.2 *The Linear Stage of Pulse Formation (Region I)*

By pumping with a flashlamp the laser active ions are excited into the upper laser level and begin to fluoresce. If the gain outweighs the losses, the radiation in the resonator begins to increase ($a_{Th} = \varkappa_0 + \gamma$). At the beginning of the linear region the radiation intensity is still small, and therefore the change in the occupation numbers due to the laser radiation in the amplifier and absorber can be ignored. The amplification increases linearly with the resonator round trip number, and from (7.12), it is given by

$$a = a_{Th} + PK. \tag{7.20}$$

The origin of time is fixed here such that at $K = 0$ the laser threshold is exceeded. In (7.11) we can assume $I_L/I_S^b \ll 1$. With that (7.11) becomes a linear differential equation whose solution is found using the Fourier transformation, and after backtransformation into the time domain, is given by

$$A_L(K,\eta) = \frac{C}{2\pi}\int_0^K dK' \int_{-\infty}^{\infty} d\omega \exp\left\{\frac{P}{4}(K^2-K'^2)\,\mathscr{L}(\omega) + \frac{(\varkappa+\gamma)}{2}(K-K')\left(\mathscr{L}(\omega)-1\right)\right\}\int_{-\infty}^{\infty} d\eta'\, e^{-i\omega(\eta-\eta')}F(K',\eta') \tag{7.21}$$

where $\mathscr{L}(\omega) = 1 - i\tau_{32}^a\omega - (\tau_{32}^a)^2\,\omega^2$ was introduced. Since the statistical characteristics of the noise term $F(K,\eta)$ can be described according to (1.58) by a white Gaussian noise process with the correlator

$$\langle F(K,\eta)\,F^*(K',\eta')\rangle = \frac{2}{\tau_{32}^a}\,\delta(K-K')\,\delta(\eta-\eta') \tag{7.22}$$

we obtain from (7.21) under the assumptions $PK \ll a_{Th}$, $K \gg 1$ and $\tau \gg \tau_{32}^a$

$$\begin{aligned}\langle A_L(K,\eta)\,A_L^*(K,\eta+\tau)\rangle &= \frac{4\,|C|^2}{\tau_{32}^a}\int_0^K dK'\int_0^{\infty} d\omega\cos\omega\tau \\ &\quad\times\exp\left\{\frac{1}{2}P(K^2-K'^2)\,\mathscr{L}(\omega) + a_{Th}(K-K')\left(\mathscr{L}(\omega)-1\right)\right\} \\ &\approx \frac{|C|^2}{\sqrt{2}\,(\tau_{32}^a)^2}\,\frac{1}{\sqrt{Pa_{Th}K}}\exp\left\{\frac{P}{2}K^2 - \left(\frac{\tau}{\tau_c{}^L(K)}\right)^2 4\ln 2\right\}\end{aligned} \tag{7.23}$$

for the correlation function of the field strength. The correlation time τ_c used here is related to the round trip number K by

$$\tau_c{}^L(K) = 2\tau_{32}^a(a_{Th}K\ln 2)^{1/2}, \tag{7.24}$$

and can also be regarded as the mean duration of the radiation fluctuations. The statistical mean value of the intensity is given by

$$\begin{aligned}\langle I_L(K)\rangle &= \sqrt{\frac{\varepsilon_0}{\mu_0}}\,\frac{2\,|C|^2}{(\tau_{32}^a)^2}\int_0^K dK'\exp\left\{\frac{P}{2}(K^2-K'^2) - a_{Th}(K-K')\right\} \\ &\quad\times I_0\left[\frac{1}{4}P(K^2-K'^2) + \frac{a_{Th}}{2}(K-K')\right] \approx I_{Th}\frac{1}{\sqrt{2Pa_{Th}K}}\exp\left\{\frac{P}{2}K^2\right\},\end{aligned} \tag{7.25}$$

where $I_{Th} = |C|^2\sqrt{\varepsilon_0/\mu_0}/(\tau_{32}^a)^2$ is the intensity at the laser threshold and $I_0(x)$ is the modified Bessel function of the 0th order. The increase of the correlation time with $\sqrt{K}$ indicates the spectral narrowing of the radiation during the linear phase; which results from the selection of the modes at the edge of the amplification profile. These modes experience an amplification that lies below the threshold value. Accordingly, for the normalized correlation function we obtain

$$\chi(K,\tau) = \frac{\langle A_L(K,\eta)\,A_L^*(K,\eta+\tau)\rangle}{\langle|A_L(K,\eta)|^2\rangle} = \exp\left\{-4\ln 2\left(\frac{\tau}{\tau_c{}^L(K)}\right)^2\right\}. \tag{7.26}$$

The linear phase is finished, when the nonlinearity of the amplifier or absorber becomes important. According to (7.12) the amplification depletion begins to play a role, when the mean intensity has the value

$$\langle I_L(K_0)\rangle = \frac{P}{u\sigma_{32}^{a} a_{Th} C_1}. \tag{7.27}$$

C_1 is here a factor that characterizes the saturation ($C_1 \gtrsim 5$); K_0 is the round trip number at the end of the linear phase. If the nonlinearity of the absorber goes into action beforehand, the end of the linear phase is reached, when the maximum pulse possesses an intensity of

$$I_L^{\max}(K_0) = \frac{I_S^{b}}{sC_2}, \tag{7.28}$$

where $s = (a_{Th} + PK_0 - \gamma)/(PK_0)$ (compare (7.39)) and $C_2 \gtrsim 5$ was again introduced to characterize the nonlinear term. If we introduce additionally a factor β, which provides the number of times the intensity of the largest pulse I_1 exceeds the mean level $\langle I_L(K_0)\rangle$, then using $\langle I_L(K_0)\rangle = I_1(K_0)/\beta$ we can calculate from (7.25) the round trip number K_0 at the end of the linear phase. Hence, we have

$$K_0 \approx \left(\frac{2}{P} \ln \frac{\langle I_L(K_0)\rangle}{I_{Th}}\right)^{1/2}. \tag{7.29}$$

Depending on the saturation mechanism, $\langle I_L(K_0)\rangle$ can be calculated either from (7.27) or using (7.28). Since $\langle I_L(K_0)\rangle/I_{Th}$ lies in the order of 10^9 and, according to (7.29), at a change in this value by two orders of magnitude K_0 changes only by 5%, we can evaluate

$$K_0 \approx 2\sqrt{\frac{1}{P}} \tag{7.30}$$

whereby K_0 becomes independent from the saturation mechanism. From (7.24) the mean duration of an intensity fluctuation at the end of the linear phase is

$$\tau_c^{L}(K_0) \approx 2{,}4\tau_{32}^{a} \sqrt{\frac{a_{Th}}{\sqrt{P}}}. \tag{7.31}$$

At the end of this section, we shall deal briefly with the statistical properties of the radiation field at the end of the linear phase. According to (7.21) the field strength is linked to the characteristics of the noise term $F(K, \eta)$, which is described by a Gaussian noise process. From the factorizability of the $2n^{\text{th}}$ order correlation function of $F(K, \eta)$ the factorizability of the corresponding correlation function for $A_L(K, \eta)$ can easily be shown. Thus, during the linear phase $A_L(K, \eta)$ behaves like a Gaussian noise process, with vanishing mean value whose second order correlation function (7.23) contains all information about the statistical properties of this process. In the complex amplitude $A_L(K, \eta) = X_1(K, \eta) + iX_2(K, \eta)$ the random quantities X_1 and X_2 are statistically independent of one another and have the same correlation function. Given the correlation function of the second order, the joint probability density of the nth order can be easily found (see e.g. [7.57]). Proceeding from the real and imaginary parts X_1 and

X_2 to the amplitude $\mathcal{A}_L = \sqrt{X_1^2 + X_2^2}$ and phase $\varphi_L = \arctan(X_2/X_1)$, the joint probability of the fourth order for phase and amplitude can be written as ([7.57])

$$p_4(\mathcal{A}, \mathcal{A}_\tau, \varphi, \varphi_\tau) = \frac{\mathcal{A}\mathcal{A}_\tau}{(2\pi\sigma^2)^2 (1-\chi^2)} \exp\left\{-\frac{\mathcal{A}^2 + \mathcal{A}_\tau^2 - 2\chi\mathcal{A}\mathcal{A}_\tau \cos(\varphi - \varphi_\tau)}{2\sigma^2(1-\chi^2)}\right\}. \tag{7.32}$$

Here, we have used the abbreviations $\mathcal{A} = \mathcal{A}_L(K, \eta)$, $\mathcal{A}_\tau = \mathcal{A}_L(K, \eta + \tau)$, $\varphi = \varphi_L(K, \eta)$, $\varphi_\tau = \varphi_L(K, \eta + \tau)$, $\sigma^2 = \langle|A_L(K, \eta)|^2\rangle/2$ and $\chi = \chi(K_0, \eta)$ according to (7.26). The probability densities of lower order can be derived from (7.32) through integration. That of the amplitude $\mathcal{A}$, for example, is described by a Rayleigh distribution $\big(p_1(\mathcal{A}) = \mathcal{A}\sigma^{-2} \exp(-\mathcal{A}^2/2\sigma^2)\big)$, whereas the phase is distributed equally in the interval $(0, 2\pi)$ $\big(p_1(\varphi) = 1/(2\pi)\big)$.

For further consideration we require the mean number $\overline{N}(\mathcal{A}, u)$ of spikes from a steady noise process that exceeds a given amplitude level $\mathcal{A}$ within the time interval u [7.55]. We subdivide the time range u into n intervals ($n \gg 1$), whose duration $\Delta\eta$ is smaller than the intensity correlation time. If $\mathcal{A} = \mathcal{A}_0$ holds at $\eta = \eta_0$ and $\mathcal{A}$ exceeds the value $\mathcal{A}_N$ at η ($\eta_0 < \eta < \eta_0 + \Delta\eta$), the derivative of $\mathcal{A}$ at $\eta = \eta_0$ is given by $\dot{\mathcal{A}}_0 = (\mathcal{A}_N - \mathcal{A}_0)/(\eta - \eta_0)$, and therefore the inequation $0 < \mathcal{A}_N - \mathcal{A}_0 < \dot{\mathcal{A}}_0\Delta\eta$ is satisfied, if $\mathcal{A}$ exceeds level $\mathcal{A}_N$ within the time interval $(\eta_0, \eta_0 + \Delta\eta)$. The probability that level $\mathcal{A}_N$ is exceeded within the time $\Delta\eta$ is given by

$$W_{\Delta\eta}^{\mathrm{ex}} = \int_0^\infty \mathrm{d}\dot{\mathcal{A}} \int_{\mathcal{A}_N - \dot{\mathcal{A}}\Delta\eta}^{\mathcal{A}_N} \mathrm{d}\mathcal{A}\, p_2(\mathcal{A}, \dot{\mathcal{A}}). \tag{7.33}$$

$p_2(\mathcal{A}, \dot{\mathcal{A}})$ is the joint probability density for the amplitude and its time derivative $\dot{\mathcal{A}}$. On the other hand, this probability (7.33) is exactly equal to the ratio of the mean number $\overline{N}(\mathcal{A}_N, u)$ of intervals in which the level $\mathcal{A}_N$ is exceeded to the total number of intervals n: $W_{\Delta\eta}^{\mathrm{ex}} = \overline{N}(\mathcal{A}_N, u)/n$. Since the total time interval u (resonator round trip time) is given by $u = n\Delta\eta$, after carrying out the limitting process $\Delta\eta \to 0$ (7.33) leads to

$$\overline{N}(\mathcal{A}_N, u) = u \int_0^\infty \mathrm{d}\dot{\mathcal{A}}\dot{\mathcal{A}}\, p_2(\mathcal{A}_N, \dot{\mathcal{A}}). \tag{7.34}$$

Using the fact that the temporal derivative of a Gaussian process is also a Gaussian process, the joint probability $p_2(\mathcal{A}, \dot{\mathcal{A}})$ can be expressed as

$$p_2(\mathcal{A}, \dot{\mathcal{A}}) = \frac{\mathcal{A}}{\sigma^2} \exp\left\{-\frac{\mathcal{A}^2}{2\sigma^2}\right\} \frac{\exp\left\{\dfrac{\dot{\mathcal{A}}^2}{2\sigma^2\ddot{\chi}(0)}\right\}}{\sqrt{-2\pi\sigma^2\ddot{\chi}(0)}} \tag{7.35}$$

where $\ddot{\chi}(0) = d^2\chi(K, \eta)/d\eta^2|_{\eta=0}$. If we introduce (7.35), together with (7.26), into (7.34), the mean number of fluctuations that exceed a given intensity level $\beta\langle I_L(K_0)\rangle$ during time u is given by [7.58]

$$\overline{N}(\beta, u) = \frac{u}{\tau_c^L(K_0)} \sqrt{\frac{4 \ln 2\beta}{\pi}}\, \mathrm{e}^{-\beta}. \tag{7.36}$$

From the requirement that the highest level is exceeded exactly one time by the largest pulse $(\overline{N}(\beta^{\max}, u) = 1)$, the intensity of the largest relative maximum averaged over many laser shots can be evaluated in relation to the mean intensity in the resonator at $K = K_0$: hence,

$$\beta^{\max} = \ln \left[\frac{u}{\tau_c^{\mathrm{L}}(K_0)} \sqrt{\frac{4 \ln 2}{\pi}} \sqrt{\ln \left(\frac{u}{\tau_c^{\mathrm{L}}(K_0)} \sqrt{\frac{4 \ln 2}{\pi}} \right)} \right]. \tag{7.37}$$

Assuming typical laser parameters we find that $\beta^{\max}$ lies in the range from 4 to 8.

7.2.3 *The Nonlinear Stage of Pulse Formation (Region II)*

A complete analysis of modelocking requires the simultaneous consideration of the saturation behavior of the absorber and amplifier. Under certain conditions, which depend on the parameters of the laser system, it can however be assumed that the depletion of the occupation inversion of the amplifier only comes into play when the absorber is already completely saturated. Such an approach provides a good description of the parameters of the maximum pulse, though it cannot be used, for describing the behavior of the next largest pulse. For the present, let us neglect the influence of the amplification depletion in the nonlinear phase and discuss this influence afterwards in detail.

7.2.3.1 *Neglect of the Occupation Change in the Amplifier*

As we shall show later, the duration of the nonlinear phase is very short compared to the linear phase, and the amplification change within this region is small compared with the threshold value amplification. Therefore, we assume that the amplification in this range remains constant ($a_0 = a_{\mathrm{Th}} + PK_0$). Furthermore because of the small number of round trips during this relatively short phase, we can neglect the frequency selective action of the amplifier, that is, we assume a_0 to be frequency independent. Consequently, (7.11) reduces to

$$\frac{\partial A_{\mathrm{L}}(K, \eta)}{\partial K} = \frac{1}{2} A_{\mathrm{L}}(K, \eta) \left[a_0 - \gamma - \frac{\varkappa_0}{1 + I_{\mathrm{L}}(K, \eta)/I_{\mathrm{S}}^{\mathrm{b}}} \right]. \tag{7.38}$$

An implicit solution of this equation for the amplitude and phase of the complex field strength $A_{\mathrm{L}} = \mathcal{A} e^{i\varphi}$ can be found by separating the variables. Thus, we have

$$(a_0 - \gamma)(K - K_0) = s \ln \frac{\mathcal{A}}{\mathcal{A}_0} - (s - 1) \ln \frac{1 + s\mu\mathcal{A}^2}{1 + s\mu\mathcal{A}_0^2} \cdot \quad \varphi = \varphi_0 \tag{7.39}$$

where $s = (a_0 - \gamma)/(a_0 - \gamma - \varkappa_0)$, $\mu = \sqrt{\varepsilon/\mu_0}/2I_{\mathrm{S}}^{\mathrm{b}}$. As we see from (7.39) nonlinearities do not play a role, as long as $s\mu\mathcal{A}^2 = sI/I_{\mathrm{s}}^{\mathrm{b}} \ll 1$ holds, from which the end of the linear phase according to (7.28) can be defined. If we assume $s\mu\mathcal{A}^2 \ll 1$ (where $s\mu\mathcal{A}_0^2 \ll 1$) for the end of the nonlinear phase (that means that the maximum pulse completely depletes the amplification), we obtain from (7.39) the closed-form solution

$$A_{\mathrm{L}} = (\mu s)^{\frac{s-1}{2}} \mathcal{A}_0^{s} \exp \left[\frac{1}{2} (a_0 - \gamma)(K - K_0) + i\varphi_0 \right]. \tag{7.40}$$

As previously, the index 0 marks here the end of the linear phase. Since the statistical distribution is known for the quantities $\mathcal{A}_0$ and φ_0, the correlation function of the field strength in the nonlinear phase can be calculated using the joint probability (7.32) Hence,

$$\begin{aligned}\langle A_L(K,\eta)\,A_L^*(K,\eta+h)\rangle &= (\mu s)^{s-1}\,e^{(a_0-\gamma)(K-K_0)}\int_0^\infty d\mathcal{A}\int_0^\infty d\mathcal{A}_\tau\int_{-\pi}^{\pi} d\varphi \\ &\quad\times\int_{-\pi}^{\pi} d\varphi_\tau \mathcal{A}^s\mathcal{A}_\tau^s\,e^{i(\varphi-\varphi_\tau)}\,p_4(\mathcal{A},\mathcal{A}_\tau,\varphi,\varphi_\tau) \\ &= \frac{(\mu s)^{s-1}}{\sigma_0^4(1-\chi_0^2)}\,e^{(a_1-\gamma)(K-K_0)}\int_0\int_0 dy\,dx\,\frac{x^{s+1}}{\sqrt{y^2+4x}} \\ &\quad\times\exp\left\{-\frac{y^2+2x}{1-\chi_0^2}\right\}I_1\left(\frac{2\chi_0 x}{1-\chi_0^2}\right) \\ &\approx (\mu s)^{s-1}\,(2\sigma_0^2)^s\,\Gamma(s+1)\left(\frac{1+\chi_0}{2}\right)^{s+1} e^{(a_1-\gamma)(K-K_0)}, \end{aligned} \tag{7.41}$$

whereby the integration variables $y=(2\sigma_0^2)^{-1/2}(\mathcal{A}-\mathcal{A}_\tau)$ and $x=(2\sigma_0^2)\,\mathcal{A}\mathcal{A}_\tau$ are introduced. In addition, $1-\chi_0^2\ll 1$ was assumed which is satisfied well inside the correlation duration of the nonlinear phase $\tau\lesssim\tau_c^{NL}\simeq\tau_L/2$ due to $\tau_c^L\gg\tau_c^{NL}$. The normalized correlation function

$$\chi^{NL}(K,\tau)=\left[\frac{1+\chi_0}{2}\right]^{s+1} \tag{7.42}$$

is only a function of $\chi_0=\chi(K_0,\tau)$, which is independent from K and determined by (7.26).

With respect to the statistical behavior of the pulses, there results a steady state that develops within the nonlinear phase. The correlation duration of the intensity τ_c^{NL} and the pulse duration τ_L, respectively, can be evaluated from (7.42) for $s\gg 1$ by

$$\tau_L=\frac{2}{\sqrt{s}}\sqrt{2\ln 2}\,\tau_{32}^{a}\sqrt{a_0K_0} \tag{7.43}$$

Thus, within the nonlinear phase a shortening of the pulses by the factor $\sqrt{2/s}$ occurs due to the action of the saturable absorption.

Until now we have left open the question of which statistical ensemble the calculated statistical mean values refer to. Due to the pulse selective action of the nonlinear absorption, only few pulses — in favorable cases only one pulse — remain in the resonator at the end of the nonlinear phase. For this reason, the time average over a resonator round trip cannot be identified with the statistical mean and is itself a statistically fluctuating value. The statistical mean in the phase of the nonlinear absorption can, however, be interpreted as a mean over a series of experiments carried out under the same conditions (i.e. of laser shots at fixed laser parameters). Time averages over a resonator round trip and statistical average are only nearly equal in the linear phase, which shows that the relative statistical square fluctuation of the time average per round trip around its statical mean $\langle\bar{I}\rangle=\langle I_0\rangle$ is very small, whereas it is about one in the nonlinear phase [7.12].

The duration of the nonlinear phase $(K_2 - K_0)u$ can be evaluated from the requirement that the intensity of the largest pulse I_1 reaches a multiple of the absorber saturation intensity I_s^b ($I_1(K_2) \approx C_3 I_s^b$ where $C_3 \geqq 5$). Since $I_1(K_1)$ is coupled with $I_1(K_0) = I_s^b/(sC_2)$ by the relation (7.40), the number $K_2 - K_0$ of resonator round trips in the nonlinear phase can be evaluated by

$$K_2 - K_0 \approx \frac{1}{a_0 - \gamma} [\ln (sC_3) + s \ln C_2] \approx \frac{\ln C_2}{a_0 - \gamma - \varkappa_0}. \tag{7.44}$$

Finally, let us provide a numerical example of the results we have obtained. If we set the laser threshold $a_{Th} = 0.5$, $\varkappa_0 = 0.4$, $\gamma = 0.1$, the resonator round trip time $u = 6$ ns and the pump rate $P = 10^{-5}$, then using the laser and absorber parameters given in table 7.1 we obtain the following results for the ruby laser: $K_0 = 1930$, $K_2 - K_0 = 120$, $s = 21.7$, $\tau_c^L = 70.6$ ps, $\tau_L = 21.4$ ps and $\beta^{max} = 5.1$. For the neodymium glass laser the results are: $K_0 = 1900$, $K_2 - K_0 = 120$, $s = 22.1$,

Table 7.1. Parameters of solid state lasers and typical saturable absorbers

Laser type	σ_{32}^a cm²	$\Delta\nu_{32}^a$ Hz	T_{31}^a s
Ruby laser	2.5×10^{-20}	3.3×10^{11}	3×10^{-3}
Neodymium glass laser	3.0×10^{-20}	7.5×10^{12}	3×10^{-4}
Nd:YAG laser	5×10^{-19}	1.2×10^{11}	2.3×10^{-4}

Absorber	Laser type	σ_{13}^b cm²	T_{21}^b s	$I_s^b h\nu$ Wcm⁻²
DDI	Ruby laser	10^{-15}	1.4×10^{-11}	2×10^7
Eastman	Nd:glass laser	5.7×10^{-16}	8.3×10^{-12}	4×10^7
No. 9740	Nd:YAG laser			

$\tau_c^L = 3.08$ ps, $\tau_L = 0.93$ ps and $\beta^{max} = 8.5$. Whereas the pulse duration found for the ruby laser corresponds well with the experimental values, the pulse duration calculated for the neodymium glass laser is about 2 to 10 times shorter than the values measured in the peak of the pulse train. This discrepancy is caused by the action of additional nonlinear effects in the glass rods of the neodymium glass laser (inhomogenous amplification depletion, intensity dependent refractive index effects and dispersion), which lead to a phase modulation of the pulse and to a pulse broadening not taken into account here. Measurements of the pulse durations at the beginning of the pulse train (where the self-phase modulation still does not play a role) show, however, better agreement between the measured pulse durations of 2 to 3 ps and the calculated ones.

7.2.3.2 *Influence of the Gain Depletion in the Amplifier*

Under certain conditions for the laser parameters, the relations given in the last section provide a good description of the characteristics of the maximum pulse. If, however, we want to investigate more closely the conditions of modelocking — that is, the formation of a single pulse and the suppression of all other pulses —, then the consideration of the depletion of the occupation inversion of the active medium is already necessary in the nonlinear phase. In a series of papers, the generation process in passively mode-locked solid state lasers was simulated on computers, whereby the influence of the amplification depletion was considered in various ways [7.7 to 7.10, 7.41 to 7.44][1]). Due to the statistical character of the pulse generation, however, one complication presents itself in a description of this kind. For the depletion of the occupation inversion the total noise intensity in the linear phase plays a role. Depending on the random intensity distribution in the linear phase, one or several pulses can be generated. Therefore, to determine the probability for the appearance of double pulses the statistic for the initial distribution of the intensity values of the most intensive fluctuation peaks must be simulated according to their statistical weight, and every laser shot must be calculated individually. Therefore a variation and with it an optimization of the laser parameters requires extreme effort. Therefore we will undertake an approximative treatment of the problem which allows an analytical solution [7.13]. For this purpose we subdivide the nonlinear phase in two parts. During the first part, which immediately follows the linear phase (region IIa in Fig. 7.1), the intensity is still small compared to the absorber intensity I_s^b, that is, we can describe the absorber losses by $\varkappa(J) \approx \varkappa_0(1 - J)$ where $J = I_L/I_s^b$. In the second part (region IIb in Fig. 7.1) of the nonlinear phase we have to use the exact expression for the nonlinear absorption $\varkappa(J) = \varkappa_0/(1 + J)$. As we shall see, the decrease of the amplification is in both parts small compared to the threshold amplification a_{Th}. However, since the laser is pumped only slightly over the threshold, this small decrease alone can exert a considerable influence on the process of modelocking. Under the assumptions mentioned, we can describe the first part of the nonlinear phase using the equations

$$\frac{\partial J(K, \eta)}{\partial K} = \tilde{a}(K)\, J(K, \eta) + \varkappa_0 J^2(K, \eta), \tag{7.45}$$

$$\frac{d\tilde{a}(K)}{dK} = -\lambda a_{Th} \bar{J}(K), \tag{7.46}$$

where we have introduced $\tilde{a} = a - a_{Th}$ and $\lambda = \sigma_{32}^a u/(\sigma_{13}^b T_{21}^b)$ and used $\tilde{a} \ll a_{Th}$. In (7.46) we have neglected the pump term P. Thus, the end of the linear phase is deter-

---

[1]) Wilbrand and Weber investigated the influence of background noise on the modelocking [7.9]. They found that above the actual laser threshold a second threshold exists for the modelocking which is linked directly to the depletion of the active medium. In their analysis, however, they proceeded on the assumption of a slow saturable absorber ($T_{21}^b \gg \tau_L$), which is not a suitable model for solid state lasers; furthermore, no satellite pulses were considered. Glenn [7.8] investigated the evolution of five pulses taking into account the amplification saturation for a fast saturable absorber. However, in this paper the depletion of the occupation inversion in the active medium was not correctly accounted for and was greatly overestimated (in equation (19) in [7.8] the pulse duration must be substituted for the resonator round trip time). The influence of the inversion reduction on the modelocking process was also investigated by New [7.11], who simulated the development of the five most intense pulses having statistically distributed initial intensities for 100 laser shots at a constant set of Nd:YAG laser parameters. The probability, evaluated on the basis of this calculation, for the occurrence of double pulses or for the breakdown of the pulse build-up agrees with our results in section 7.2.3.3.

mined by the relation $\bar{J}(K_0) = P/\lambda a_{\text{Th}}$, and therefore $C_1 = 1$ must be set due to (7.27). Using the new variable

$$z(K) = \int_{K_0}^{K} \mathrm{d}K' \exp\left\{\int_{K_0}^{K'} \mathrm{d}K'' \tilde{a}(K'')\right\}$$

we can write the solution of (7.45) as

$$J(K, \eta) = \frac{J(K_0, \eta)\,\dfrac{\mathrm{d}z}{\mathrm{d}K}}{1 - \varkappa_0 z J(K_0, \eta)}. \tag{7.47}$$

This solution only remains meaningful if also for the intensity $J_1(K_0, \eta)$ of the most intensive fluctuation peak the relation $\varkappa_0 J_1(K_0, \eta)\, z \lesssim 0.9$ holds (which determines the end of the first part of the nonlinear phase). However, since the intensity of the maximum pulse exceeds the mean intensity $\bar{J}(K_0)$ by a factor $\beta^{\max} = 4$ to 8, we can expand $J(K)$ into a powers series to solve (7.46). In calculating the general expansion term $\bar{J}^n(K_0)$ we have to take into consideration that $J(K_0, \eta)$ is defined only within the finite time interval $0 \leqq \eta \leqq u$, because the statistical mean and the time average of the variables $J^n(K_0, \eta)$ deviate greatly from one another at $n > \beta^{\max}$. Up to a value of $n = 3$, however, the two values coincide well, and therefore we can set $\bar{J}^2(K) = 2(\bar{J}(K))^2$, $\bar{J}^3(K) = 6(\bar{J}(K))^3$ due to the Gaussian distributed field intensity. The relation for $\bar{J}(K)$ obtained from (7.47) in this way is inserted in (7.46). Considering the relation between $z(K)$ and $a(K)$, we obtain the following equation for z:

$$\frac{\mathrm{d}^2}{\mathrm{d}K^2} \ln \frac{\mathrm{d}z}{\mathrm{d}K} = -\lambda a_{\text{Th}} a \bar{J}(K_0) \frac{\mathrm{d}z}{\mathrm{d}K} \left\{1 + 2z\varkappa_0(\bar{J}(K_0)) + 6(\varkappa_0 \bar{J}(K_0)\, z)^2\right\}. \tag{7.48}$$

With regard to the initial condition $z(K_0) = 0$, $\mathrm{d}z/\mathrm{d}K|_{K_0} = 1$ and $\mathrm{d}^2z/\mathrm{d}K^2|_{K_0} = \tilde{a}(K_0)$, (7.48) can be integrated, and hence

$$\frac{\mathrm{d}z}{\mathrm{d}K} = -\frac{\lambda}{2} a_{\text{Th}} \bar{J}(K_0)\, z^2 \left\{1 + \frac{2}{3} \varkappa_0 \bar{J}(K_0)\, z + (\varkappa_0 \bar{J}(K_0)\, z)^2\right\} + \tilde{a}(K_0)\, z + 1. \tag{7.49}$$

This equation can be solved by separating the variables, however we only require $\mathrm{d}z/\mathrm{d}K$ as a function of z in the calculations that follow.

If the depletion of the occupation inversion in the active medium leads to $\tilde{a} < 0$, then intensity fluctuations with $J < -\tilde{a}/\varkappa_0$, are attenuated (since here $\mathrm{d}J/\mathrm{d}K < 0$) and those with $J > -\tilde{a}/\varkappa_0$ are amplified. If $\mathrm{d}J/\mathrm{d}K < 0$ applies for a fluctuation peak at a certain round trip this condition is preserved at later round trips as well. Thus, $\mathrm{d}J/\mathrm{d}K = 0$ not only characterizes the achievement of the maximum intensity, but also the end of the amplification range for any chosen fluctuation peak. Since for all K the inequality $\mathrm{d}z/\mathrm{d}K > 0$ holds $\mathrm{d}J/\mathrm{d}K = 0$ is equivalent to $\mathrm{d}J/\mathrm{d}z = 0$.

Although the amplification in the time range considered here falls below the threshold amplification, the most intensive fluctuations are nevertheless further amplified due to the likewise reduced absorption (as a result of the saturation of the absorption that begins), while the fluctuations of weaker intensity are absorbed. The small depletion of the occupation inversion in the active medium combined with the saturation of the absorption of the absorber thus leads to a greater pulse discrimination than the saturable absorption alone could produce.

Let us now investigate more closely the condition under which fluctuation peaks are amplified and absorbed. For this purpose we calculate $\mathrm{d}J(K, \eta)/\mathrm{d}z = 0$ from (7.47), where equation (7.49) is used for $\mathrm{d}z/\mathrm{d}K$. The intensity of the fluctuation maxima at the end of the linear phase $J(K_0, \eta)$ is normalized to the mean intensity: $\beta = J(K_0, \eta)/\bar{J}(K_0)$. The equation $\mathrm{d}J/\mathrm{d}z = 0$ can than be solved with respect to $\beta = \beta_\mathrm{D}$

$$\beta_\mathrm{D} = -\frac{\tilde{a}(z)}{\varkappa_0 \bar{J}(K_0)\left[\dfrac{\mathrm{d}z}{\mathrm{d}K}(z) - \tilde{a}(z)\, z\right]} \tag{7.50}$$

with

$$\tilde{a}(z) = \frac{\mathrm{d}}{\mathrm{d}z}\left[\frac{\mathrm{d}z}{\mathrm{d}K}(z)\right] = -\lambda a_\mathrm{Th} \bar{J}(K_0)\, z\{1 + \varkappa_0 \bar{J}(K_0)\, z + 2(\varkappa_0 \bar{J}(K_0)\, z)^2\} + \tilde{a}(K_0). \tag{7.51}$$

Equation (7.50) can be interpreted here such that all pulses whose intensity is $\beta_\mathrm{D}\bar{J}(K_0)$ at the end of the linear phase reach their maximum at $z = z(K)$. The graphs of $a(z)$ and $\beta_\mathrm{D}(z)$ are shown in Fig. 7.2. All pulses whose relative intensity at the end of the

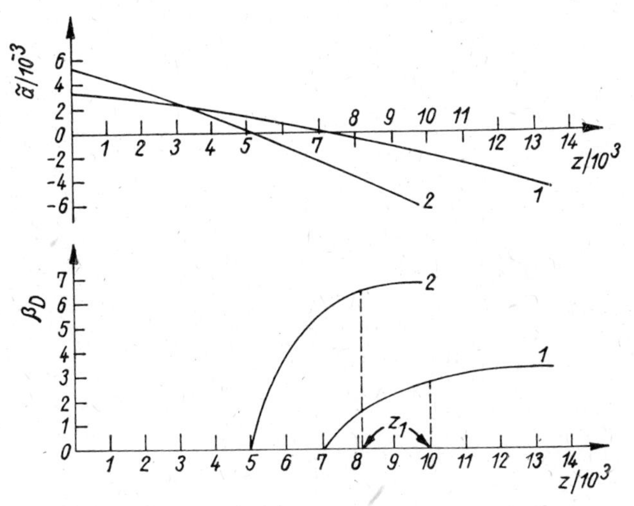

Fig. 7.2. Critical relative intensity level β_D and the net gain $\tilde{a} = a - a_\mathrm{Th}$ as functions of

$$z = \int_{K_0}^{K} \mathrm{d}K' \exp\left\{\int_{K_0}^{K'} \mathrm{d}K''\tilde{a}(K'')\right\}$$

Parameters: Ruby laser (curve 1): $u = 6$ ns, $P = 4 \times 10^{-7}$, $a_\mathrm{Th} = 0.5$, $\varkappa_0 = 0.3$; Nd: glass laser (curve 2): $u = 6$ ns, $P = 10^{-6}$, $a_\mathrm{Th} = 0.5$, $\varkappa_0 = 0.3$. The corresponding laser and absorber parameters are given in table 7.1.
(from Herrmann, Weidner and Wilhelmi [7.13])

linear phase is greater than β_D are amplified, though pulses with $\beta < \beta_\mathrm{D}$ are absorbed. The function $\beta_\mathrm{D}(z)$ achieves its maximum $\beta_\mathrm{D}^{\max}$ at a certain $z^{\max}$. Due to the assumed approximations, however, the validity range of the curves ends at smaller $z = z_1$ ($z_1 < z^{\max}$) and can be determined from the limit of validity of the expansion with respect to $zJ(K_0)\,\varkappa_0$, whereby we set $J_1(z_1) = 0.2$ in (7.47) (see Fig. 7.2); z_1 and the resonator round trip number K_1 belonging to it, respectively, mark the end of the first part of the nonlinear phase. For $K > K_1$, we must take into consideration the exact functional dependence of the absorption coefficient on the intensity, which in this range delivers higher absorber losses than the approximation $\varkappa(J) = \varkappa_0(1 - J)$. Consequently, the rise of the curve $\beta_D(z)$ in Fig. 7.2 is steeper than the rise of the curve that would result with the exact $\varkappa(J)$. Because $\beta_D(z)$ changes only little in the range $z > z_1$, the maximum values of the two curves calculated with the various absorber losses differ only slightly from one another, so that we can assume $\beta_D{}^{\max}$ as the upper limit for the pulse discrimination. From (7.50) we see that the condition $d\beta_D/dz = 0$ is identical with $dz/dK = 0$, where we obtain $\beta_D{}^{\max} = \left(\varkappa_0 \bar{J}(K_0)\, z^{\max}\right)^{-1}$. From (7.49) $\beta_D{}^{\max}$ can then be written as

$$(\beta_D{}^{\max})^2 + \frac{\tilde{a}(K_0)}{\varkappa_0 \bar{J}(K_0)}\,\beta_D{}^{\max} - \frac{1}{2}\,\frac{\lambda a_{\mathrm{Th}}}{\varkappa_0{}^2 \bar{J}(K_0)}\left\{1 + \frac{2}{3\beta_D{}^{\max}} + \frac{1}{(\beta_D{}^{\max})^2}\right\} = 0. \tag{7.52}$$

The parameters $a(K_0) = PK_0$ and $\bar{J}(K_0) = P/(\lambda a_{\mathrm{Th}})$ were evaluated in section 7.2.2, with which the equation (7.52) still depends only on the laser parameters.

7.2.3.3 *Criteria for Good Modelocking: The Probability for the Breakdown of the Pulse Build up and the Occurrence of Double Pulses*

The existence of a critical intensity level, below which all pulses are absorbed, is linked with a criterion for the occurrence of modelocking. If none of the intensity fluctuations at the end of the linear phase exceed the critical level $\beta_D{}^{\max}\bar{J}(K_0)$, all maxima are absorbed, and the pulse build up breaks down. In this case, the laser remains in the non-modelocked regime. The occurrence of this case can only be described by a certain probability, which we want to determine in the following.

In section 7.2.2, we calculated the mean number $\bar{N}(\beta, u)$ of fluctuations exceeding the relative intensity level β during the resonator round trip time u, which is determined by (7.36). The probability that level β is exceeded within an arbitrarily chosen (very small) time interval $\Delta\eta$ was given by $W_{\Delta\eta}^{\mathrm{ex}}(\beta) = \bar{N}(\beta, u)/n$, where n is the total number of time intervals ($n \gg 1$). Accordingly, the probability that level β is not exceeded during the n time intervals considered is given by

$$W(\beta, u) = [1 - W_{\Delta\eta}^{\mathrm{ex}}(\beta)]^n = \exp\{-\bar{N}(\beta, u)\}. \tag{7.53}$$

Inserting $\beta = \beta_D{}^{\max}$ in (7.36) and (7.53), we obtain the probability that a laser shot does not result in the formation of modelocked laser pulses.

The probability $W(\beta_D{}^{\max}, u)$ shows an interesting dependence on the resonator length. For fixed laser parameters there exists an optimal resonator round trip time u_{opt} (and resonator length, respectively) at which the probability for the breakdown of the pulse build up is minimal. Taking into account the dependence of λ, $a(K_0) = PK_0$ and $P = pu$ on the resonator round trip time u, (7.52) provides a relation between u und $\beta_D{}^{\max}$ in accordance with (7.29). Therefore we can eliminate u in (7.53) and replace the minimum condition $dW(\beta, u)/du = 0$ by $d\bar{N}(\beta_D{}^{\max})/d\beta_D^{\max} = 0$. The evaluation of this condition leads approximately to $(\beta_D{}^{\max})_{\mathrm{opt}} = 3.659$, which is completely independent of the laser parameters. With that the optimal resonator length is

$$u_{\mathrm{opt}} = 67.8p \left[\frac{\sigma_{13}^{\mathrm{b}} T_{21}^{\mathrm{b}} \varkappa_0}{\sigma_{32}^{\mathrm{a}} a_{\mathrm{Th}}}\right]^2 Q, \tag{7.54}$$

$$\overline{N}_{\mathrm{opt}} = 10.1 p^{3/2} \Delta\nu_{32}^{\mathrm{a}} \frac{1}{(a_{\mathrm{Th}})^3} \left[\frac{\sigma_{13}^{\mathrm{b}} T_{21}^{\mathrm{b}} \varkappa_0}{\sigma_{32}^{\mathrm{a}}}\right]^{5/2} \tag{7.55}$$

where

$$Q = \ln\left[4.83 \frac{p^{3/2}\sqrt{\varkappa_0}}{a_{\mathrm{Th}} I_{\mathrm{Th}}} \left(\frac{\sigma_{13}^{\mathrm{b}} T_{21}^{\mathrm{b}}}{\sigma_{32}^{\mathrm{a}}}\right)^{3/2}\right] + \frac{1}{4} \ln\ln\left[\frac{p\sigma_{13}^{\mathrm{b}} T_{21}^{\mathrm{b}} \sqrt{2}}{I_{\mathrm{Th}} \sigma_{32}^{\mathrm{a}} \sqrt{a_{\mathrm{Th}}}}\right].$$

Beside a smallest possible probability for the breakdown of the pulse build up, a smallest possible probability for the appearance of double pulses is another feature suitable to characterize the quality of the modelocking in the laser. For this characterization, the probability distribution must be calculated for the ratio Z of the intensity of the maximum pulse to that of the second largest pulse, $Z = J_1(K)/J_2(K)$. First, we calculate this distribution function at the end of the linear phase. In this phase, the radiation field contains $M = u/\tau_c(K_0)$ fluctuation peaks. The probability that the relative intensity of the largest peak lies in the interval $(\beta_1, \beta_1 + \mathrm{d}\beta_1)$ $(\beta_1 = J_1(K_0)/\langle J(K_0)\rangle)$ and the relative intensity of the next largest in the interval $(\beta_2, \beta_2 + \mathrm{d}\beta_2)$, while the intensity of the remaining $(M - 2)$ maxima are located in the interval $(0, \beta_2)$ is given as follows [7.11]:

$$p_2(\beta_1, \beta_2)\, \mathrm{d}\beta_1\, \mathrm{d}\beta_2 = M(M-1)\, \mathrm{e}^{-\beta_1}\, \mathrm{e}^{-\beta_2}[1 - \mathrm{e}^{-\beta_2}]^{M-2}\, \mathrm{d}\beta_1\, \mathrm{d}\beta_2. \tag{7.56}$$

The random variables β_1 and β_2 are transformed into the new quantities $X = \beta_1/\beta_2$ and β_1. For the distribution function $F(X) = \int_1^\infty \mathrm{d}\beta_1 \int_1^X \mathrm{d}X' p_2(\beta_1, X')$ we obtain

$$F(X) = 1 - M \int_1^\infty \mathrm{d}\beta_1\, \mathrm{e}^{-\beta_1} \left[1 - \exp\left(-\frac{\beta_1}{X}\right)\right]^{M-1}. \tag{7.57}$$

The ratio of the intensities $Y = J_1(K_1)/J_2(K_1)$ at the end of the first part of the nonlinear phase can be calculated from (7.47), where $z(K_1)$ must be eliminated by solving (7.47) with respect to z_1. If we generate the inverse function $X = X(Y, \beta_1)$ (which, however, can only be done by approximation) and substitute it into the integrands in (7.57), then we obtain the distribution function $F(Y)$ at the end of the first part of the nonlinear phase. In the second part of the nonlinear phase the enlargement of the net amplification increases rapidly due to the saturation of the absorption, which is the dominating effect within this stage. In view of that, we can evaluate the change in the amplification due to the pumping and depletion of the amplification as very small. For this reason, we may set approximately $a \approx a_{\mathrm{Th}}$ in (7.11), from which we obtain

$$\frac{\partial J(K, \eta)}{\partial K} = \frac{\varkappa_0 J(K, \eta)}{1 + J(K, \eta)}. \tag{7.58}$$

Accordingly, we can calculate the intensity ratio of the two most intensive pulses $Z = J_1(K_2)/J_2(K_2)$ at the end of the second part of the nonlinear phase ($K = K_2$, region IIb), which results in

$$Z = Y \exp\left\{\frac{1}{J_1(K_1)}(Y-1)\right\}, \tag{7.59}$$

where $J_1(K_2) \gtrapprox 3$ and $J_2(K_2) \gtrapprox 3$ was assumed. The intensity at the end of the first part of the nonlinear phase occuring in (7.59) was again assumed to be $J_1(K_1) = 0.2$.

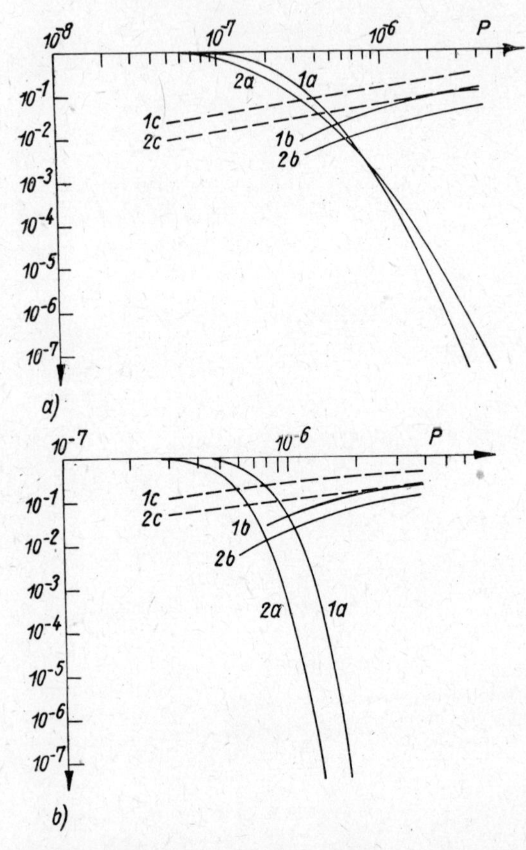

Fig. 7.3. Probability for the breakdown of the pulse build up (index a) and for the occurrence of double pulses (index b) taking into account amplification depletion as functions of the pump rate P. The dotted curves (index c) indicate the double pulse probability neglecting the amplification depletion.

Parameters: $u = 6$ ns; curve (1): $a_{\text{Th}} = 0.5$, $\varkappa_0 = 0.3$, curve (2): $a_{\text{Th}} = 0.9$, $\varkappa_0 = 0.7$. In the calculation ruby laser parameters were used in a) and Nd: glass laser paraméters in b). The laser and absorber parameters are given in table 7.1.
(from Herrmann, Weidner, Wilhelmi [7.13])

If we now generate the inverse function from (7.59) $Y = Y(Z)$ and substitute $Y(Z)$ into the distribution function $F(Y)$, we obtain finally the distribution function $F(Z)$ at the end of the nonlinear phase. In the final amplification stage (region III), in which the absorber is already saturated and the amplifier reaches the saturation range, the intensity ratio between two pulses remains constant (see 7.2.4), so that the distribution function $F(Z)$ characterizes the double pulse probability at the end of the amplification process. We define the two most intense pulses as double pulses, if the ratio Z of the intensities of both pulses lies in the interval $1 < Z < 10$ at the end of the amplification process. The double pulse probability is then given by $F(Z = 10)$.

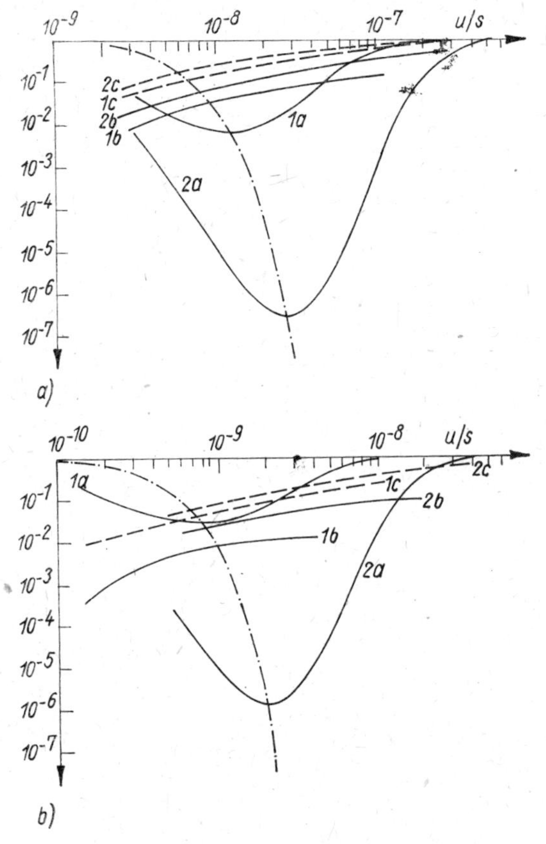

Fig. 7.4. Probability for the breakdown of the pulse build up (index a) and for the occurrence of double pulses (index b) as functions of the resonator round trip time u. The dotted curves (index c) describe the double pulse probability neglecting the amplification depletion and the dot-dashed curves indicate the position of the minima of the probability for the breakdown of the pulse evolution.

Parameter: $a_{\text{Th}} = 0.5$, $\varkappa_0 = 0.3$;
a) Ruby parameters where $p = P/u = 100\ \text{s}^{-1}$ (curve 1) and $p = P/u = 200\ \text{s}^{-1}$ (curve 2);
b) Neodymium glass parameters where $p = 100\ \text{s}^{-1}$ (curve 1) and $p = 230\ \text{s}^{-1}$ (curve 2);
the corresponding laser and absorber parameters are given in table 7.1.
(from Herrmann, Weidner, and Wilhelmi [7.13])

The double pulse probability calculated in this manner for neodymium glass and ruby laser parameters is shown in Fig. 7.3 with respect to the pump rate P and in Fig. 7.4 with respect to the resonator round trip time u [7.13]. In the figures the probability for the breakdown of the pulse build up is additionally plotted according to (7.53). The curves for the breakdown probability are designated with (a) in the diagram, and the curves for the double pulse probability with (b). For a comparison the double pulse probability is plotted without considering the depletion of the amplification (as calculated in [7.11]) (dotted, where (c) designates the curves).

Fig. 7.3 shows that the breakdown probability depends considerably more on the pump rate P than the double pulse probability. The very large decrease in the breakdown probability particularly at greater gain and smaller absorption cross sections (compare Fig. 7.3) in a certain way warrants speaking of the existence of a second threshold. If the pump intensity exceeds the first threshold, the laser begins to oscillate, but remains initially in a non-modelocked regime. Only with the exceeding of the second threshold — denoted by the pump rate P at which the breakdown probability falls greatly — ultrashort pulses can build up. As we have seen, however, in contrast to the first threshold the second threshold is not uniquely determined, but rather each pump rate P is related solely to a probability for the build up of a pulse. At smaller absorber losses the drop in this probability is not as pronounced (compare (Fig. 7.3a)), so that the concept of the second threshold is problematic here.

In contrast to the breakdown probability the double pulse probability increases with increasing pump rate. Since both probabilities should be as small as possible, P must be chosen such that a lowest possible value for both quantities results. Increasing absorber losses lead to a decrease in the double pulse probability, and in the neodymium glass laser to a smaller breakdown probability as well.

A comparison between curves (b) and (c) in Fig. 7.3 shows that the probability for the selection of the most intense pulses from noise increases at the end of the linear phase due to the action of the depletion of the occupation inversion, whereby the double pulse probability decreases in the optimal range by about an order of magnitude. Thus, the influence of the inversion depletion in the active medium favorably affects the modelocking in solid state lasers, too, though their operating mode is completely different from that of dye lasers.

The active medium in the solid state laser is far from being saturated in the time range, in which the pulse evolves. In Fig. 7.4 the resonator round trip time u varies. The parameter of the curve is $p = P/u$. The breakdown probability shows a significant minimum whose depth is enhanced with increasing pump intensity. The formation of this minimum can be understood by the fact that at longer resonator round trip times the number of fluctuation peaks at the end of the linear phase increases, whereby a rapid depletion of the occupation inversion and with it an increased breakdown probability is produced. At resonators that are too short the number of fluctuation maxima at $K = K_0$ is reduced, so that the probability increases that none of these maxima will exceed the level $\beta_D^{\max} \bar{J}(K_0)$. The position of the minima of the breakdown probability resulting from the formula (7.54) is depicted in Fig. 7.4 by a dash-dot line and corresponds to the minima of the numerically calculated curve. The double pulse probability increases monotonically with increasing resonator round trip time, so that considering both criteria it is more favorable to choose smaller resonator lengths u than the length u_{opt} that results from equation (7.54).

In the numerical examples previously carried out we referred only to the neodymium glass laser and ruby laser, however we consciously excluded the neodymium YAG laser. As evident from table 7.1 the Nd:YAG laser has a cross section for stimulated emission, which is by about a factor of 20 greater than that of the other two laser types. The result is a considerably faster depletion of the occupation inversion, whereby the assumption for the equation (7.46) ($\tilde{a} \ll a_{\text{Th}}$) is no longer justified and the approximations undertaken here do not apply. Let us at least qualitatively discuss how the greater cross section affects the process of modelocking. The more rapid depletion of the occupation caused leads of course to an increased breakdown probability, which requires higher pump rates for the modelocking. On the other hand, the faster depletion of the occupation inversion affects very favorably the reduction of the double pulse probability, which should now become considerably smaller at pump rates that are not too high. As we have seen previously, it is more difficult to decrease the double pulse probability than the breakdown probability. Thus, under optimal conditions, the greater emission cross section should produce a higher stability of the generation of ultrashort pulses, which experiments have confirmed.

Instead of an increased emission cross section the same effect can be achieved using a telescope inside the resonator. If the beam cross section is different in the absorber and amplifier, then in equation (7.46) $\lambda' = \sigma_{32}^{a} u q^{b} / (\sigma_{13}^{b} T_{31}^{b} q^{a})$ must be substituted for the factor λ, where q^{a} and q^{b} respresent the beam cross sections in the amplifier and absorber. An increase in λ' can thus be achieved by increasing the ratio q^{b}/q^{a}. On the other hand, the value of λ' must not be too large, because in this case the critical level β_{D} becomes too large and thus the laser remains in the free-running regime.

7.2.4 *The Gain Depletion (Region III)*

In the phase of the complete depletion of amplification the nonlinear absorption in the pulse center no longer plays a part, since the absorber was saturated beforehand. We can describe accordingly the radiation field and the amplification by the equations

$$\frac{\partial J(K, \eta)}{\partial K} = [a(K) - \gamma]\, J(K), \tag{7.60}$$

$$\frac{\mathrm{d}a(K)}{\mathrm{d}K} = -\lambda a(K)\, \bar{J}(K). \tag{7.61}$$

However, they provide only a relatively rough description because the evaluation of the amplification a_2 at the round trip number $K = K_2$ is faulty due to the assumptions made, and furthermore nonlinear optical processes in the laser crystal at high intensities, as they appear particularly in the Nd:glass laser, are neglected. In the validity range of the equations (7.60) and (7.61) the statistical properties of the radiation field (e.g. shape of the pulses or possible satellite pulses) do not change any more. Since at the end of region II only one pulse circulates under optimal conditions, we need only consider the intensity of the maximum pulse $J_1(K, \eta)$, and therefore we can substitute $\tau_{L} J_1(K, 0)/u$ for $\bar{J}(K)$ in (7.61) (using τ_{L} from (7.43)). If we solve (7.61) with respect to $J_1(K, 0) = J_1(K)$, substitute this into (7.60) and integrate, we obtain

$$J_1(K) - J_1(K_2) = -(u/\tau_{L}\lambda)\big(a(K) - a(K_2)\big) + \gamma(u/\tau_{L}\lambda) \ln \frac{a(K)}{a(K_2)}. \tag{7.62}$$

Combining (7.62) and (7.61), we find the relation

$$K - K_2 = -\int\limits_{a(K_2)}^{a(K)} \frac{\mathrm{d}a'}{a'\left[\frac{\lambda\tau_L}{u} J_1(K_2) - \left(a' - a(K_2)\right) + \gamma \ln \frac{a'(K)}{a(K_2)}\right]}, \tag{7.63}$$

from which $a(K)$ can be calculated numerically.

The pulse has achieved its maximum at $K = K_{\max}$, if according to (7.60) $a(K_{\max}) = \gamma$ holds. With $a(K_2) \approx a_{\mathrm{Th}} = \varkappa_0 + \gamma$ and $I_1(K_{\max}) \gg I(K_2)$ the intensity of the pulse at the maximum of the pulse train is

$$J_1(K_{\max}) = \frac{u}{\tau_L \lambda}\left[\varkappa_0 + \gamma \ln\left(1 + \frac{\varkappa_0}{\gamma}\right)\right]. \tag{7.64}$$

For an approximate calculation of the halfwidth of the pulse train ΔK we solve (7.61) with respect to the intensity of the maximum pulse $J_1(K) = u\bar{J}(K)/\tau_L$ and substitute it into (7.60). The integration of the equation from $K = K_2$ to $K = \infty$ yields

$$a_2 - a_\infty + \frac{u}{\tau_L \lambda}\left(J_1(K_2) - J_1(K = \infty)\right) = \gamma \frac{u}{\tau_L \lambda} \int\limits_{K_2}^{\infty} \mathrm{d}K J_1(K). \tag{7.65}$$

Since the pulse energy at the beginning and at the end of the pulse train is neglibly small compared to the "saturation energy" of the amplifier, the second term in (7.65) can be neglected. In the same approach we obtain from (7.62)

$$a_\infty = a_2 - \gamma \ln \frac{a_2}{a_\infty}. \tag{7.66}$$

If we assume approximately a Gaussian shaped envelope of the pulse train

$$J_1(K) = J_1(K_{\max}) \exp\left\{-4 \ln 2 \left(\frac{K - K_{\max}}{\Delta K}\right)^2\right\} \tag{7.67}$$

then from (7.65) and (7.64) we obtain

$$\Delta K = \frac{a_2 - a_\infty}{\gamma\left[\varkappa_0 - \gamma \ln\left(1 + \frac{\varkappa_0}{\gamma}\right)\right]} \tag{7.68}$$

for the halfwidth of the pulse train ΔK. Assuming $\varkappa_0 = 0.4$, $\gamma = 0.1$ and $a_{\mathrm{Th}} = 0.5$ we obtain $a_\infty = 0.0035$ from (7.66), from which $\Delta K \approx 20$ follows. With an outcoupling mirror transmission of 10%, a maximum output intensity of $(h\nu_L\gamma)\, I_1(K_{\max}) = 6 \times 10^9$ W cm^{-2} results for the ruby laser and $(h\nu_L\gamma)\, I_1(K_{\max}) = 10^{11}$ W cm^{-2} for the neodymium glass laser.

These values lie about one order of magnitude above typical experimental data. The cause for this discrepancy, besides the reasons already mentioned, is mainly the influence of the inhomogeneous broadening. Neodymium glass, for example, has an inhomogenous linewidth of $\Delta\nu_{\mathrm{inhom}} \approx 10^{13}$ Hz and a homogenous line width of $\Delta\nu \approx 5 \times 10^{11}$ Hz, whereas the cross relaxation time is $T_{\mathrm{inhom}} \approx 70$ µs. The characteristic times in the linear phase are sufficiently large, so that transfer processes between the various ions have already occurred. Therefore it is justified to neglect the inhomogeneous character of the line broadening in this stage as presented in the theory. The saturation phase of the laser is, however, short compared to the cross relaxation time. As a result, only those spectral ranges in the center of the amplification profile, that lie within the homogeneous line width will contribute to the amplification in the amplification depletion phase. This depletion accordingly produces a hole burning in the amplification profile, as described in section 1.1.6. Due to the reduction of the number N of participating ions by the factor $N\Delta\nu/\Delta\nu_{\text{inhom}}$ thus produced, the maximum intensity of the pulses decreases, while the pulse duration remains unaffected. More precise experimental and theoretical investigations of hole burning and other nonlinear optical processes have been conducted by Penzkofer and Weinhardt [28].

7.3 Experimental Set-ups and Results

7.3.1 Set-ups and Features of Passively Modelocked Solid State Lasers

A typical experimental arrangement for a modelocked solid state laser is shown in Fig. 7.5. The most important components here are the optical resonator, the laser rod and the dye cell, which contains the saturable absorber. Usually, a single pulse is selected from the pulse train after leaving the laser oscillator (see. 7.3.3). To generate a reproducible modelocked pulse train of high pulse quality that contains no subpulses, a careful optimization of the individual components, such as the resonator configuration, the dye, its solvent and its concentration as well as the thickness of the dye cell and its position, is necessary. Independent of this, however, double pulses evolve or the laser remains in the free-running regime with a certain probability that results from the process of modelocking itself according to the fluctuation model, as shown in section 7.2. Only by varying the effective cross sections of absorber and amplifier in order to balance carefully the action of the amplification depletion with respect to the absorber saturation can this probability become a minimum.

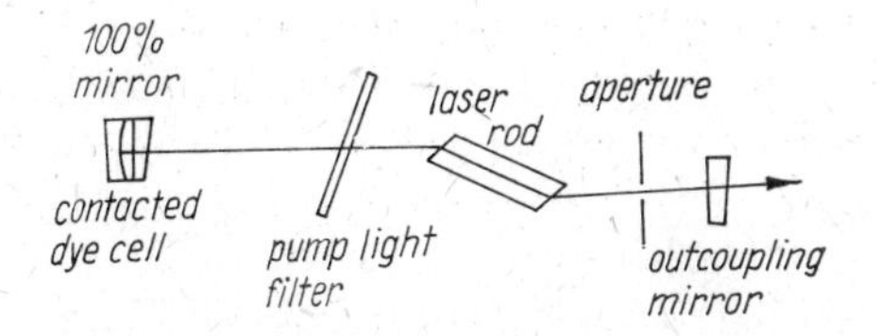

Fig. 7.5. Scheme of a resonator configuration for a passively modelocked solid state laser

A primary requirement for a modelocked system is the elimination of reflections that intracavity and extracavity components may produce. The reflection at optical surfaces arranged parallel to the resonator mirrors leads to sub-resonators, whereby the modelocking process is considerably disturbed, the main pulse broadened, and the formation of a large number of stochastically superimposed pulses is caused. By grinding the laser rod and inclining the dye cell at the Brewster angle or through dielectric antireflection coatings, such resonances can be avoided.

Another condition for a good reproducibility of the ps pulses is the suppression of higher transverse modes using an aperture diaphragm. Frequently, additional correction lenses are used to compensate for the thermal lenses that form in the active medium. Moreover, in the configuration in Fig. 7.5 a pump light filter is used that protects the dye from the possible photodisintegration due to the scattered light from the flashlamp.

The length of the resonator is usually chosen to be of the order of 1 m, and accordingly the pulse spacing is about 10 ns. The properties of the dye and the position of the dye cell inside the resonator are other important factors that influence the modelocking process. As in the passive modelocking of dye lasers, a short saturable absorber that is in contact with a resonator mirror provides the optimal condition for stable modelocking and the formation of the shortest pulses [7.17, 7.18]. According to the theory, in contrast to the dye laser, the relaxation time of the saturable absorber in a solid state laser must be smaller than the pulse duration or at least of the same order of magnitude. The absorption band of the absorber must lie in the frequency range of the stimulated emission of the active medium and have a larger halfwidth than the emission line. For Nd:glass and Nd:YAG lasers polymethine dyes are suitable as saturable absorbers. For the modelocking of the ruby laser, for example, cryptocyanine (dissolved in acetone) or DDI (dissolved in methanol or ethanol) can be used [7.20]. Reproducible and especially short pulses could be obtained using a mixture of DDI and rhodamine 6G dissolved in methanol [7.21]. Because of the possible transitions to the long-lived triplet state, it is necessary periodically to replace the dye with a fresh solution. This can be achieved by pumping the dye from a large reservoir through the cell, whereby a uniform mixture of the dye as well as a constant temperature and concentration are maintained over a long time. The thickness of the dye cell usually varies between 0.02 cm and 1 cm, where for an optimal utilization of the coherent effects caused by contacted absorbers it must be smaller than the optical length of the pulse. In order to protect the dye from photodisintegration, cell windows made of UV-absorbing glass can be used instead of pump light filters.

Another condition for achieving good modelocking is the careful adjusting of the pump intensity to a regime slightly above the second laser threshold. This requirement follows likewise directly from the mechanisms of modelocking according to the theoretical investigations in section 7.2. A higher pump intensity causes a rapid rise in the probability for the development of satellite pulses. The saturable absorber must possess here the lowest possible value of the small signal transmission (50 to 80%). In order to decrease the double pulse probability, it is often preferable to use a telescopic beam expansion inside the resonator to reduce the intensity in the dye relative to that in the amplifier. This is equivalent to increasing the effective ratio of the cross sections $\sigma_{13}^{b}q^{a}/(\sigma_{32}^{a}q^{b})$ (q^{a}, q^{b} — beam cross section of the pulses in the amplifier and absorber, respectively).

7.3.2 *Properties of the ps-Pulses of Passively Modelocked Solid State Lasers*

The main advantages of modelocked solid state lasers are that the maximum intensity and the energy of the pulses are considerably higher than in dye lasers. For example, in solid state lasers the pulse energy lies in the mJ range, whereas in dye lasers it is only some nJ. This occurs because the cross sections σ_{32}^{a} of the solid materials used are considerably smaller ($\sigma_{32}^{a} \simeq 10^{-20}$ cm^2) and therefore the saturation intensities considerably greater than in dyes. As opposed to these advantages there are several disadvantages, particularly in short-time spectroscopic applications, such as longer pulse durations, no direct tunability of the frequency and less reproducibility. Another important difference is that flashlamp pumped solid state lasers work in a nonsteady-state pulse regime, and therefore the emitted radiation of a modelocked solid state laser consists of a train of ultrashort pulses with a length of 50 ns to 200 ns (see Fig. 7.6). The pulse duration in the Nd : YAG laser varies between 20 and 40 ps, while the total energy of the pulse train runs from 1 mJ to 10 mJ, and the energy of a single pulse lies between 0.1 mJ and 1 mJ [7.22]. Nd : YAG lasers can usually be operated with a pulse repetition rate of several Hz, in extreme cases up to 100 Hz. Particularly short Nd:YAG pulses of 15 ps duration could be attained in [7.60] using a special ring laser arrangement in which counterpropagating pulses meet in the absorber, as in the dye ring lasers already described. In modelocked ruby lasers pulses having a duration of 15 to 30 ps are commonly generated. The energy per pulse in the center of the train is of the order of 0.1 mJ to 1 mJ.

The development and experimental investigation of modelocked Nd:glass lasers seems especially interesting because this laser material has a considerably greater linewidth than the other two materials. The reciprocal value of the line width is about 0.2 ps.

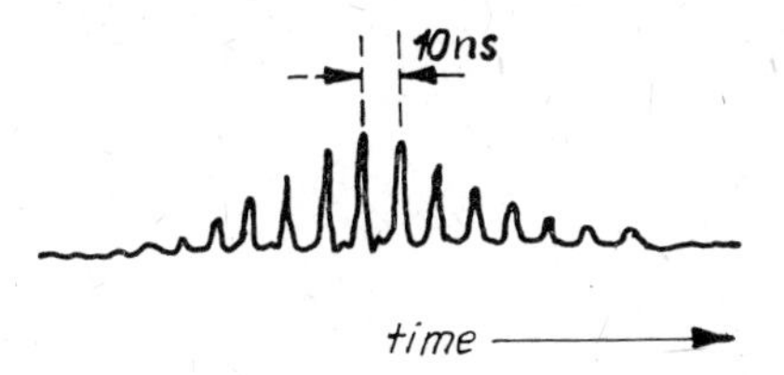

Fig. 7.6. Pulse train from a passively modelocked ruby laser (from [7.59])
Saturable absorber: DDI dissolved in methanol ($u = 10$ ns, $\tau_L \approx 20$ ps). The pulse train was recorded using a vacuum photodiode (rise time 0.5 ns) and the GHz oscilloscope I 2—7. The pulse duration was determined by means of two photon fluorescence (TPF).

Typical Nd:glass lasers generate pulses that have a duration of 2 to 20 ps, a pulse energy in the maximum of the train of 1 to 10 mJ and a halfwidth of the pulse train of 50 to 200 ns. A comparison between experimental results and the theoretical values calculated in section 7.2 for the pulse duration in Nd :glass lasers shows a satisfactory agreement only for pulses at the beginning of the pulse train. The pulse duration in the maximum of the pulse train is considerably greater than the one calculated theoretically, and the pulses possess a complicated temporal structure. Careful investigations of the temporal and spectral structure of the radiation emitted from modelocked Nd :glass lasers [7.14 to 7.18, 7.25 to 7.30] led essentially to the following explanation of the complicated structure of high-intense pulses in Nd:glass lasers. At the beginning of the pulse train the pulses have a duration of 2 to 5 ps and a spectral halfwidth that corresponds to the reciprocal pulse duration [7.16, 7.18] ($\Delta\nu_L\tau_L \approx 0.5$). TPF-measurements of the pulses provide a contrast ratio of 1:3, as expected in the case of good modelocking (compare 3.). Therefore, we adjust the pulse selection (compare section 7.3.3) after the laser resonator such that a single pulse is selected from the leading part of this train for further amplification and for the subsequent experiments. In the further evolution of the pulse train the spectral bandwidth of the pulses increases greatly, and the pulses exhibit distinctly formed substructures in the pulse spectrum as well as in the temporal intensity profile. The spectral broadening is caused by the spectrally inhomogeneous depletion of the amplification and the self-phase modulation of the radiation, which arise due to the nonlinear interaction of the intense radiation with the host material (glass) (compare 7.2.4). At relatively high laser intensities, changes in the refractive index, which depend on the intensity I_L of the pulse, $n = n_0 + n_2 I_L$, occur in the laser rod.

In this manner nonlinear effects arise, which lead, on the one hand, to a self-phase modulation and on the other hand to self-focussing of the radiation. The self-phase modulation produces a positive frequency sweep with increasing time and a spectral broadening, and may lead under certain conditions to a fissuring of the pulse into many components [7.27]. The self-focussing deflects portions of the radiation out of the resonator, whereby the pulse is similarly fissured and obtains a complicated temporal and spatial structure [7.29, 7.30].

It should be noted that in ruby and Nd:YAG lasers the effects linked to the self-phase modulation and self-focussing have a considerable influence only at intensities that lie an order of magnitude above the values usually achieved in the laser operation, whereby they have a smaller influence on the pulse structure.

7.3.3 *Single Pulse Selection and Amplification*

Instead of a train of pulses of various intensity and duration, in most experiments a single pulse with favorable parameters is required. For this purpose, an appropriate pulse is separated from the train and further amplified. Fig. 7.7 shows the experimental

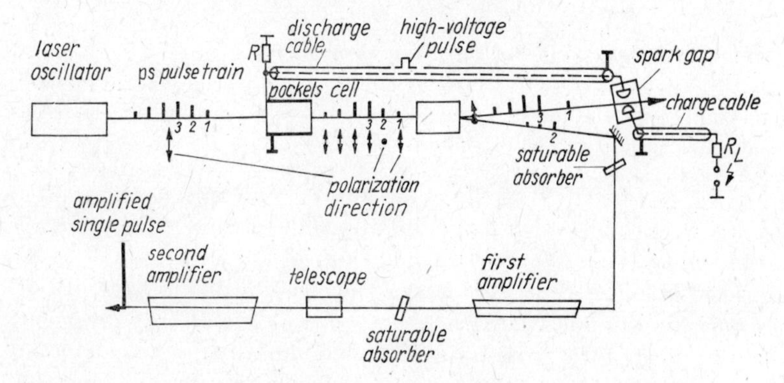

Fig. 7.7. Scheme for the selection and amplification of single pulses

arrangement for the selection and amplification of single pulses (see e.g. [7.16, 7.23, 7.36]). The laser oscillator emits a picosecond pulse train whose light is linearly polarized. This train passes through a Pockels cell and subsequently a polarizing beam splitter. Without supplying a voltage at the Pockels cell, the entire train is deflected at the polarizing beam splitter in the direction that corresponds to the polarization of the laser oscillator. If a high-voltage is switched on, the charging cable is charged and the voltage applied at the spark gap. If the discharge threshold is suitably set the pulse, designated by 1 in Fig. 7.7, that lies in the leading part of the train now causes an ignition of the spark gap, and consequently the high-voltage cable is discharged in a short time. In this manner a voltage pulse is applied at the Pockels cell for several nanoseconds, which at a suitable transit time influences precisely the polarization of pulse 2. The high-voltage is selected such that the polarization direction is rotated 90°. In order to avoid reflections at the cable end, the load resistance R must have the same magnitude as the characteristic impedance of the charge and discharge cables. A sufficiently large charging resistor R_L prevents a too rapid charging of the cable as well as a triggering of the spark gap due to the subsequent light pulses of the train. Instead of the spark gap discharge, a fast response photodetector can also be used for switching the Pockels cell in connection with high-voltage pulse electronics. By rotating the polarization of the second pulse, this pulse can be separated spatially from the remaining pulse train in the subsequent polarizing beam splitter. Before the selected single pulse reaches the amplifier, incompletely selected pulses — especially those at the maximum of the train — are supressed in a saturable absorber of small relaxation time and very low small signal transmission ($\lesssim 10^{-9}$), which altogether improves the pulse to background ratio. In general, the amplifier consists of the same active material as the oscillator. Through a suitable inclining of the amplifier rods or antireflection coatings optical feedback can be avoided. Saturable absorbers between the individual amplification stages again serve to improve the pulse to background ratio, to suppress weak feedback pulses and to shorten the pulses. Before the amplifier stages (in the figure before the second stage) the light beam must be suitably expanded to avoid undesired nonlinear optical effects — such as the self-phase modulation and self-focussing already mentioned.

In order to achieve a high pulse quality it is useful to begin with a pulse of low energy from the oscillator ($\lesssim 0.1$ mJ) and amplify it in several stages isolated from one another by saturable absorbers (see e.g. [7.37]). With two amplification stages pulses having an energy of about 10 mJ can thus be generated, which at a pulse length of 2 ps corresponds to a power of 5 GW. With suitable dimensioning of the amplification stages and the saturable absorbers arranged between them, whose relaxation time in this case must be greater than τ_L, a further pulse shortening can be achieved [7.38]. The reason for this effect is similar to those for the pulse shaping in the dye laser: Due to the depletion of the amplification the trailing edge is suppressed during the pulse passage, whereas the slowly relaxing saturable absorber attenuates mainly the leading edge. An arrangement for shortening the high power pulses through the action of multi-photon absorbers was discussed in [7.39] (compare 8.3).

7.3.4 *Investigations of the Evolution Process of Ultrashort Pulses*

An experimental verification of the selection mechanism in the evolution of ultrashort pulses due to the action of the nonlinear absorber was reported in [7.4, 7.31]. A direct investigation of the evolution process of ultrashort pulses and an immediate verification of the fluctuation mechanism in the pulse evolution was conducted by Kriukov et al. [7.30, 7.32]. In this experiment, the temporal intensity profile of the pulses from an Nd:glass ring laser was recorded at various moments of the pulse development using an electrooptical streak camera. Fig. 7.8 shows the temporal structure of the radiation 1200, 900, 600 and 300 ns before the maximum of the pulse train. The formation of a periodically recurring pulse out of the stochastic radiation fluctuations can be clearly seen from these recordings. In particular, it was found that in passively modelocked solid

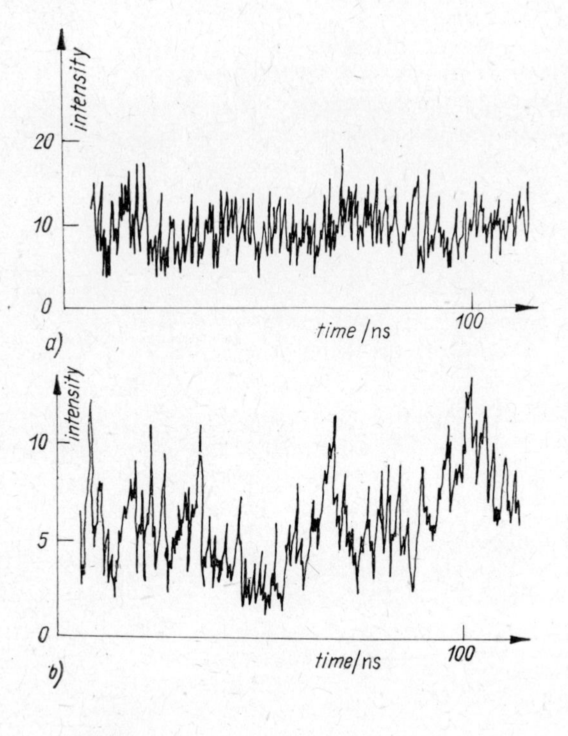

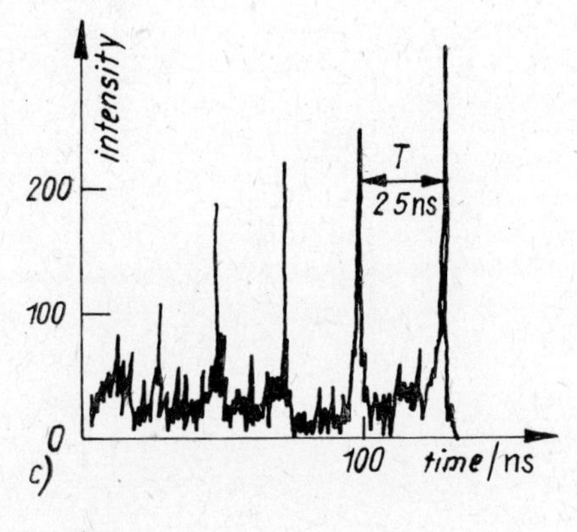

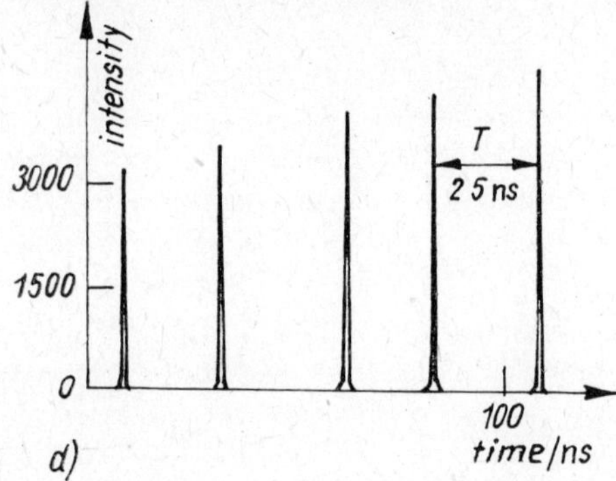

Fig. 7.8. Successive build up of an ultrashort pulse in a Nd:glass laser
The radiation was recorded a) 1200 ns, b) 900 ns, c) 600 ns, d) 300 ns before the peak of the pulse train.
(from Sakharov et al. [7.32])

state lasers two different thresholds exist. If the first threshold (laser threshold) is exceeded, the laser begins to oscillate in the free-running regime, which results in stochastically distributed intensity fluctuations. Only if the pump intensity exceeds a certain value above the first threshold, i.e. a second threshold (modelocking threshold), does a train of ultrashort pulses develop. In order to investigate more accurately the entire generation process, the dependence on time of the peak intensities of different fluctuation peaks was measured (Fig. 7.9). The hatched areas (1, 2, 3, 4) correspond here to the

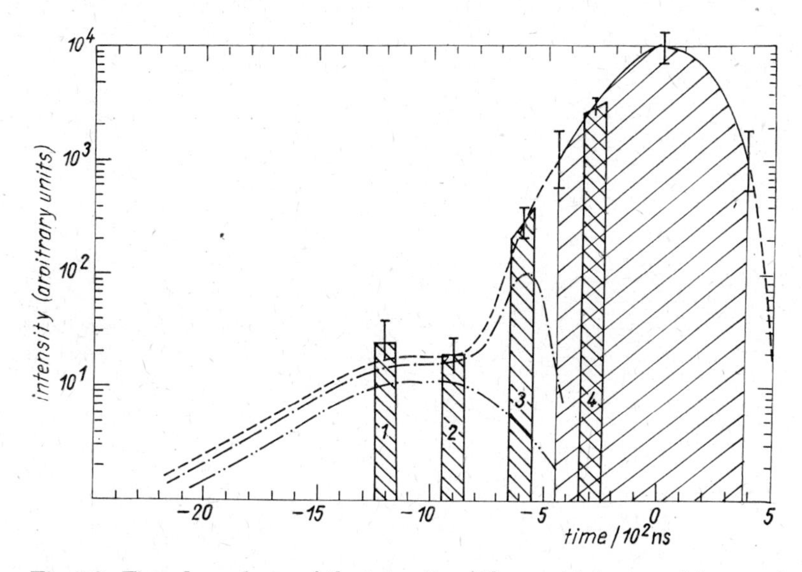

Fig. 7.9. Time dependence of the intensity of the most intense and two weaker fluctuations (from Sakharov et al. [7.32])

stages of the evolution processes (a, b, c, d) in Fig. 7.8. In the same figure the temporal evolution of pulses having lower intensities is also depicted. These pulses reach their maximum earlier than the maximum pulse and are then suppressed. Weaker pulses are absorbed even faster. The result is that in the last stages of the pulse evolution (in which the intensity of the main pulse is at a maximum) the intensity of the remaining pulses is at least 50 times smaller which means that the secondary pulses are again absorbed. This behavior of the pulses can be explained by the fact that between stages (a) and (b) in Fig. 7.8 the decrease of the occupation inversion due to laser radiation begins, itself, to play a role, as described in detail in section 7.3.2.3. Although the cross section of the amplifier is four orders of magnitude smaller than that of the absorber, the amplification depletion may exert an influence earlier than the absorber saturation for two reasons. First, the absorber is saturated by the intensity of a single pulse, whereas the amplifier is saturated by the total energy from the beginning of the laser oscillation. Secondly, the laser operates close to the threshold, so that the regime falls below the threshold due to only a small reduction of the amplification. For these reasons the selection of the most intense pulse is improved considerably. Since this pulse experiences a smaller loss from the absorber than the other pulses, it is further amplified despite the reduction of the amplification. The remaining pulses not only experience a smaller net amplification but, due to the decrease of the occupation inversion in the active medium, they undergo a net loss.

7.3.5 *Influence of the Effective Cross Sections of the Absorber and Amplifier*

The influence of the laser gain depletion on the pulse evolution process is determined largely by the ratio of the effective cross sections of the amplifier and absorber $\sigma_{32}^{a}/\sigma_{13}^{b}$. A ratio that is too small should lead theoretically to a higher double pulse probability and a smaller breakdown probability, since in this case the amplification depletion has no influence. At too large a ratio, however, the breakdown probability greatly increases, so that no pulses exceed the critical intensity level β_D, that is, all pulses are again damped, and the system remains in the state of free-running oscillations.

The influence that the effective cross sections have on the evolution process of ultrashort pulses in a Nd:YAG laser were studied experimentally in [7.30, 7.33]. Because the cross sections of the absorber and amplifier cannot be altered easily, a telescope was used to vary the ratio of the geometric cross sections q^b/q^a in the absorber and amplifier, which leads to the same effect, since the modelocking depends on the parameter $\sigma_{32}^{a}q^b/(\sigma_{13}^{b}q^a)$. Using the telescope the beam cross section could be adjusted with the ratios 4:1, 2:1, 1:1. Here again the pulse development was studied using a streak camera.

At a ratio of the cross sections of 4:1, no regime of ultrashort pulses developed even at higher pump energies, but rather only a regime of free-running oscillations. The number of fluctuations increases with increasing pump energy. Thus, the increase of the effective cross section of the amplifier leads in this case to a critical intensity level β so high that at high probability no fluctuation will exceed it.

At a ratio of the beam cross sections of 1:1 a train of ultrashort pulses was generated in the laser. However the main pulse was accompanied by satellites whose intensity amounted to more than 1% of the main pulse. A second threshold could not be observed here, although the pump energy was controlled with an accuracy of better than 1%. According to the theory in section 7.2.3.2, this behavior can be explained by the fact that the action of the amplification depletion is too small. The critical intensity level β_D is now so small that with a probability of nearly 1 the maximum pulse peak exceeds this level even at pump energies immediately above the first threshold. For this reason, however, the selecting action of the amplification depletion also decreases, and accordingly the double pulse probability increases.

At a beam cross section ratio of 2:1 finally only satellites of very low intensity were observed. In this case two thresholds could be clearly distinguished. Evidently, the action of the amplification depletion here leads to a favorable regime that is distinguished by a small probability for the breakdown of the pulse development as well as by a small double pulse probability. In this manner, the critical influence of the ratio of the cross sections on the modelocking process and the favorable action of gain depletion with respect to the main pulse selection could be directly demonstrated by experiment. The regime with a maximum separation of the two thresholds proves here to be optimal for a high reproducibility of the ps pulses and for a high contrast. Such a regime can only be achieved if the losses for the TEM_{00}-mode are considerably lower than those for the higher transverse modes.

7.3.6 *The Influence of the Relaxation Time of the Absorber on the Formation of Ultrashort Pulses*

An important result of the fluctuation model is that the selection of a main pulse is greatly reduced, if the relaxation time of the saturable absorber is greater than the spacing between adjacent fluctuations (which lies in the order of the mean duration of a fluctuation given by the reciprocal emission bandwidth):

$$\tau_{\mathrm{Fl}} \approx \frac{1}{\varDelta\nu_{32}^{\mathrm{a}}},$$

where $\varDelta\nu_{32}^{\mathrm{a}}$ is the emission linewidth of the laser. For a verification of this assertion the relation between the relaxation time of the absorber T_{21}^{b}, the duration of a fluctuation and the properties of the emitted radiation were investigated experimentally using a ruby laser in [7.30, 7.34].

In order to observe directly the temporal structure of the radiation and the pulse duration an absorber (vanadiumphthalocyanine dissolved in nitrobenzol) of relatively long relaxation time $T_{21}^{\mathrm{b}} = 1.2 \pm 0.6$ ns was used. Since it is difficult in experiment to change the absorber recovery time T_{21}^{b} arbitrarily the effective bandwidth of the radiation $\varDelta\nu_{\mathrm{eff}}$ in the resonator was changed, which became possible after introducing in the resonator a frequency selective element in the form of plane-parallel plates of variable thickness. In this way τ_{Fl} could be varied. Three ranges were investigated:

$$(1)\quad \tau_{\mathrm{Fl}} = \frac{1}{\varDelta\nu_{\mathrm{eff}}^{\mathrm{a}}} \ll T_{21}^{\mathrm{b}}:$$

Here $\varDelta\nu_{\mathrm{eff}}^{\mathrm{a}}/c = 0.1 \cdots 0.5\ \mathrm{cm}^{-1}$ was chosen. The radiation consists of a superposition of very many pulses, i.e., the absorber is not able to select a single pulse.

$$(2)\quad \tau_{\mathrm{Fl}} = \frac{1}{\varDelta\nu_{\mathrm{eff}}^{\mathrm{a}}} \lessapprox T_{21}^{\mathrm{b}}:$$

In this case the radiation spectrum is narrowed to a range of $\varDelta\nu_{\mathrm{eff}}^{\mathrm{a}}/c = 0.04$—$0.08\ \mathrm{cm}^{-1}$. The result of a streak camera recording is shown in Fig. 7.10a. A pulse having a duration of 1 ns, and possessing a temporal substructure, develops. The trailing edge of the pulse has an exponential shape.

$$(3)\quad \tau_{\mathrm{Fl}} = \frac{1}{\varDelta\nu_{\mathrm{eff}}^{\mathrm{a}}} > T_{21}^{\mathrm{b}}:$$

In this case, the spectrum is narrower than $0.02\ \mathrm{cm}^{-1}$, and the stability reaches its maximum. The pulse has no temporal substructure (see Fig. 7.10b), and its duration is 2 ns. Usually, the leading edge is not as steep as the trailing edge.

The influence of the absorber recovery time on the pulse generation was investigated in [7.35] by comparing the temporal profiles of the pulses resulting from the use of two different absorbers (DDCI and DTDCI) with different relaxation times (13 ps and 120 ps). The pulse duration of the pulses generated depends heavily on the absorber used and comes to 15 to 30 ps with DDCI as absorber and 90 to 105 ps with DTDCI. This result also confirms the theoretical predictions of the fluctuation model, according to which the lower limit for the pulse duration is given by the relaxation time of the absorber.

Kolmeder and Zinth [7.42] also conducted theoretical and experimental investigations into the dependence of the pulse durations on various laser parameters particularly the relaxation time of the absorber. They employed three different absorbers for the modelocking of a Nd:glass laser. Using the Kodak dye no. 9860 (dissolved in 1–2 dichlorethane) with a relaxation time of 7 ps, they attained pulse lengths of about 3 ps,

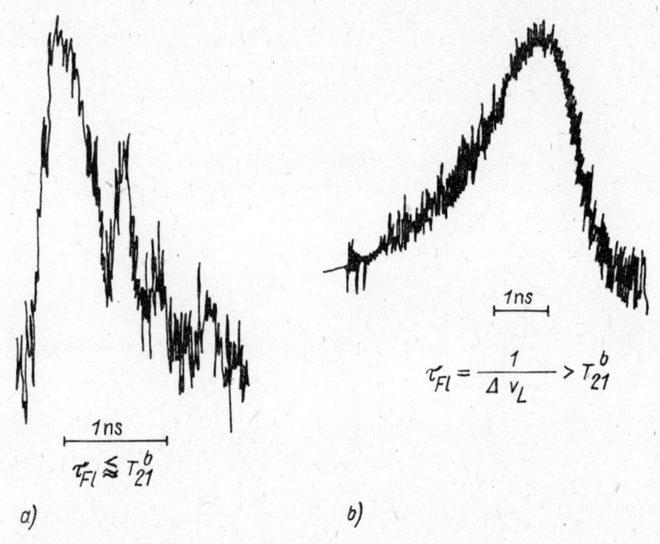

Fig. 7.10. Microdensitometer trace of streak camera records that illustrate the influence of the absorber relaxation time T_{21}^{b} on the pulse shape
a) $\tau_{\mathrm{Fl}} \lessapprox T_{21}^{\mathrm{b}}$, b) $\tau_{\mathrm{Fl}} > T_{21}^{\mathrm{b}}$
(from Kriukov et al. [7.34])

whereas the same dye dissolved in 1–2 dichlorethane with tetrabutylammonium iodide which has a relaxation time of 4.7 ps yielded 2.5 ps pulses. The shortest pulses, having a duration of 1.7 ps, occur using perylium dye no. 5, which has a relaxation time of 2.7 ps. The numerical calculation of a theoretical model, in which the temporal behavior of the absorber and amplification depletion were considered, provided corresponding results.

7.4 Semiconductor Lasers

As already pointed out in chapter 2, the semiconductor diode lasers are of great interest, particularly because they offer important advantages for many applications due to their small dimensions and their low power consumption. The generation of ultrashort pulses plays an important part here in many possible applications — such as in fast transmission and processing of information. In contrast to other types of lasers, even without applying modelocking very short pulses can be generated with laser diodes in the regime of self-excited pulsations [7.45] and by means of gain or Q-switching [7.46, 7.47, 7.48]. The switching of resonator Q or gain need not occur with a fixed given repetition frequency as in active modelocking but can, for instance, consist of a single switching operation, which leads to the formation of a short single pulse (compare 2.5). The minimum obtainable pulse duration is related here to the transient response of the resonator (response time τ_R), which amounts to at least several resonator round trips $2L/c$. Using typical laser diodes we can build up short resonators ($L \simeq 1$ mm) with round trip times in the order of 10^{-11} s, which makes it possible to generate very short single pulses. With short injection current pulses laser pulses of $\tau_L \approx 50$ ps duration can be achieved. With special resonator arrangements and optical gain switching it is possible to advance even into the picosecond and subpicosecond range. In this way a resonator with an optical length of only 12 µm corresponding to a round trip time of 0.08 ps was built up in [7.49].

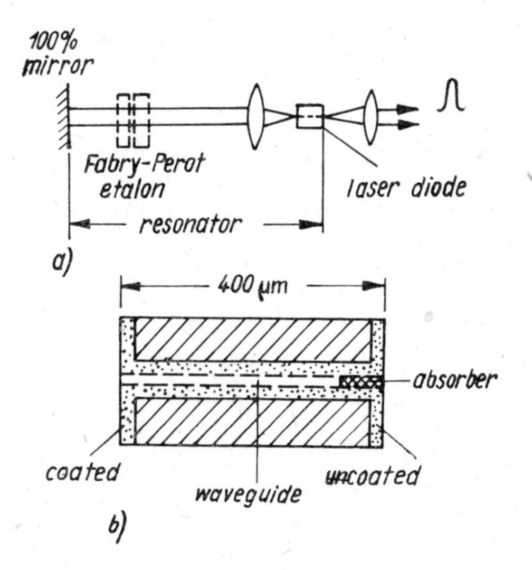

Fig. 7.11. Passively modelocked semiconductor diode laser according to [7.52] and [7.53]
a) Arrangement with the laser diode in an external resonator
b) Plan view of the laser diode
Shown here is a modified strip buried heterostructure GaAlAs diode (see e.g. [7.54]).

Here the active semiconductor material, a GaAs layer, had a thickness of merely 1 to 2 µm. This device was pumped by a single pulse of a modelocked dye laser ($\tau_L = 1$ ps). By means of this extremely fast gain switching, pulses having a duration of somewhat less than 1 ps could be generated. The generation of such short pulses using this method, however, seems to be possible at present only by pumping with modelocked lasers.

As evident from this presentation, the pulse lengths achievable in this way are restricted to the range above the resonator round trip time. As with other types of lasers, shorter pulses can only be generated through modelocking. In section 5.3.1, it was briefly pointed out that it is possible to modelock actively [7.50, 7.51] or pump synchronously semiconductor lasers by means of an injection current modulation whose frequency coincides with the reciprocal resonator round trip time. As with other laser types described, extremely short pulses are also obtained by means of passive modelocking. Passive modelocking in semiconductor lasers was first observed by Ippen, Eilenberger and Dixon [7.52]. The shortest pulses attained up to now were measured in [7.53]. In both papers similar arrangements were used, the schemes of which are depicted in Fig. 7.11. The laser diode is operated as an active element in an external resonator. The outcoupling mirror of the external laser resonator is formed by one of the laser diode facets. From the other, well antireflection-coated facet of the diode the radiation passes through a microscope objective with a large numerical aperture and from there to the 100% mirror. The laser resonator has a length of about 10 cm, which corresponds to a pulse repetition frequency in the GHz range. A Fabry-Perot etalon can be placed in the beam path to narrow the bandwidth. The diode is continuously (cw) pumped using a constant injection current. In the case of GaAs- and GaAlAs-diodes the active material can act simultaneously as saturable absorber due to the presence of defects, especially in unpumped regions [7.52]. In [7.53] the saturable absorption of the semiconductor material was increased through the implantation of ions (600 keV protons at a flux density of 3×10^{15} cm^{-2}). In this manner the saturable absorber can be concentrated in a small region at the outcoupling mirror or output facet. As already described in chapter 6, the saturable absorber in contact with a resonator mirror ensures a particularly favorable effect. (Note, however, that the optimum is only achieved when the corresponding mirror is nearly 100% reflective which, due to the arrangement, is not the case here.) The pulse duration is determined from the measurement of the autocorrelation function by generating the second harmonic in $LiIO_3$-crystals, where in [7.53] $\tau_L = 0.65$ ps was obtained. The spectral width was determined at $\Delta\nu = 5.2 \times 10^{11}$ Hz (according to $\Delta\lambda = 1.2$ nm). Using these results the pulse-duration-bandwidth-product was found to be $\Delta\nu \cdot \tau_L = 0.34$, which is near to the value for bandwidth-limited sech^2-pulses ($\Delta\nu \cdot \tau_L = 0.315$). The shape of the autocorrelation function is also in agreement with a pulse profile of this kind, which also occurs approximately in passively modelocked dye lasers (compare 6.).

Very high stability of modelocked laser operation was achieved in [7.61] by using a GaAs/GaAlAs multiple-quantum well (MQW) structure (compare 2.4.4) as external saturable absorber, which consisted of 47 periods of 9.8 nm GaAs layers alternated with 9.9 nm of $Ga_{0.71}Al_{0.29}As$ layers grown by molecular-beam epitaxy. The commercial GaAs laser diode Hitachi HLP-1400 modified by an antireflection coating on one facet served as the active element. The MQW structure provides the advantage of very low saturation intensity. It is known that the optical saturation of the excitonic absorption is caused by the screening effect of the laser-created carriers and that this saturation takes place at intensities more than one order of magnitude lower than those required to saturate the band-to-band transition. Thus the effective cross section of saturable absorption is large compared to that of the amplifier ($\sigma^b \gg \sigma^a$). This is of advantage for stable moedelocking with homogeneously broadened laser transitions (compare 7.3.5).

8. Nonstationary Nonlinear Optical Processes

The modelocked lasers described in chapters 4. to 7. can be employed as primary light sources for the generation of ultrashort excitations and suitable probe pulses. For many applications, however, light pulses are needed that differ in their parameters from the original laser pulses, for instance with respect to the wavelength or pulse duration.

Since in many cases in the excitation, as well as for the measuring, resonant interactions should take place between the light and the sample, one must be able to choose pulses of appropriate wavelength. In many lasers — e.g. in the ruby laser, the Nd :glass and Nd : YAG laser and the gas lasers — the emission wavelengths are, however, restricted within narrow limits. On the other hand, due to the broad fluorescence profiles of suitable organic molecules the dye lasers can be tuned over a large wavelength range of about 100 nm. Through the selection of several dyes, which can be used interchangeably as active material, the entire visible spectral range can be covered (compare 2.). For the exciting of electronic, vibrational and rotational levels of different substances, however, excitation radiation from the ultraviolet to the infrared spectral range is required. For this purpose various methods of frequency conversion are applied, in which a pulse of the mid-frequency ω_0 generates one having the changed mid-frequency ω. A special conversion process has already been described when we dealt with the pulse generation in synchronously pumped dye lasers. The frequency conversion between primary and secondary radiation occurs here by means of a two quanta process, which proceeds in temporally successive stages: after the absorption of a high-energy photon a low-energy photon is emitted and the difference in energy passes in the form of heat into the environment of the fluorescent molecules. Simultaneously, the pulse duration is considerably decreased.

Using methods of nonlinear optics [11] a variety of different frequency conversions can be carried out, which have also been applied successfully with ultrashort light pulses. The basic principle of nonlinear optical frequency conversion should first be explained before we describe in the following sections specific methods, such as the generation of harmonics, of waves with sum and difference frequencies and of Raman and parametric signals.

Included in the field of "nonlinear optics" are all electromagnetic phenomena at high frequencies ($\nu \gtrsim 10^{11}$ Hz), which are due to the nonlinearity of the material equations in the Maxwell theory of electromagnetic waves. These nonlinearities may lead to the generation of harmonic waves and frequency mixing in the optical range as in the well-known processes in the radio wave range. With strong electric fields, which can be produced in a material by high-power lasers, one must generally account for nonlinear dependences of the induced atomic dipole moments on the electric field $\vec{E}$ and thus of the polarization $\vec{P}$ on $\vec{E}$. As long as these fields still lie well below the damage threshold of the materials, the general nonlinear relation $\vec{P}[\vec{E}]$ can be expanded in a rapidly converging series:

$$\vec{P} = \sum_{n=1} \vec{P}^{(n)}, \tag{8.1}$$

where $\vec{P}^{(n)}$ is related to the nth power of the electric field. Symbolically we can write

$$\vec{P}^{(n)} = \varkappa^{(n)}(\vec{E})^n. \tag{8.2}$$

It should be noted that $\varkappa^{(n)}$ in this relation is generally a tensor of the $(n+1)^{\text{th}}$ order, which relates the vector components of the field to those of $\vec{P}$. With respect to the influence on the field strength in the time domain $\varkappa^{(n)}$ represents, in general, an integral operator by which the polarization at time t can be calculated using an n-fold integral from $-\infty$ to t over the n time coordinates of the field strength product. The time interval before the moment t in which the field strength contributes substantially to the polarization, hence the memory time of the material, depends on the frequency of the incident light as well as the transition frequencies and relaxation times of the atomic systems. If the frequencies of the incident light and their sums and differences are relatively far away from the resonance frequencies of the material, then the memory effect can be neglected. Under these conditions, the temporally slowly varying polarization amplitude $\overline{P}_1^{(n)}(t, \vec{r})$ of the nth order in the frequency range around the mid-frequency ω_1 is given approximately by:

$$\overline{P}_1^{(n)}(t, \vec{r})\, e_\alpha^{(1)\prime}\, e^{-i\vec{k}_1[\alpha]\vec{r}} = \varepsilon_0 \sum_{\beta,\ldots,\mu} \chi^{(n)}_{\alpha\beta\ldots\mu}(\omega_1; \omega_1, \ldots, \omega_n)\, e_\beta^{(1)} \cdots e_\mu^{(n)}$$
$$\times A_1(t, \vec{r}) \cdots A_n(t, \vec{r})\, e^{-i(\vec{k}_1[\beta] + \cdots + \vec{k}_n[\mu])\vec{r}}$$

where

$$\omega_1 = \sum_{m=1}^{n} \omega_m. \tag{8.3}$$

$A_m(t, \vec{r})$ is the slowly varying amplitude of the electric field component in the frequency range around ω_m; $\vec{e}^{(m)}$ and $\vec{e}^{(1)\prime}$, are the polarization unit vectors of the field component at ω_m and the polarization component at ω_1, respectively, whose vector components are designated by small Greak letters; $\vec{k}_m^{[\gamma]}$ represents the wave number vector for the vector component γ of the field at the frequency ω_m. The quantity $\chi^{(n)}$ represents a nonlinear susceptibility of the nth order whose argument contains the mid-frequencies of the emerging polarization and those of the generating fields. (From this formula the approximation described above regarding a memory-free media becomes clear: We use the nonlinear susceptibility at the $(n+1)$ mid-frequencies $\omega_1, \omega_1, \ldots, \omega_m$ to join fields whose Fourier components lie in small ranges of the order of $1/\tau_L$ around these frequencies; thus, for nonresonant, nonlinear interaction processes we neglect the dispersion of the nonlinear susceptibilities within these frequency bands.)

The convergence in the series expansion according to (8.1) and in the corresponding approximate relations is determined by an expansion parameter ζ that is equal to the quotient from the laser field strength and the strength of the intraatomic electric field

($\simeq 10^{11}$ V/m), which for example is given in a hydrogen atom by the field of the proton at the position of the electron. For the most part, we do not know the general relation $\vec{P}[\vec{E}]$ at all, but determine the lowest orders of the given expansion with a perturbation calculation — beginning at $\vec{P}^{(1)}$. (A perturbation calculation of this kind can, for example, be carried out on the basis of the semi-classical treatment of light-matter interaction, as described in section 1.3.) All phenomena of common optics, i.e., linear optics can be explained by taking into account the relation between $\vec{P}$ and $\vec{E}$ in the first order. The large range of validity of the linear theory in nearly all experiments conducted with classical light sources results from the smallness of the expansion parameter ζ under these conditions.

We now insert the calculated nonlinear optical polarization in the wave equation as a source term. Under the conditions discussed in chapter 1, we can immediately use the corresponding nonlinear polarization amplitude $\bar{P}^{(n)}$ from (8.3) in the reduced wave equation (1.50) for the slowly varying field amplitude at frequency ω_l. Since we restrict ourselves to nonresonant interactions, the term $\bar{P}_l^{\mathrm{LR}}(t, z)$ in (1.50) disappears and we obtain

$$\left(\frac{\partial}{\partial z} + \frac{1}{v_l}\frac{\partial}{\partial t}\right) A_l(t, z) = \sum_{\alpha,\beta,\ldots,\mu} \delta^{(n)}_{\alpha\beta\ldots\mu}(\omega_l; \omega_1, \ldots, \omega_n) \times A_1(t, z) \cdots A_n(t, z)\, \mathrm{e}^{-\mathrm{i}(k_1^{[\beta]} + \cdots + k_n^{[\mu]} - k_l^{[\alpha]})z} \tag{8.4a}$$

with the effective susceptibility tensor

$$\delta^{(n)}_{\alpha\beta\ldots\mu}(\omega_l; \omega_1, \ldots, \omega_n) = -\frac{\mathrm{i}\omega_l^2}{2c^2 k_l^{[\alpha]}}\, e_\alpha^{(l)'} \chi^{(n)}_{\alpha\beta\ldots\mu}(\omega_l; \omega_1, \ldots, \omega_n)\, e_\beta^{(1)} \cdots e_\mu^{(n)}\,. \tag{8.4b}$$

The relation (8.4) can be further simplified, if we assume that for every frequency component of the field there is only one value of the wave number. This condition is always satisfied in isotropic media; in crystals, however, it is only satisfied in experimental arrangements where the electromagnetic field at a certain frequency either propagates only as an ordinary or only as an extraordinary wave with the wave numbers k_j^{o} or k_j^{e}. (Due to the uniqueness of the relationship between polarization direction and frequency it suffices then to denote the wave number only by means of the frequency.) The summation in (8.4) now refers only to the susceptibility and polarization unit vectors and we can write

$$\left(\frac{\partial}{\partial z} + \frac{1}{v_l}\frac{\partial}{\partial t}\right) A_l(t, z) = \Delta^{(n)}(\omega_l; \omega_1, \ldots, \omega_n)\, A_1(t, z) \cdots A_n(t, z)\, e^{-\mathrm{i}(k_1 + \cdots + k_n - k_l)z} \tag{8.5a}$$

with the effective susceptibility

$$\Delta^{(n)}(\omega_l; \omega_1, \ldots, \omega_n) = \frac{-\mathrm{i}\omega_l^2}{2c^2 k_l} \sum_{\alpha,\beta,\ldots,\mu} e_\alpha^{(l)'} \chi^{(n)}_{\alpha\beta\ldots\mu}(\omega_l; \omega_1, \ldots, \omega_n)\, e_\beta^{(1)} \cdots e_\mu^{(n)}\,. \tag{8.5b}$$

From (8.4), (8.5) it is obvious that in a nonlinear optical material a field at frequency ω_l can build up from fields at other frequencies $\omega_1, \ldots, \omega_n$ where $\omega_l = \omega_1 + \cdots + \omega_n$, which is impossible in linear optics. Furthermore, it is obvious that such a process will occur with particular efficiency, if the period of the spatially oscillating factor $\exp\left[-\mathrm{i}(k_1^{[\beta]} + \cdots + k_n^{[\mu]} - k_l^{[\alpha]})\, z\right]$ on the right side of (8.4a) approaches infinity, because then the amplitude A_1 changes monotonically over long distances. This leads to the condition

$$\Delta k = k_1^{[\alpha]} - (k_1^{[\beta]} + \cdots + k_n^{[\mu]}) = 0\,. \tag{8.6a}$$

If not all waves propagate in the z-direction, as assumed here, then instead of (8.6a) we have the vector relation

$$\Delta \vec{k} = \vec{k}_1^{[\alpha]} - (\vec{k}_1^{[\beta]} + \cdots + \vec{k}_n^{[\mu]}) = 0\,. \tag{8.6b}$$

The relations (8.6a, b) are designated as phase matching relations. The satisfying of these relations means that the nonlinear polarization and the electric field have the same wave number at the frequency ω_1. Thus, a certain phase relation between both quantities remains over a long distance, from which again a monotonic increase or drop of a certain wave follows when passing through the nonlinear material. We shall come back to this discussion in greater detail in our treatment of second harmonic generation.

After the following description of some frequency conversion techniques we will briefly discuss methods of influencing the pulse shape, particularly with respect to pulse shortening in the last section of this chapter.

8.1 Generation of the Second Harmonic

Only shortly after the first lasers were constructed, second harmonic generation (SHG) was discovered by P. A. Franken et al. [8.1]. In this experiment the light of a ruby laser was incident on a quartz crystal. At the output of the crystal, in addition to the laser light at $\lambda_1 = 694$ nm, radiation at half the wavelength ($\lambda_2 = 347$ nm) could be detected. Beginning with this experiment the experimental techniques of nonlinear optics evolved. Second harmonic generation, especially, gained considerable practical importance, because it provides a simple method of converting light into light of shorter wavelength.

The generation of the second harmonic is caused by the nonlinear polarization of the second order, which is that component of the polarization that varies with the square of the electric field. We can describe this process using the general relation (8.4) with suitable modification, where a field with the frequency ω_2 is built up from fields at the frequency ω_1 where $\omega_2 = \omega_1 + \omega_1$.

For an effective generation of the second harmonic it is necessary according to (8.6) to satisfy the phase matching condition $\Delta k = 0$. For this purpose the birefringence in the nonlinear optical crystal can be utilized. In uniaxial birefringent crystals the refractive index $n^e(\theta, \omega)$ for the wave of extraordinary polarization, which propagates at an angle θ to the optical crystal axis, is given by

$$\frac{1}{n^e(\theta, \omega)^2} = \frac{\cos^2\theta}{[n^o(\omega)]^2} + \frac{\sin^2\theta}{[n^e(\omega)]^2}, \tag{8.7}$$

where $n^e(\omega)$ is the refractive index of an extraordinary wave, which propagates perpendicularly to the optical crystal axis. In Fig. 8.1, $n^e(\theta, \omega_1)$, $n^o(\omega_1)$ as well as $n^e(\theta, 2\omega_1)$ and $n^o(2\omega_1)$ are depicted for a negatively birefringent crystal. There are two possible ways of achieving phase matching. In type I phase matching, pump light propagates in negatively birefringent crystals as an ordinary wave, where it generates a second harmonic wave of extraordinary polarization. To satisfy $\Delta k = 0$ we obtain

$$n^{e}(\theta, 2\omega_1) = n^{o}(\omega_1). \tag{8.8a}$$

In positively birefringent crystals ($n^e > n^o$), however,

$$n^{o}(2\omega_1) = n^{e}(\theta, \omega_1) \tag{8.8b}$$

must be satisfied. In type II phase matching pump light that contains a component with ordinary as well as one with extraordinary polarization is used. The second harmonic develops by coupling both components and exhibits in negatively and positively

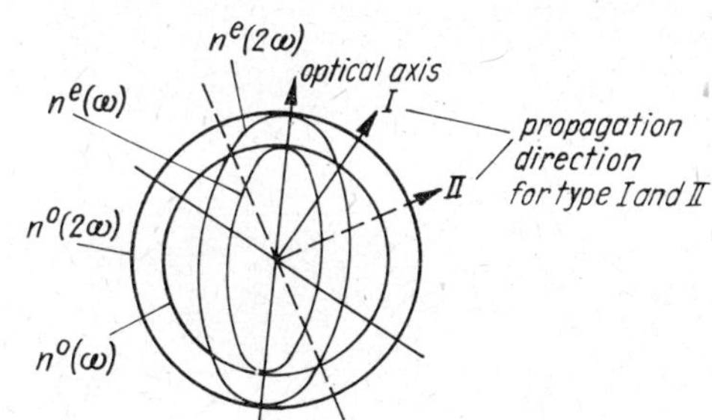

Fig. 8.1. Refractive index for the ordinarily and extraordinarily polarized wave in a negatively birefringent crystal

birefringent crystals extraordinary and ordinary polarization, respectively. The phase matching condition $\Delta k = 0$ leads to

$$n^{e}(\theta, 2\omega_1) = \frac{1}{2}\,[n^{e}(\theta, \omega_1) + n^{o}(\omega_1)] \quad \text{for} \quad n^e < n^o \tag{8.8c}$$

and

$$n^{o}(2\omega_1) = \frac{1}{2}\,[n^{e}(\theta, \omega_1) + n^{o}(\omega_1)] \quad \text{for} \quad n^e > n^o. \tag{8.8d}$$

In the further treatment let us restrict ourselves to type I phase matching. In this case, we can begin with the simplified basic relation (8.5) and obtain the following equations for the temporal and spatial change of the fundamental wave A_1 and its harmonic A_2:

$$\left(\frac{\partial}{\partial z} + \frac{1}{v_1}\frac{\partial}{\partial t}\right) A_1(t, z) = \Delta^{(2)}(\omega_1; 2\omega_1, -\omega_1)\, A_2(t, z)\, A_1^*(t, z)\, e^{-i\Delta kz}, \tag{8.9a}$$

$$\left(\frac{\partial}{\partial z} + \frac{1}{v_2}\frac{\partial}{\partial t}\right) A_2(t, z) = \Delta^{(2)}(2\omega_1; \omega_1, \omega_1)\, [A_1(t, z)]^2\, e^{i\Delta kz} \tag{8.9b}$$

where $\Delta k = k_2 - 2k_1$.

The solution of the differential equations (8.9a) and (8.9b) permits a complete treatment of the generation of the second harmonic from coherent monochromatic laser radiation, accounting for the weakening of the fundamental wave. Let us consider only the case of small conversion rates in which the spatial change of the fundamental wave can be neglected and the problem reduced to the integration of (8.9b). If the amplitude of the second harmonic at the input of the crystal vanishes at $z = 0$, then after introducing the new variables $\eta = t - z/v_2$ and $z = \xi$ (where we write z instead ξ in the following) we obtain the solution of (8.9b) by simple integration:

$$A_2\left(t - \frac{z}{v_2}, z\right) = \Delta^{(2)}(2\omega_1; \omega_1, \omega_1) \int\limits_0^z dz'\, A_1{}^2 \left(t - \frac{z}{v_2} + D\frac{z'}{c}\right) e^{i\Delta k z'}, \tag{8.10}$$

where $D = c\left(\frac{1}{v_2} - \frac{1}{v_1}\right)$ is the dispersion parameter, which is proportional to the difference of the reciprocal group velocities of the fundamental wave and the harmonic.

Let us discuss the solution (8.10) first for the case of a monochromatic fundamental wave from a stationary radiation source or of a very long pulse of duration $\tau_L \gg Dz/c$ (z — path in the nonlinear crystal through which the light passes). In this case we can replace $D = 0$ in (8.10) and obtain for the absolute value of the amplitude

$$|A_2(t, z)|^2 = |\Delta^{(2)}(2\omega_1; \omega_1, \omega_1)|^2 \frac{\sin^2 \frac{\Delta k z}{2}}{\left(\frac{\Delta k}{2}\right)^2} |A_1(t, z)|^4 . \tag{8.11}$$

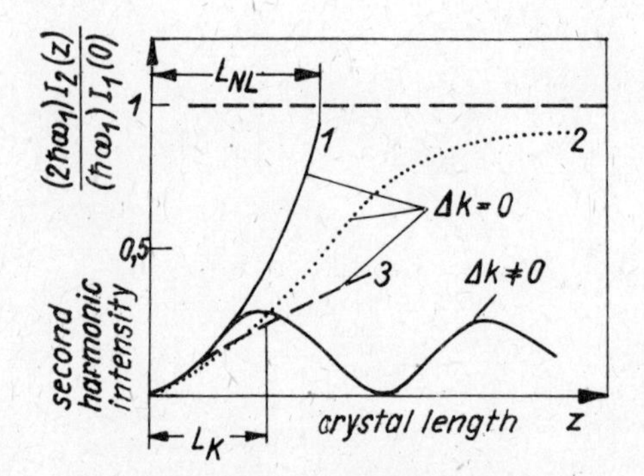

Fig. 8.2. Intensity of the second harmonic as a function of the traversed crystal length

Accordingly, the intensity of the second harmonic varies with the square of the intensity of the fundamental wave. For $\Delta k = 0$, I_2 increases with the square of the crystal length z that the light passes through. (This dependence, of course, can only hold until the requirement of small conversion rates is violated.) The condition $\Delta k = 0$ means that the waves of the nonlinear polarization and the electric field at the frequency $2\omega_1$ propagate with equal phase velocities, so that between them a fixed phase relation can be maintained. For $\Delta k \neq 0$, the intensity of the second harmonic varies periodically with z (see Fig. 8.2); it increases to its maximum after a length $L_K = (\pi/\Delta k)$, which is called the phase coherence length. Subsequently, due to the altered phase relations between the polarization and the electric field the sign of the amplitude change reverses, so that energy from the harmonic is pumped back into the fundamental wave. Beyond the length $2L_K$ the intensity of the harmonic falls again to zero. For comparison, the increase of the intensity $(2\hbar\omega_1)\, I_2$ for $\Delta k = 0$ is also plotted in figure 8.2 (curve 1). This function, increasing monotonically with z^2, reaches the value of the input intensity of the fundamental wave $(\hbar\omega_1)\, I_1(z = 0)$ at $z = L_{NL}$. The nonlinear conversion length L_{NL} is given here by

$$L_{\mathrm{NL}} = \frac{1}{|\Delta^{(2)}(2\omega_1; \omega_1, \omega_1) \, A_1(z = 0)|}. \tag{8.12}$$

As already mentioned, the attenuation of the fundamental wave must, of course, be considered before these high intensity values are achieved; the corresponding curve (calculated exactly) increases likewise monotonically with z, but not as rapidly as z^2 and reaches the value $(\hbar\omega_1) \, I_1(z = 0)$ only asymptotically (curve 2 in figure 8.2) — compare, for example, [21, 22]. At $z = L_{\mathrm{NL}}$ the intensity of the second harmonic is 0.58 $(\hbar\omega_1) \, I_1(z = 0)$.

In the generation of the second harmonic with ultrashort light pulses further complications occur in contrast to the excitation with monochromatic light. In particular, the phase matching condition $\Delta k = 0$ can still be precisely fulfilled for the mid-frequencies of the pulses, but no longer for the total frequency spectrum. In the solution (8.10) this is expressed by the dependence of the integrands on the dispersion parameter D that contains the difference of the group velocities. The integration in (8.10) can be carried out analytically for specific amplitude functions $A_1(t)$. For a Gauss function the result can be expressed, for example, by the error integral with complex arguments. From this the intensity as well as its time integral, the pulse energy per unit area, can be determined. It becomes evident that the pulse energy for short pump pulses also increases for $z \ll L_{\mathrm{NL}}$ more slowly than with z^2. This is due to the violation of the phase matching condition for certain spectral components of the pulses. As a characteristic measure we can introduce the length

$$L_{\mathrm{D}} = \tau_{\mathrm{L}} \cdot c/D \tag{8.13a}$$

or more generally for non-bandwidth-limited pulses

$$L_{\mathrm{D}}' = \frac{c}{\Delta\omega \cdot D}. \tag{8.13b}$$

For $z \ll L_{\mathrm{D}}, L_{\mathrm{D}}'$ the steady-state relations (curves 1 and 2) hold, whereas for $z \gg L_{\mathrm{D}}, L_{\mathrm{D}}'$ the signal increases only linearly with z (see curve 3 in Fig. 8.2).

At this point, the spectrum of the second harmonic from ultrashort light pulses should be more closely considered. The Fourier transform of $E_2(t)$ using (8.10), is given by (see [8.2])

$$\underset{\sim}{E}_2(\omega, z) = \Delta^{(2)}(2\omega_1; \omega_1, \omega_1) \, \frac{\sin \xi}{\xi} \, H(\omega) \cdot z, \tag{8.14}$$

where $H(\omega)$ is the convolution integral of the fundamental wave, i.e.

$$H(\omega) = \int_{-\infty}^{\infty} \mathrm{d}t \, A_1^2(t) \, \mathrm{e}^{-\mathrm{i}(\omega - 2\omega_1)t} = \frac{1}{2\pi} \int_{-\infty}^{\infty} \mathrm{d}\omega' \underset{\sim}{A}_1(\omega') \, \underset{\sim}{A}_1(\omega - \omega') \tag{8.15}$$

and

$$\xi = \left[\Delta k - (\omega - 2\omega_1) \frac{D}{c}\right] \frac{z}{2}.$$

Δk is the difference of the wave numbers at the mid-frequencies. For the square of the electric field $|\underset{\sim}{E}_2(\omega, z)|^2 \equiv |\underset{\sim}{A}_2(\omega, z)|^2$ of the second harmonic (proportional to the intensity), we obtain

$$|\underset{\sim}{E}_2(\omega, z)|^2 = |\Delta^{(2)}(2\omega_1; \omega_1, \omega_1)|^2 \left(\frac{\sin \xi}{\xi}\right)^2 |H(\omega)|^2 z^2. \tag{8.16}$$

As we see from this equation, the frequency dependence of the intensity of the second harmonic is determined essentially by the product of $H(\omega)$ and $(\sin \xi)/\xi$. Generally, spectral measurements do not resolve the individual modes. Therefore it is sufficient to consider the envelopes of the mode spectrum. For an estimation we assume a Gaussian-shaped envelope of halfwidth $\Delta\omega$ and $H(\omega)$ of halfwidth $2\Delta\omega$, where the latter follows precisely for real amplitudes.

For the picosecond Nd:glass laser, for example, the frequency dependence of $H(\omega)$ can be neglected up to deviations of some 10^{11} s^{-1} from the center frequency. Whether the factor

$$\left(\frac{\sin \xi}{\xi}\right)^2 = \frac{\sin^2\left[\frac{1}{2c} D(\omega - 2\omega_1)\, z\right]}{\left[\frac{1}{2c} D(\omega - 2\omega_1)\, z\right]^2} \tag{8.17}$$

changes significantly within this range ($\Delta k = 0$ was chosen) depends on the crystal parameters and on z. At a given crystal length the dispersion parameter D of the crystal can be determined from the intensity spectrum of the second harmonic. If $\delta\lambda$ at a certain crystal length z is the separation of two minima in the intensity spectrum of the second harmonic, then we have

$$|D| = \frac{\lambda^2}{z\delta\lambda}. \tag{8.18}$$

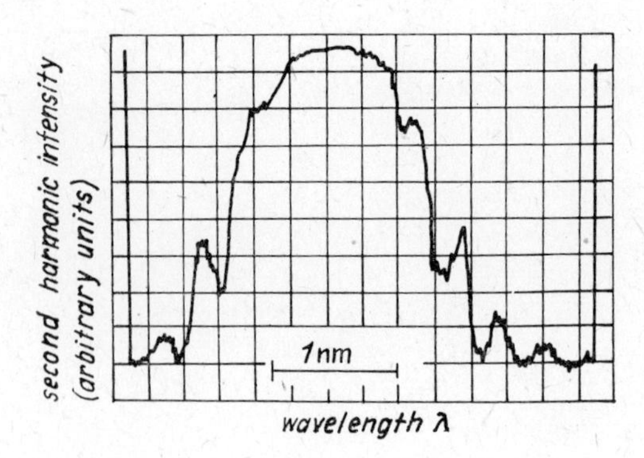

Fig. 8.3. Recorded spectrum of the second harmonic of a picosecond Nd:glass laser using a 10.4 mm thick $LiJO_3$ crystal [8.2]

Fig. 8.3 shows a recorded photometer curve of the spectrum of the second harmonic generated in a $LiJO_3$ crystal of 10.4 mm length. (According to (8.17) the curve is expected to be completely modulated. The observed deviation from this case results from the violation of the conditions formulated at the outset, particularly as a result of a finite divergence of the laser beam.) From the separation of the intensity minima in Fig. 8.3 we obtain with (8.18) $|D| = 0.0775 \pm 0.012$. From the refractive index behavior we can calculate the dispersion parameter and obtain

$D = 0.084 \pm 0.08$. The agreement within 10% of both values is good, since even small errors in the refractive indices ($\simeq 10^{-4}$) strongly influence the derivatives and differences necessary for the calculation of D.

Using KDP and ADP, however, no spectral modulations of the intensity of the second harmonic were observed under radiation with Nd:laser light (compare also [8.3, 8.4, 8.5]). In these crystals the dispersion parameter $|D|$ is so small (compare table 8.1) that for typical crystal lengths of several centimeters the frequency dependence of the intensity of the second harmonic is essentially determined by the factor $|H(\omega)|^2$.

Table 8.1. Dispersion parameter $D = c\left(\frac{1}{v_2} - \frac{1}{v_1}\right)$ for several crystals

Crystal	λ_1 µm	D
KDP	1.06	0.003
	0.53	0.075
$LiNbO_3$	1.06	0.156
$LiJO_3$	1.06	0.084

Let us draw several conclusions for the experimental work. The violation of the phase matching condition for certain components of the frequency spectrum of the exciting radiation leads to a strong reduction in the conversion rates. In order to avoid this, the condition

$$\Delta\lambda_1 \lessapprox \frac{0{,}2 \cdot \lambda_1^2}{D \cdot l} = \frac{0{,}2\lambda_1}{l}\left[\left(\frac{\partial n^o}{\partial \lambda}\right)_1 - \frac{1}{2}\left(\frac{\partial n^e}{\partial \lambda}\right)_2\right]^{-1} \tag{8.19}$$

must be maintained between the bandwidth $\Delta\lambda_1$ of the exciting radiation at the wavelength λ_1, the crystal length l and the dispersion parameters of the crystal, where $l \lessapprox L_{NL}$. If we again take as an example the frequency doubling of Nd laser light the relation (8.19) indicates no important limitation in KDP (at $l = 2$ cm, $\Delta\lambda_1 \lessapprox 13.5$ nm follows), whereas in a $LiNbO_3$ crystal of length $l = 2$ cm $\Delta\lambda_1 \lessapprox 0.35$ nm has to be required. This corresponds to the requirement $\tau_L > 4$ ps for the duration of bandwidth limited pulses. Therefore KDP is preferred to $LiNbO_3$ and $LiJO_3$ for the conversion of very short pulses, although $LiNbO_3$ and $LiJO_3$ possess higher nonlinear coefficients. Furthermore, another effect occurs which lowers the efficiency. Due to the different group velocities of two pulses at the wavelengths λ_1 and λ_2, which enter a crystal simultaneously at $z = 0$, the overlapping decreases as the crystal length that the pulses pass through increases; the one pulse hastens away the other. In order to avoid this effect the relation

$$\tau_L \gtrapprox \frac{l \cdot D}{c} = \frac{\lambda_1 l}{c}\left[\left(\frac{\partial n^o}{\partial \lambda}\right)_1 - \frac{1}{2}\left(\frac{\partial n^e}{\partial \lambda}\right)_2\right] \tag{8.20}$$

must be satisfied [8.3, 8.6]. (For bandwidth-limited pulses the requirements (8.19) and (8.20) are identical.) The estimations mentioned referred to extended plane waves. If the light is focussed into the crystal, then not all angular parts of the beams satisfy the phase matching condition simultaneously and therefore a further loss in the conversion rates occurs. Despite these complications, by heeding the conditions mentioned conversion efficiencies of more than 90% have been achieved with picosecond pulses.

Through repeated second harmonic generation higher harmonics of the original frequency ω_1 can be generated. The absorption of the crystals beginning in the ultraviolet spectral region restricts the process to a minimum wavelength $\lambda \simeq 200$ nm. Even shorter wavelengths are obtained by generating harmonics in gases and metal vapours, where the absorption ranges are very narrow, and thus a resonance of the light pulses with atomic transitions can be avoided. In all gases, vapours and liquids (more generally, in all systems with inversion symmetry), the nonlinear optical coupling coefficients of the lowest, that is, the second order as well as all other coupling coefficients of even order disappear [11]. Therefore, in the gases and vapors mentioned, of course not the second, but the third, fifth, ... harmonics with the frequencies $3\omega_1$, $5\omega_1$, ... can be produced. At the present, beginning with the ps pulses of the neodymium glass laser ($\lambda = 1.06$ µm) or of a dye laser ($\lambda = 0.579$ µm), one can generate radiation up to the shortwave ultraviolet ($\lambda = 38$ nm) using these methods. Due to the application of nonlinear processes of higher order the efficiency is small, though.

8.2 Application of Other Nonlinear Optical Processes for Frequency Conversion

8.2.1 *Frequency Mixing*

The generation of pulses at the sum and difference frequency $\omega_1 \pm \omega_2$ occurs at the simultaneous passage of two light pulses of the frequencies ω_1 and ω_2 through a nonlinear optical crystal as in the generation of harmonics. According to the general relation (8.5a) the amplitude of the wave A_3 having the difference or sum frequency $\omega_3 = \omega_1 \pm \omega_2$ can be calculated by the differential equation

$$\left(\frac{\partial}{\partial z} + \frac{1}{v_3}\frac{\partial}{\partial t}\right) A_3(z, t) = \Delta^{(2)}(\omega_3; \omega_1, \omega_2)\, A_1(z, t)\, A_2(z, t)\, \mathrm{e}^{-\mathrm{i}(k_1+k_2-k_3)z}. \tag{8.21}$$

If we again consider only small conversion rates, the change of the two incident waves inside the crystal can be neglected $\left(A_1(t, z) \approx A_{10}\left(t - \frac{z}{v_1}\right),\ A_2(t, z) \approx A_{20}\left(t - \frac{z}{v_2}\right)\right)$. Substituting the new variables $\eta = t - \frac{z}{v_3}$, $\xi = z$ in (8.21), the solution can easily be found and after backtransforming to the original variables t and z it can be written as

$$A_3\left(z, t - \frac{z}{v_3}\right) = \Delta^{(2)}(\omega_3; \omega_1, \omega_2) \int_0^z \mathrm{d}z' A_{10}\left(t - \frac{z}{v_3} + D_1 \frac{z'}{c}\right) \times A_{20}\left(t - \frac{z}{v_3} + D_2 \frac{z'}{c}\right) \mathrm{e}^{\mathrm{i}\Delta k z'} \tag{8.22}$$

where $D_{1,2} = c\left(\frac{1}{v_3} - \frac{1}{v_{1,2}}\right)$ and $\Delta k = k_3 - k_1 - k_2$. Thus, the nonsteady-state behavior is caused by the various group velocities of the participating waves as in the case of the second harmonic, and relations similar to those result. The efficiencies achieved in the frequency mixing also correspond to those of the generation of the second harmonic. Using at least one pulse of tunable frequency the sum and difference frequencies can also be continuously changed. In particular, the radiation in the infrared spectral range can be tuned over wide ranges by only slightly altering the wavelength in the visible range (for example, the wavelength of a dye laser). In the crystal proustite, for instance, the difference frequency could be tuned between the long-wave limit of the visible spectral range and about 15 μm. The considerations for phase matching can be carried out similarly to those for the case of harmonics generation.

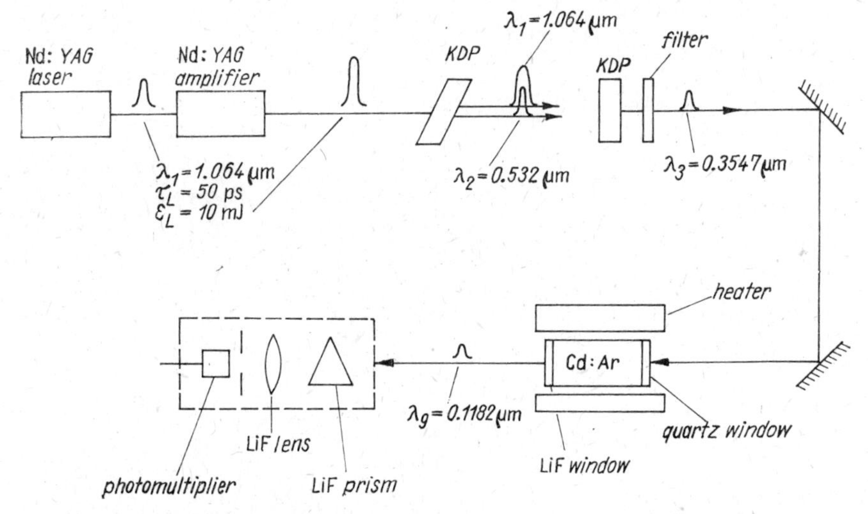

Fig. 8.4. Set-up for generating the ninth harmonic of a Nd:YAG laser [8.7]

In the first KDP crystal the second harmonic of the laser radiation is generated; in the second crystal the second harmonic is mixed with the fundamental wave to generate the sum frequency $3\omega_1 = 2\omega_1 + \omega_1$. Subsequently, starting with radiation at 0.3547 μm, the third harmonic (i.e. the ninth harmonic in regard to the laser radiation) is generated in cadmium vapor, which is distinguished by its high nonlinear coefficients. Through a suitable admixture of argon, which exhibits normal dispersion in this wavelength range, to the cadmium vapor, in which an anomalous dispersion occurs, phase matching $\Delta k = 3k_3 - k_9 = 0$ can be achieved. The radiation at 0.1182 μm is detected using a spectrometer with components made of lithium fluoride (due to the good UV transmission).

As an example for the generation of harmonics and sum frequencies an experimental set-up is presented in Fig. 8.4, with which the light from a Nd:YAG laser ($\lambda_1 = 1.064$ μm) can be converted into radiation at $\lambda_2 = 0.532$ μm, $\lambda_3 = 0.3574$ μm and $\lambda_9 = 0.1182$ μm. As already mentioned, radiation up to $\lambda = 38$ nm was generated by means of similar cascade processes [8.8, 8.29].

In [8.29] two nonlinear optical conversion processes of higher order were used to achieve very small wavelengths. As already mentioned, the efficiency in such processes is very low. In order to balance out at least partially the losses due to the low efficiency, not only the fundamental wave but also the radiation after the first frequency conversion was suitably amplified in this experiment. The initial light source was a dye laser that emitted 6 ps pulses at 579 nm. After passing through three amplifier stages these pulses produced the third harmonic at 193 nm in strontium vapor. The low-power ultraviolet pulses were then amplified in three consecutive ArF^+ excimer laser amplifiers from about 3×10^2 W to 4×10^9 W. Due to the large bandwidth of the excimer amplifier only a small increase in the pulse duration occurs. With these high power ultraviolet pulses the third harmonic ($\lambda = 64$ nm) and fifth harmonic ($\lambda = 38$ nm) were subsequently generated in hydrogen with an admixture of argon as buffer gas, with which the dispersion relations and the phase matching conditions could be set. At the third harmonic a power of 20 kW and a pulse duration of 10 ps were achieved.

8.2.2 Optical Parametric Generation

Beside dye lasers optical parametric generators represent the most important light source for generating frequency tunable ultrashort light pulses. By parametric amplification and generation we understand the amplification and the generation of two light waves, respectively, at frequencies ω_2 and ω_3 in a suitable material if a strong light wave called the pump wave is incident at frequency ω_1. The parametric interaction can be considered as the reversal of the frequency mixing. Under continuous monochromatic radiation the frequencies of the amplified or generated waves ω_2, ω_3 and the pump wave ω_1 obey the relation

$$\omega_1 = \omega_2 + \omega_3 . \tag{8.23}$$

This relation can also be interpreted as the energy conservation equation. The waves of the frequencies ω_2, ω_3 emerge with particular efficiency if the corresponding phase matching condition

$$\Delta\vec{k} = \vec{k}(\omega_1) - \vec{k}(\omega_2) - \vec{k}(\omega_3) = 0 \tag{8.24a}$$

between them and the pump wave of the frequency ω_1 is satisfied, which in the case of a colinear interaction reduces to the scalar relation

$$\Delta k = k(\omega_1) - k(\omega_2) - k(\omega_3) = 0 . \tag{8.24b}$$

This relation can be interpreted in the photon case as the momentum conservation equation. The parametric light generation can be understood similarly to the parametric amplification or parametric oscillation of radio-frequency oscillations. The term "parametric process" derives from the fact that this process is caused by a periodic change of a parameter of a resonant circuit, usually of the capacity of the circuit. In this manner certain frequencies that are related to the change of the system parameters are generated and amplified, respectively. In optical parametric amplification or generation the nonlinear optical crystal replaces the oscillator circuit. By pumping with an intense wave the susceptibility is forced to change with the radiated frequency, which corresponds to the periodic change of the capacity in the above mentioned oscillator circuit. The parametric interaction effect can be used for a large number of purposes in the optical range, too.

The device operates as a parametric amplifier, when in addition to the strong pump wave of frequency ω_1 a signal wave of frequency ω_2 is radiated on the crystal. During the amplification process a third wave develops, called the auxiliary wave or idler wave, with the frequency ω_3 and the wave number vector $\vec{k}_3 = \vec{k}_1 - \vec{k}_2$. With a very strong pump wave the parametric process can also take place without a signal being applied at the input, that is; the photon noise can act as an input signal, and the amplifier becomes a generator. The frequencies ω_2, ω_3 at which light favorably emerges are determined by the phase matching condition and the interaction geometry. By changing the effective refractive indices in the nonlinear optical crystal (e.g. by rotating the crystal or altering its temperature) the frequencies ω_2 and ω_3 can be tuned. In Fig. 8.5 the tuning range is given for some crystals and pump wavelengths for ps operation [8.16].

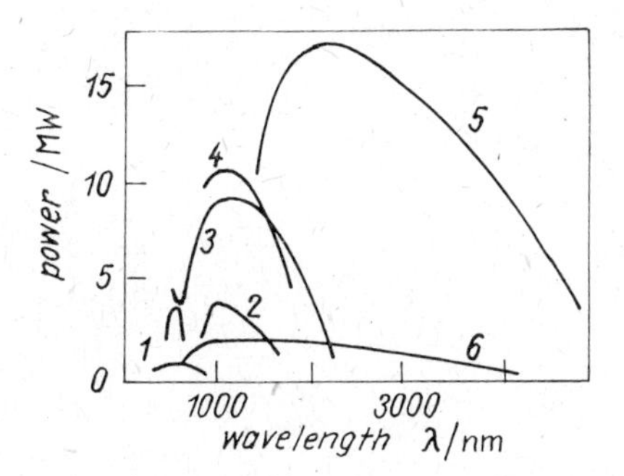

Fig. 8.5. Tuning range of a parametric generator [8.16]
The radiation of a modelocked Nd laser or its harmonics serves as the pump light source: KDP, α-HJO_3 and $LiNbO_3$ were used as nonlinear optical crystals.

Curve	Pump Wave Length	Crystal
1	266 mm	KDP
2	354 nm	KDP
3	532 nm	α-HJO_2
4	532 nm	KDP
5	1064 nm	$LiNbO_3$
6	532 nm	$LiJO_3$

Similar to the generation of the harmonics, in parametric generation certain requirements must be placed on the pulse parameters and on the crystal parameters, in order that the violation of the phase matching condition resulting from the broad spectrum and the spreading of pulses having different mid-frequencies (due to the dispersion of the group velocity) do not lead to a considerable decrease of the conversion rates or to a pulse broadening. A review of experimental and theoretical investigations on this set of problems is presented in [8.16], in which particular attention is also given to the attaining of high conversion rates ($\gtrsim 80\%$). Extremely short pulses were generated by Kaiser et al. (see e.g. [8.17]). Here, a parametric generator is followed by a second crystal serving as parametric amplifier. Minimum signal pulse lengths could be achieved through suitable adjustment of the delay between signal and pump pulse in the amplifier. A two crystal arrangement of this kind furthermore permits a better wavelength adjustment.

8.2.3 Parametric Four Photon Interaction

In the parametric processes described above we dealt with a three photon interaction ($\omega_1 = \omega_2 + \omega_3$), which is caused by the nonlinear susceptibility of lowest, that is second order. Furthermore, parametric effects of higher order also occur. The same susceptibilities that characterize the generation of the third harmonic ($\omega_1 + \omega_1 + \omega_1 \rightarrow 3\omega_1$) are, for instance, also responsible for the parametric four photon interaction, in which two photons are annihilated and two other photons generated ($\omega_1 + \omega_2 \rightarrow \omega_3 + \omega_4$ or specifically $2\omega_1 \rightarrow \omega_3 + \omega_4$). The process is again very efficient, if the corresponding phase matching relation $\Delta\vec{k} = \vec{k}_1 + \vec{k}_2 - \vec{k}_3 - \vec{k}_4 = 0$ is satisfied (see Fig. 8.6). At very

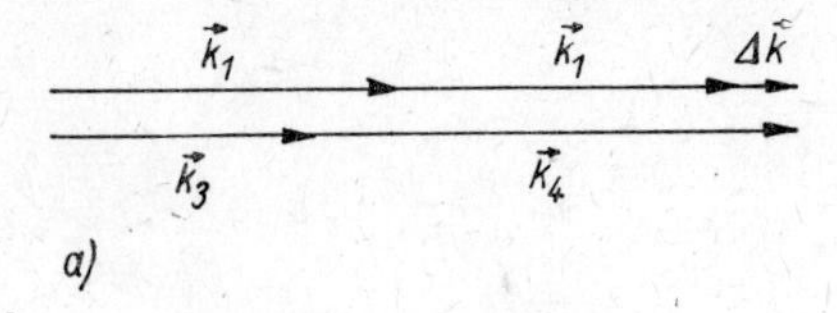

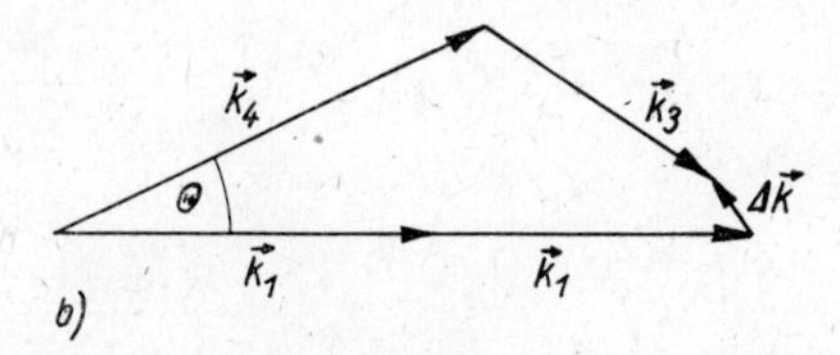

Fig. 8.6. Phase matching condition of the parametric four photon interaction

a) Colinear interaction

$\Delta\vec{k} = \vec{k}_3 + \vec{k}_4 - 2\vec{k}_1$, $\omega_3 + \omega_4 = 2\omega_1$

$\Delta\vec{k} = 0$ is only achievable with anomalous dispersion.

b) Noncolinear interaction

$\Delta\vec{k} = \vec{k}_3 + \vec{k}_4 - 2\vec{k}_1$

$$\Delta\vec{k} = 0 \text{ for } \cos\Theta = \frac{4k_1^2 + k_4^2 - k_3^2}{4k_4k_1}$$

high laser intensities a cascade-like process can be produced due to the parametric four photon interaction: out of the laser radiation strong waves at ω_3 and ω_4 develop, which in turn produce parametric effects and so forth. Using this method white, that is spectrally very expanded, ultrashort pulses are generated. Alfano and Shapiro [8.18] conducted the first experiments in the picosecond region, in which they focussed the pulses of a modelocked Nd:glass laser in a glass sample. At intensities of about 10^{15} W/m^2, for instance, the pulse of a Nd laser produces in water (cell length of some 10^{-2} m) a spectral continuum from the ultraviolet to the near infrared spectral range, where about 10% of the laser power is converted and the conversion efficiency per wave number is several 10^{-7} cm (see [8.19] and [8.20] and references there, compare Fig. 8.7).

Using this method white fs pulses of large spectral width can also be generated. For this purpose pulses of a passively modelocked or synchronously pumped dye laser having a duration of 60—100 fs are amplified up to powers in the Gigawatt range and subsequently focussed into a suitable sample. Fork, Shank, Yen and Hirliman [8.29] used a jet-stream of ethylene glycol with a thickness of only 1 mm. The spectrum of the white pulses generated comprises the range from 0.19 μm to 1.6 μm.

These white light pulses are used as very suitable probe pulses in laser spectroscopy (see 9.2.). Subpicosecond IR continua (3—14 μm) are described in [8.58].

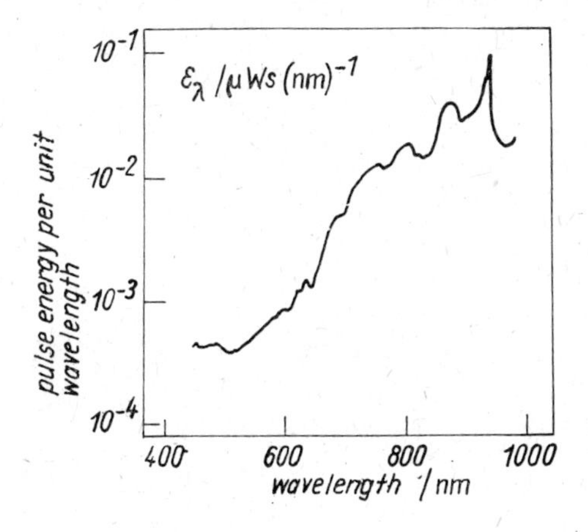

Fig 8.7. Spectral energy distribution of a ps-continuum generated in D_2O according to [8.20] (excitation by means of a Nd:glass laser with a pulse duration of 10 ps and an intensity of 4×10^{11} W/cm²). The divergence of the continuum is about 10 mrad.

8.2.4 *Optical Rectification and Generation of Čerenkov Radiation*

If monochromatic light at frequency ω_1 is incident on a nonlinear optical crystal without inversion symmetry (an electrooptical crystal) then besides the polarization at the sum frequency $\omega_1 + \omega_1 = 2\omega_1$ leading to second harmonic generation, a component of the polarization of second order at the difference frequency $\omega_1 - \omega_1 = 0$ develops. The latter causes accordingly the formation of a d.c. field and a d.c. voltage. This effect, which is termed optical rectification and can be interpreted as an inverse electrooptical effect, was observed in 1962 by Bass, Franken, Ward and Weinrich [8.53] after the generation of the second harmonic. It can be used to detect optical signals where, however, the sensitivity is not very high.

If we use light pulses instead of monochromatic cw light, then instead of the temporally constant polarization and the d.c. voltage, polarization pulses and, subsequently, field strength pulses and voltage pulses develop due to the nonlinear optical effect. Since in typical electrooptical crystals the nonlinearity results from a pure electronic effect, far from the resonances the polarization follows the electric field of the light pulses almost inertialessly; the corresponding response times are only a few femtoseconds. This means that an ultrashort light pulse ($\tau_L \gtrsim 10$ fs) causes the formation of a polarization pulse of equal duration at a certain point in the crystal. If we assume an inertialess nonlinear optical effect and use the parametric approximation (i.e. neglecting the weakening of the laser wave as it passes through the sample) the space-time structure of the polarization is given approximately by

$$P(t, z) \propto \left| A_L \left(t - \frac{z}{v_L} \right) \right|^2 .$$

Thus, in the crystal a dipole excitation develops, which propagates in the transmission direction with the group velocity of the laser wave v_L. Similar to a moving charge, the fast propagation of a dipole leads to the emission of a transient electromagnetic field that contains frequencies up to the order of $1/\tau_L$. Due to the contribution that the lattice vibrations yield to the linear optical susceptibility at low frequencies in the crystals mentioned, the propagation velocity v_F of these electromagnetic fields is smaller than the group velocity v_L of the light and, hence, smaller than the velocity of the source of the field, which is identical to v_L. Thus, as in the usual Čerenkov effect a wave front develops which forms a cone. Auston [8.54] was the first to point out this effect. A theoretical treatment can be found in [8.55]. Note that as opposed to the Čerenkov effect the

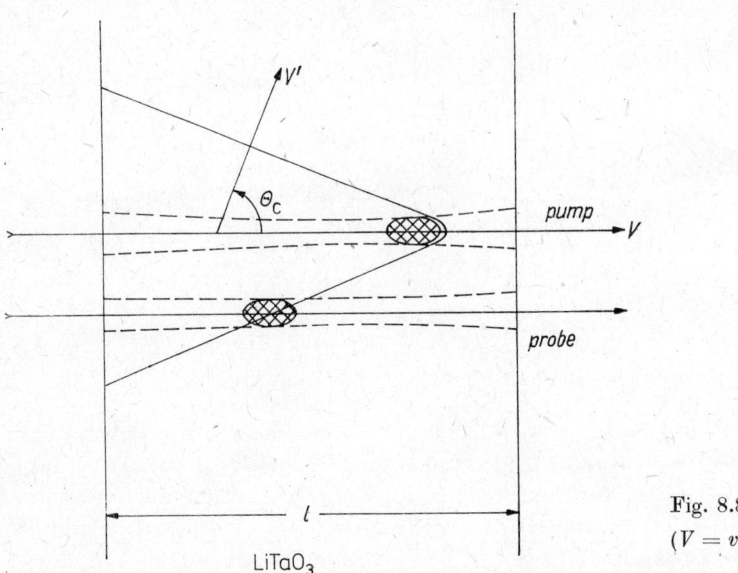

Fig. 8.8a)
($V = v_L$, $V' = v_F$)

radiation source is extended spatially; according to the relation given above, the dimension in the propagation direction z is determined by the pulse duration, and the lateral dimensions by the diameter of the laser beam. Thus, the properties of the radiation characteristics depend on the pulse duration and pulse shape as well as on the beam waist of the exciting laser radiation. A precise experimental investigation of the optical Čerenkov effect was undertaken in [8.56]. An excitation pulse with a duration of about 100 fs and a wavelength of $\lambda = 625$ nm passes through a lithium tantalate crystal of 1 mm thickness, in which due to the inverse electrooptical effect described it generates a field whose wave front propagates on the lateral area of a cone. Due to the common electrooptical effect, the electric field causes birefringence in the crystal. This birefringence is measured with a probe pulse by means of the electrooptical sampling technique (compare 3.1.2). First, the laser pulse is divided at a beam splitter into a strong excitation pulse and a weak probe pulse. After a variable delay line the probe pulse passes through the crystal parallel to the excitation beam at a variable distance to it (see Fig. 8.8a). In this way the beam moves in precise synchronism to the Čerenkov radiation, whereby at a low delay relatively large interaction lengths and, consequently, high sensitivities can be achieved. The delay time, at which the Čerenkov wave and the probe pulse have the maximum overlap, depends on the distance of the probe beam from the excitation beam and the Čerenkov angle, and therefore the Čerenkov angle can be determined by varying this distance. In the experiment with lithium tantalate an angle of

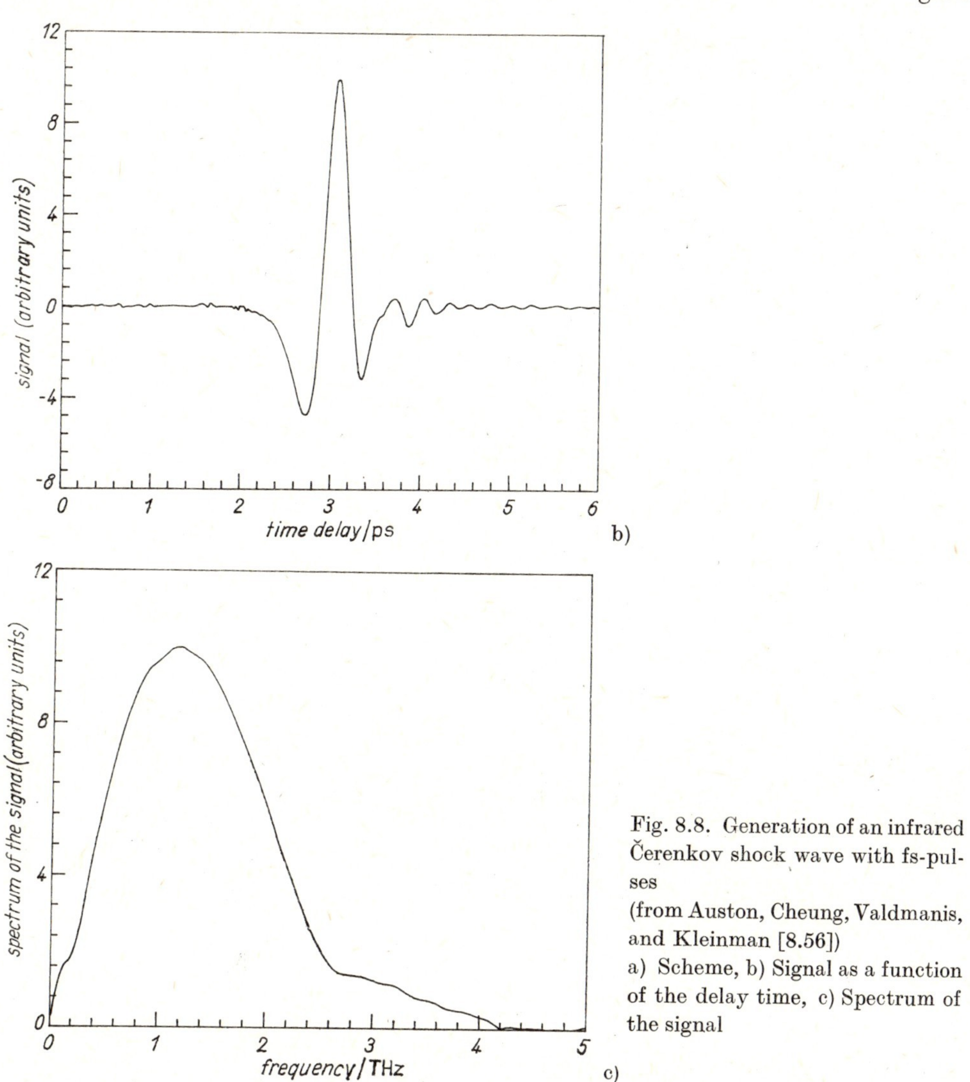

Fig. 8.8. Generation of an infrared Čerenkov shock wave with fs-pulses (from Auston, Cheung, Valdmanis, and Kleinman [8.56]) a) Scheme, b) Signal as a function of the delay time, c) Spectrum of the signal

about 70° was measured which agrees very well with the theoretically predicted value of 68° for the Čerenkov angle at given velocities $v_L = 0.428c$ and $v_F = 0.158c$. The measured temporal evolution of the shock wave is shown in Fig. 8.8b, the corresponding spectrum in Fig. 8.8c. As expected, an ultrashort excitation occurs, which represents approximately a single cycle of a 1 THz wave. Note that this frequency corresponds to a wavelength of about 300 μm. As a result single cycles of infrared radiation could for the first time be generated, which of course is of fundamental physical interest. Let us point out that the slowly varying amplitude or envelope approximation (SVEA), which we have used in the entire book and according to which the field strength is split into a factor that varies periodically with a carrier frequency and a slowly varying amplitude, loses all meaning here (compare 1.3.1).

It is expected that the technique described can be applied with very good results for transient spectroscopic investigations in the far infrared range. The advantages are due to the extremely short time excitation, the high time resolution and the good signal-to-noise ratio at the detection.

8.2.5 *Stimulated Raman Scattering*

Stimulated Raman scattering is a two photon process in which a photon of the incident laser light of frequency ω_L is annihilated, whereby the molecule is raised into an excited vibrational level that corresponds to the frequency ω_{21} and a photon is emitted with the Stokes frequency $\omega_s = \omega_L - \omega_{21}$ (compare Fig. 9.1c). The vibrational frequencies that determine the frequency changes reach maximum values of $\omega_{21}/2\pi c \approx 3000\ \mathrm{cm}^{-1}$. It is possible to describe this process with an effective two-level model, where one must additionally account for the participating virtual intermediate levels of the molecule by means of a modified interaction operator. In accordance with the two-quanta nature of the process, (1.57) no longer provides the matrix elements of the interaction operator H_{12}^{w}. Instead, they are given by

$$H_{12}^{w} = -\frac{1}{2}(\alpha_{ij})_{12} E_i E_j. \tag{8.25}$$

i and j are the tensor indices of the polarization operator $\hat{\alpha}$, whose expectation value is related to the transition moments $\vec{\mu}_{1k}$ and $\vec{\mu}_{k2}$ between the virtual intermediate levels k and the ground level 1 and end level 2 as follows:

$$(\alpha_{ij})_{12}(\omega) = \frac{1}{\hbar}\sum_k \frac{(\mu_{1k})_i (\mu_{k2})_j}{\omega_{k1} + \omega} + \frac{(\mu_{k2})_i (\mu_{1k})_j}{\omega_{k2} - \omega}. \tag{8.26}$$

The density matrix equation (1.65) of the effective two-level system can now be rewritten as

$$\frac{\partial \tilde{\varrho}_{12}}{\partial t} + \frac{1}{\tau_{21}}\tilde{\varrho}_{12} = \frac{\mathrm{i}}{2\hbar}\alpha_{12}(\varrho_{22} - \varrho_{11}) A_L A_S^* \mathrm{e}^{-\mathrm{i}(\vec{k}_L - \vec{k}_S)\vec{r}} \tag{8.27}$$

where $\alpha_{12} = (\alpha_{ij})_{12}(e_L)_i (e_S)_j$.

After introducing the nonlinear polarization $P = N(\varrho_{12}\alpha_{21} + \alpha_{12}\varrho_{21})$ in (1.50) the equation for the amplitude of the Stokes wave becomes

$$\frac{\partial A_S}{\partial z} + \frac{1}{v_S}\frac{\partial A_S}{\partial t} = \frac{\mu_0 N \omega_S^2}{4\mathrm{i}k_S}\alpha_{12}(\omega_S)\,\tilde{\varrho}_{21} A_L \mathrm{e}^{-\mathrm{i}(\vec{k}_L - \vec{k}_S)\vec{r}}. \tag{8.28}$$

Under the assumption of small amplification ($|A_S| \ll |A_L|$) and small changes in the level occupation ($\varrho_{22} \ll \varrho_{11}$), we can calculate $\tilde{\varrho}_{21}$ from (8.27) and (8.28), and obtain with the new variables $\eta = t - \dfrac{z}{v_L}$, $z = z$ for A_S the following hyperbolic differential equation:

$$\frac{\partial^2 A_S}{\partial z\,\partial \eta} + D_L\frac{\partial^2 A_S}{\partial \eta^2} + \left[\frac{1}{\tau_{21}} - \frac{\mathrm{d}}{\mathrm{d}\eta}\ln A_L(\eta)\right]\left[\frac{\partial A_S}{\partial z} + D_L\frac{\partial A_S}{\partial \eta}\right] = Q\,|A_L(\eta)|^2, \tag{8.29}$$

where $D_L = \frac{1}{v_S} - \frac{1}{v_L}$ and $Q = \frac{\mu_0 N \omega_S^2 |\alpha_{12}|^2}{8\hbar k_S}$. For the case of negligible dispersion ($D_L = 0$) an exact solution of this equation with arbitrary pump pulse amplitude $A_L(\eta)$ can be found ([8.12] to [8.15]). In the presence of dispersion ($D_L \neq 0$) we can solve equation (8.29) analytically in the special case of a rectangular pulse with constant phase [8.30]. Since, however, the solution is physically interesting particularly in the asymptotic range of large amplification, and since this is the only range where simple expressions exist, we want to restrict our consideration to this range. Following the treatment by Herrmann [8.21] the calculation can be simplified considerably by the ansatz

$$A_S = C_S(z, \eta) \exp \{u(z, \eta)\} \tag{8.30}$$

and by the requirement of large amplification $u(z, \eta) \gg 1$. If we assume further the essentially nonstationary case of a short exciting laser pulse for which $\tau_L < \tau_{21} u(z, \eta)$ is satisfied, then we can solve the differential equation (8.29) similarly to the quasi-optical approximation or the W.K.B. method of quantum mechanics [8.21] providing that

$$\frac{\partial^2 u}{\partial z\, \partial \eta} \ll \frac{\partial u}{\partial z} \frac{\partial u}{\partial \eta}, \quad \frac{\partial^2 u}{\partial \eta^2} \ll \left(\frac{\partial u}{\partial \eta}\right)^2. \tag{8.31}$$

For $u(z, \eta)$, then, we have the following differential equation of first order:

$$\frac{\partial u}{\partial z} \frac{\partial u}{\partial \eta} + D_L \left(\frac{\partial u}{\partial \eta}\right)^2 + \left(\frac{1}{\tau_{21}} - \frac{d}{d\eta} \ln A_L(\eta)\right) \left(\frac{\partial u}{\partial z} + D_L \frac{\partial u}{\partial \eta}\right) = Q\, |A_L|^2. \tag{8.32}$$

Using the Riemann method of characteristics the amplification coefficient in different physical situations can be obtained relatively simply from this equation. If we consider first the dispersionless case $D_L = 0$, then after introducing the new variable $\tau = \int_{-\infty}^{\eta} d\eta' \, |A_L(\eta')|^2$ the gain coefficient $G_T = 2u(z = L, \eta)$ can be written as

$$G_T = 4 \sqrt{QL \int_{-\infty}^{\eta} |A_L(\eta')|^2 \, d\eta' - \frac{2\eta}{\tau_{21}} + 2 \ln A_L(\eta)/\tilde{C}_S}, \tag{8.33}$$

where $\tilde{C}_S$ is a constant.

If the Raman active medium is excited with short pulses ($\tau_L < \tau_{21} G_T/2$), the gain coefficient for the Stokes radiation thus increases with the square root of the amplifier length and the pump energy. Compared with this the gain coefficient varies linearly with the amplifier length L and the pump intensity ($G_S = 2Q\, |A_{L0}|^2 L \tau_{21}$) under steady conditions ($\tau_L \gg \tau_{21}$). The smaller gain coefficient at the excitation with ultrashort pulses as opposed to the steady-state case is due to the finite transverse relaxation time that results in a delayed starting of the oscillation of the nonlinear polarization. This results also in a shortening and a spectral broadening of the Stokes pulse with respect to the laser pulse. These effects can be qualitatively estimated from (8.33) for a rectangular pump pulse by expanding the gain coefficient at its maximum, which lies at $\eta = \tau_L$:

$$G_T \approx 4 \sqrt{QL\, |A_{L0}|^2 \, \tau_L} - (\tau_L - \eta)\, 2 \sqrt{\frac{QL\, |A_{L0}|^2}{\tau_L}} + \cdots. \tag{8.34}$$

From this, we can evaluate the duration of the Stokes pulse at $\tau_S \approx \sqrt{\dfrac{\tau_L}{QL\,|A_{L0}|^2}}$ and the spectral halfwidth of the Stokes radiation at $\Delta\omega_S \approx \sqrt{\dfrac{QL\,|A_{L0}|^2}{\tau_L}}$. Besides the local non-steady states considered above, dispersion effects influence the SRS of ultrashort pulses because, due to the different group velocities, the Stokes and laser pulses separate. For a discussion of these effects, we look for the solution of the differential equation (8.32) for $D_L \neq 0$ with a rectangular pulse ($A_L = A_{L0}$ for $|\eta| < \dfrac{\tau_L}{2}$; $A_L = 0$ for $|\eta| > \dfrac{\tau_L}{2}$) as pump pulse. Using the Riemann method of characteristics we obtain

$$u(z, \eta) = 2\sqrt{Q\,|A_{L0}|^2\,L[\eta - D_L L]}\,. \tag{8.35}$$

In normal dispersion ($D_L < 0$) the Stokes pulse overhauls the laser pulse, whereby the vibrational excitation is temporally delayed with respect to the Stokes pulse. Thus, the laser pulse always interacts with molecules in which the molecular vibrational polarization is more excited than in the dispersionless case. Therefore, Stokes radiation is generated along a greater effective amplifying length. This becomes obvious from (8.35) because greater amplification lengths $z \gg L_S = \dfrac{\tau_L}{|D_L|}$ lead to a quasi-steady state gain coefficient

$$G_T = 4\sqrt{Q\,|A_{L0}|^2\,|D_L|}\;L \tag{8.36}$$

which is proportional to the amplification length L, as in the steady state case.

In anormal dispersion ($D_L > 0$) the laser pulse propagates at $z > L_S$ in a non-excited medium, and therefore at an amplification length $z = \dfrac{L_S}{2}$ the amplification is saturated. The maximum gain coefficient results from (8.32). Hence, we have

$$G_T = \sqrt{\frac{Q\,|A_{L0}|^2}{|D_L|}}\;\tau_L\,, \tag{8.37}$$

where $\eta = \tau_L$ and $z = \dfrac{L_S}{2}$.

If the laser pulse is phase modulated the gain coefficient of the SRS in a dispersive medium decreases and saturation may occur even in the case of normal dispersion, if the phase modulation is very strong. As an illustration we consider a constant chirp and expand the pulse amplitude in the vicinity of its maximum. Thus $A_L(\eta)$ can be written as

$$A_L(\eta) = A_{L0}\left(1 - \frac{\eta^2}{\tau_1^2}\right)\exp\left(\mathrm{i}\beta\eta^2\right). \tag{8.38}$$

For a phase modulation that is not too strong $\left(\beta^2 < \dfrac{Q\,|A_{L0}|^2}{D_L\tau_1^2}\right)$ the solution of equation (8.32) again provides a quasi-steady state gain coefficient [8.21]

$$G_T = 4L\sqrt{Q\,|A_{L0}|^2\,D_L - \tau_1^2\beta^2 D_L^2}\,, \tag{8.39}$$

which is however smaller than those given in (8.36). If the pulse is strongly phase modulated $\beta^2 > \frac{Q\,|A_{L0}|^2}{D_L \tau_1^2}$, however, the amplification becomes saturated and the maximum gain coefficient is given by

$$G_T = \frac{Q\,|A_{L0}|^2\,\pi\tau_1}{\sqrt{Q\,|A_{L0}|^2\,D_L - \tau_1^2 D_L^2 \beta^2}}. \tag{8.40}$$

In addition to the generation of a Stokes pulse of the frequency $\omega_S = \omega_L - \omega_{21}$, an anti-Stokes pulse may also be generated in a Raman active medium. Similar to the 3-wave coupling in the optical parametric generation the pulse matching condition $\Delta k = 2k_L - k_A - k_S \approx 0$ must be satisfied. The gain coefficient for the anti-Stokes component in the nonstationary case (i.e. under the excitation with ultrashort light pulses with $\tau_L < \tau_{21}G_T/2$) was calculated in [8.21] for dispersive and non-dispersive media in asymptotic approximation. In both cases, the maximum of the anti-Stokes radiation propagates in the direction determined by $\Delta k \approx G_T/L$, where depending on the actual conditions G_T is given by (8.34) or (8.37). Thus, G_T determines the angle between the anti-Stokes radiation and the propagation direction of the laser pulse. Consequently, the anti-Stokes radiation propagates on the lateral area of a cone.

The first experiments on the stimulated Raman effect with picosecond pulses were conducted by Shapiro et al. [8.9] and by Bret and Weber [8.10]. Here, the second harmonic of a modelocked Nd:glass laser was focussed into various liquids, such as benzol, tuluol, carbon disulphide, and nitrobenzene, as well as into various liquid mixtures. In [8.10] it could be demonstrated that the conversion rate greatly decreases, if the spectral width of the laser pulse exceeds the width of the Raman vibrational transition, which corresponds to a non-steady-state regime. The shortening of the Stokes pulse compared to the laser pulse was investigated in later experiments by numerous authors ([8.31] to [8.35]), where shortenings up to a factor of four near the pump threshold and conversion rates up to 70% were found. Colles [8.32] employed this shortening of Stokes radiation in a synchronously pumped Raman oscillator, whereby Stokes pulses with one tenth of the pump pulse duration of the Nd:glass laser were generated. The application of nonstationary SRS for measuring vibrational relaxation times is discussed in detail in section 9.2.2.2.

8.3 Nonresonant Optical Processes for Controlling the Pulse Shape and Pulse Duration

Every nonlinear optical interaction can lead to a change in the shape of given pulses. Using examples let us discuss these processes in the following section and describe possible ways to shorten pulses. Pulses that are not transform-limited can also be shortened by means of linear, dispersive processes, which we shall discuss in section 8.3.2.

8.3.1 Pulse Shaping through Nonlinear Optical Interactions

If the light-matter interaction can be assumed to be inertialess (i.e. without memory) as is the case in the generation of harmonics or the parametric generation far from atomic resonances, then the amplitude of the evolving pulse at η depends only on the amplitudes of the exciting pulses at the same moment η (compare 8.1). The conversion will therefore only proceed with high efficiency at the times at which the corresponding product of the amplitudes in the expression for the nonlinear polarization is large. If bandwidth-limiting effects can be neglected, which for instance is justified for the har-

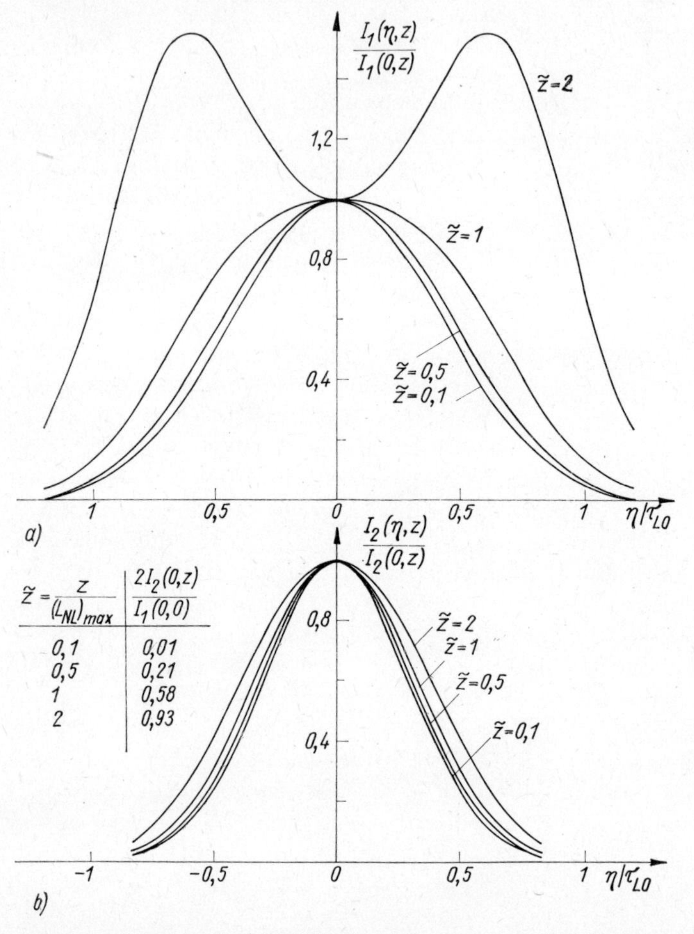

Fig. 8.9. Change in the pulse shape at the generation of the second harmonic

A Gaussian shaped input pulse of duration τ_{L0} strikes a crystal whose length is measured in units of $(L_{NL})_{max}$. $(L_{NL})_{max}$ is the nonlinear interaction length for the maximum input amplitude defined according to (8.12).

a) Output pulses at the fundamental frequency for various crystal lengths

b) Output pulses of the second harmonic for various crystal lengths

monic generation in KDP in the visible region (compare table 8.1), then at small conversion rates the intensity of the n-th harmonic $I_n(t)$ varies approximately with $(I_1(t))^n$, which implies a suppression of the pulse wings and consequently a pulse shortening (see Fig. 8.9b). If the attenuation of the fundamental wave is not negligible, the nonlinear frequency conversion causes a particularly strong damping of the pump intensity around the pulse maximum which leads to a flattening and eventually to a "hollowing out" of the pump pulse (see Fig. 8.9a). In addition the peak intensity of the pump pulse is stabilized. In this manner, the generation of the second harmonic in the laser resonator results in a broadening and amplitude stabilization of the pulses at the fundamental frequency.

In the case where a signal pulse is amplified in the field of a pump pulse the delay between the two pulses at the input of the amplifier presents another parameter, the suitable choice of which can optimize the pulse shortening. This possibility has already been mentioned in the treatment of the parametric amplification in section 8.2.3.

There are even more possible ways of shaping pulses if we also include resonant interactions. Resonant, nonlinear one photon processes alone may have very different influences on the pulse shape, which depends on the relaxation parameters of the material that is to be saturated. For example, a saturable absorber with a very short relaxation time ($T_{21}^{\mathrm{b}} \ll \tau_{\mathrm{L}}$) shortens the pulses, because it attenuates the pulses more at the leading and trailing edges than in their center (compare 7.). Accordingly, a symmetrical pulse remains symmetrical. However, a "slow", saturable absorber ($T_{21}^{\mathrm{b}} \gg \tau_{\mathrm{L}}$) weakens the pulse particularly at the leading edge, while a "slow", saturable amplifier discriminates against the trailing edge. These effects, which lead to a pulse asymmetry were discussed in detail in chapter 6 in connection with the generation of ultrashort light pulses in the dye laser. Furthermore these effects were dealt with in the discussion of the pulse amplification in section 7.3.3.

8.3.2 *Compression of Phase Modulated Pulses in Linear Optical Media*

Whereas transform-limited pulses cannot be shortened in linear optical interactions (compare section 3.2.5), with phase modulated pulses this is possible. For example, pulses with monotonically increasing carrier frequency (up chirp, positive chirp) or monotonically decreasing frequency (down chirp, negative chirp), whose slowly varying amplitude in the simplest case can be expressed by

$$A_{\mathrm{L}}(\eta, z) = |A_{\mathrm{L}}(\eta, z)|\, \mathrm{e}^{\mathrm{i}[\beta(z)\cdot\eta^2 + \varphi_{\mathrm{L}}(0)]} \tag{8.41}$$

— compare (2.93) — can be shortened by frequency filtering or more effectively by pulse compression. The principle of pulse compression, which Treacy [8.22] introduced to laser physics, is shown in Fig. 8.10. In Fig. 8.10a) a pulse with decreasing frequency (down chirp), in which the components of shorter wavelength are accordingly in the leading edge, strikes a glass plate with "normal" optical dispersion, in which the longer wavelength components thus propagate more rapidly. In this way the long-wavelength pulse trailing edge catches up with the leading edge after a glass path of suitable length, where in the optimal case a transform-limited pulse evolves. If this bandwidth-limited pulse passes further through the glass, it is again broadened and up-chirped. To calculate the pulse compression, we cannot as before neglect the dispersion of the group velocity, but must consider in (1.50') the term proportional to k_{L}''. In a linear optical medium the polarization $\overline{P}'$ is negligible, and accordingly we have to solve the equation

$$\mathrm{i}\,\frac{\partial A_{\mathrm{L}}}{\partial z}(z, \eta) + \frac{k_{\mathrm{L}}''}{2}\,\frac{\partial^2 A_{\mathrm{L}}}{\partial \eta^2}(z, \eta) = 0, \tag{8.42}$$

which leads to the Poisson integral:

$$A_L(z, \eta) = \frac{1 - ik_L''/|k_L''|}{2\sqrt{\pi |k_L''| z}} \int_{-\infty}^{\infty} d\eta' A_{L0}(\eta') e^{i\frac{(\eta-\eta')^2}{2k_L''z}}, \tag{8.43}$$

where $A_{L0}(\eta) = A_L(z = 0, \eta)$ represents the field amplitude at the input of the medium. For the incident pulse we assume a Gaussian shaped pulse with constant chirp

$$A_{L0}(\eta) = A_{max} \exp \left\{-2 \ln 2 \left(\frac{\eta}{\tau_{L0}}\right)^2 + i(\beta_0 \eta^2 + \varphi_{L0})\right\} \tag{8.44}$$

which has the time-varying frequency

$$\omega(t) = \omega_L + \frac{d}{dt} \varphi(t) = \omega_L + 2\beta_0 t \tag{8.45}$$

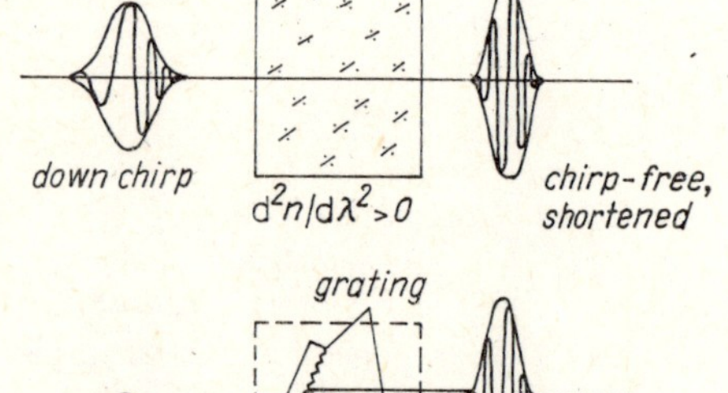

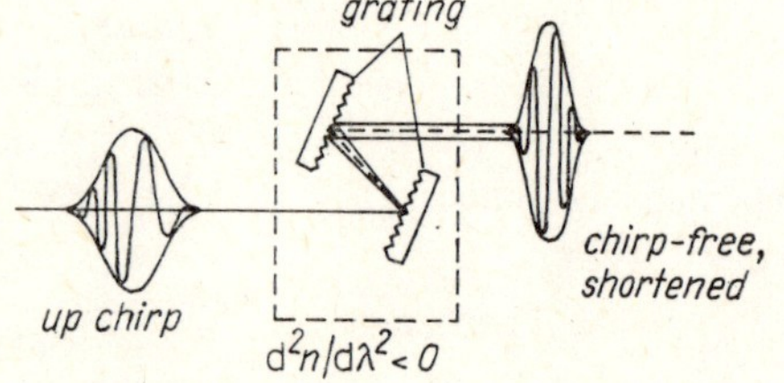

Fig. 8.10. Scheme for the compression of pulses exhibiting a monotonic frequency sweep
a) With negative frequency sweep (down chirp)
b) With positive frequency sweep (up chirp)

(compare figure 2.25). Combining (8.44) and (8.43) we obtain

$$A_L(z, \eta) = \frac{A_{max}}{\sqrt{2k_L'' iz \left(\frac{2 \ln 2}{\tau_{L0}^2} - i\beta_0\right) + 1}} \exp \left\{-2 \ln 2 \left(\frac{\eta}{\tau_L}\right)^2 - i2 \ln 2 \left(\frac{\eta}{\tau_L}\right)^2 \right.$$

$$\left. \times \left[\frac{4}{\tau_{L0}^2} \ln 2zk_L'' + \frac{\beta_0 \tau_{L0}^2}{2 \ln 2} (1 + k_L''\beta_0 z)\right]\right\}. \tag{8.46}$$

The pulse duration at z is determined by

$$\tau_L(z) = \tau_{L0} \sqrt{1 + 2\beta_0 k_L'' L_\beta \left[1 - \left(1 - \frac{z}{L_\beta}\right)^2\right]}. \tag{8.47}$$

L_β is a critical length, which is defined by

$$L_\beta = \frac{-2\beta_0 k_L'' \tau_{L0}^4}{(4 \ln 2 k_L'')^2 + (2\beta_0 \tau_{L0}^2 k_L'')^2}. \tag{8.48}$$

For $L_\beta > 0$ (i.e. $k_L''\beta_0 < 0$) the pulse is shortened on the path from $z = 0$ to $z = L_\beta$. At $z = L_\beta$ the pulse is bandwidth limited, so that L_β provides the optimal thickness of a

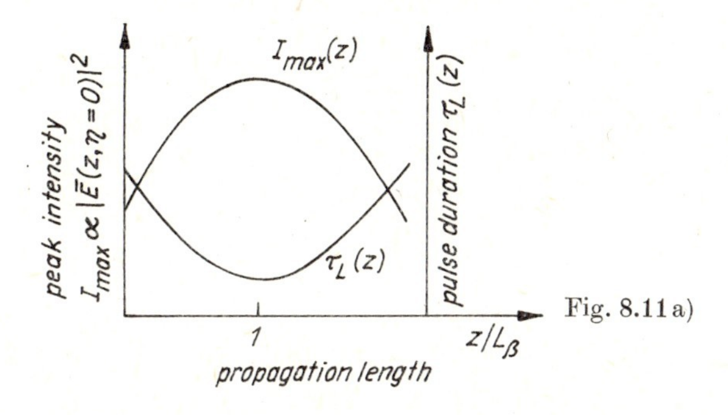

Fig. 8.11 a)

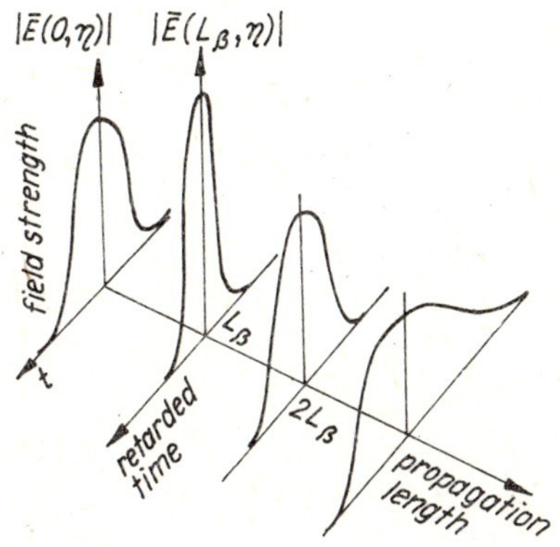

Fig. 8.11 b)

Fig. 8.11. Change in the duration of a chirped pulse as it passes through a dispersive glass plate

a, b) Theoretical results for the pulse change. c) Pulse with down chirp and variable initial length as it passes through BK5 glass with a thickness of 17 cm (from [6.21]). d) Up-and down-chirped input pulses having durations of 260 fs as they pass through SF5 glass of variable thickness (from [6.36]).

dispersive plate for an optimal pulse shortening at given input parameters. Note that L_β depends on the initial pulse length. In contrast bandwidth-limited Gaussian pulses ($\beta_0 \to 0$) double their pulse lengths after passing through the so-called dispersion length

$$L_D \approx 0.6 \frac{\tau_{L0}^2}{k_L''}. \tag{8.49}$$

Fig. 8.11a, b shows the dependence of the pulse duration $\tau_L(z)$ and the maximum intensity I_{max} on the pathlength z in the medium as well as the spatial and temporal pulse evolution according to the relation (8.46). Fig. 8.11c, d) shows, as an example, experimental results for the compression of phase modulated pulses in a glass plate. In Fig. 8.11c the shaping of a down-chirped femtosecond pulse ($\lambda = 0.61\ \mu m$) in a BK5 glass sample of 17 cm length is shown ([6.21]). The long input pulses ($\tau_{L0} > 0.17$ ps) experience a compression, whilst very short pulses are lengthened due to $z > L_\beta$. Fig. 8.11d shows the experimentally obtained shortening of a 260 fs down-chirped pulse and the behavior of an up-chirped pulse with respect to the external optical path in SF5 glass ([6.36]). For the pulse with increasing frequency (up chirp), only a broadening of the pulses is possible because $k_L'' > 0$. Instead of an element with "normal" dispersion ($k_L'' > 0$), for its shortening an optical element with $k_L'' < 0$ would have to be used[1]). For this purpose a grating can be employed, as shown in Fig. 8.10b.

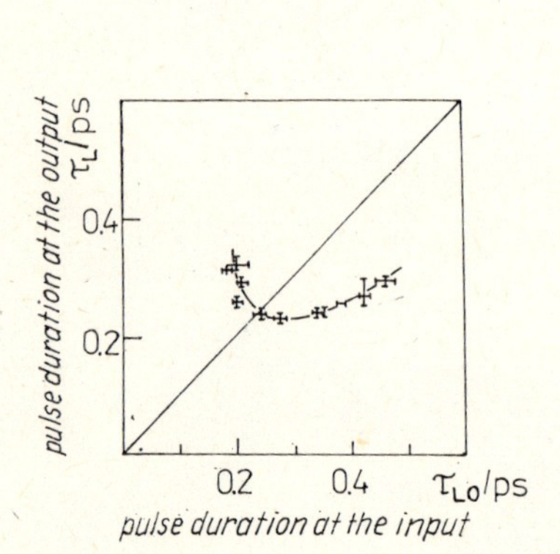

Fig. 8.11c)

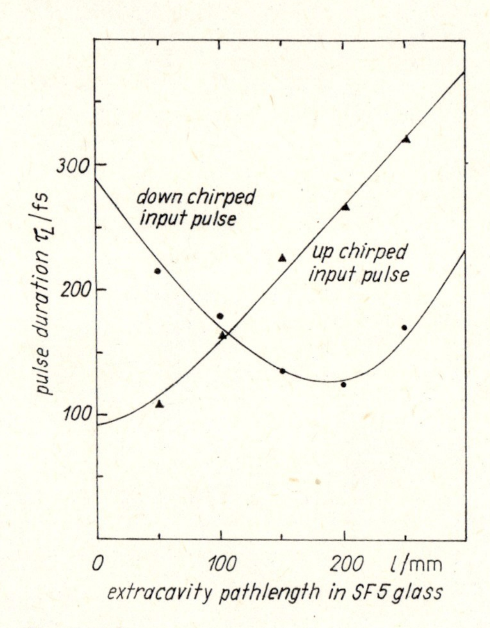

Fig. 8.11d)

---

[1]) In this connection, i.e. in problems of pulse propagation, the label "normal" dispersion, which has a different meaning from the same label in common optics, is used in the literature, and we have therefore placed it in inverted commas.
Whereas in the common normal dispersion the sign of the first derivative of the refractive index with respect to the wavelength $dn/d\lambda$ is essential ($dn/d\lambda < 0$ corresponds to "normal" dispersion), the equations for the pulse propagation contain the second derivative of the refractive index with respect to the wavelength ($d^2n/d\lambda^2 > 0$ corresponds to "normal" dispersion here). In optical materials that are in thermodynamic equilibrium $dn/d\lambda > 0$ (and hence anomalous dispersion) is only possible within absorption ranges. In contrast, values $d^2n/d\lambda^2 < 0$ and hence $k_L'' < 0$ can also appear in ranges where no absorption occurs. In optical glasses, for example, one finds $k_L'' > 0$ for most of the visible range, because the dispersion profile is determined essentially by the ultraviolet absorption bands; towards the red or near infrared spectral range k_L'' can however change its sign after a zero passage, because here the influence of the phonon transitions at large wavelengths is dominant in determining the dispersion. This means that up-chirped pulses can also be compressed in suitable optical materials. For this purpose one replaces the grating in Fig. 8.9b) with a block consisting of the corresponding material with $k_L'' < 0$. This possibility can be made use of in optical fibre communications. For the wavelength range 1.3–1.6 μm of the pulses employed here one finds weakly attenuating glasses with $k_L'' \approx 0$, in which pulses hardly spread apart, as well as glasses with $k_L'' < 0$, in which pulses with up chirp can be compressed.

8.3.3 *Pulse Propagation in a Nonlinear Optical Medium without Dispersion*

Let us now consider the propagation of a very short pulse in a medium with a nonlinear polarization $\bar{P}^{(3)}$, which is produced by the propagating pulse of frequency ω_L, itself. The nonlinear interaction can here be described by the nonlinear polarization

$$\bar{P}^{(3)}(z, t) = 2\varepsilon_0 n_0 \tilde{n}_2 \, |A(z, t)|^2 \, A(z, t) \tag{8.50}$$

which produces an intensity dependent change in the refractive index

$$n(z, t) = n_0 + \tilde{n}_2 \, |A(z, t)|^2 \tag{8.51}$$

$\tilde{n}_2$ is caused here by electronic transitions in the ultraviolet range, which provide the main contribution for a nonresonant nonlinear polarization in the visible and near infrared range. Other contributions, such as those produced by the Kerr effect, play a smaller part in solids in the visible and near infrared range.

Inserting (8.50) in (1.50) and neglecting group velocity dispersion the propagation of the pulse is described by

$$\mathrm{i} \frac{\partial A}{\partial z}(z, \eta) = \varkappa \, |A(z, \eta)|^2 \, A(z, \eta) \tag{8.52}$$

where $\varkappa = \dfrac{k_L \tilde{n}_2}{n_0}$. The solution is given by

$$A(z, \eta) = A_0(\eta) \exp\{-\mathrm{i}\varkappa \, |A_0(\eta)|^2 \, z\}. \tag{8.53}$$

Since $\varkappa$ is real in a loss-free medium, the pulse amplitude remains constant, whereas the time-varying phase changes during the passage through the medium. A significant phase change occurs beyond a propagation length of

$$L_{NL} = \frac{1}{\varkappa \, |A_{0max}|^2} = \frac{n_0}{\tilde{n}_2 k_L \, |A_{0max}|^2}. \tag{8.54}$$

If we approximate the input pulse around its peak with

$$|A_0(\eta)|^2 = |A_{0max}|^2 + \frac{1}{2} \frac{\partial^2}{\partial \eta^2} |A_0(\eta)|^2\Big|_{max} \, \eta^2 \tag{8.55}$$

we obtain an output pulse with a frequency that increases linearly with time i.e. with a constant chirp. The corresponding chirp parameter β is given by

$$\beta = \frac{1}{2} \tilde{n}_2 \left(-\frac{\partial^2}{\partial \eta^2} |A_0(\eta)|^2\Big|_{max} \right) z. \tag{8.56}$$

Thus, the nonlinearity of the medium causes the development of a chirp that increases linearly with the pathlength z in the medium (Fig. 8.12a). In most cases $\tilde{n}_2 > 0$ results from the electronic contributions to the nonlinearities. Consequently, β has a positive value which indicates an up chirp, i.e. an increasing frequency with respect to time. From the discussion in the last section it is evident that such a chirp can be compensated during a passage through a linear optical medium with group velocity dispersion ($k_L'' < 0$), whereby a pulse shortening is possible. Through an appropriate combination of linear and nonlinear optical elements, an effective pulse shortening can thus be achieved ([8.22], [8.24] to [8.27], [8.36], [8.37]). It must, however, be noted that the approximations considered here, which lead to a constant chirp, are only valid for the pulse center. At the pulse wings a time varying chirp results, which cannot be compensated in a linear optical dispersive element. Therefore the wings are not compressed like the pulse center, but lead to secondary pulses (see figure 8.12b).

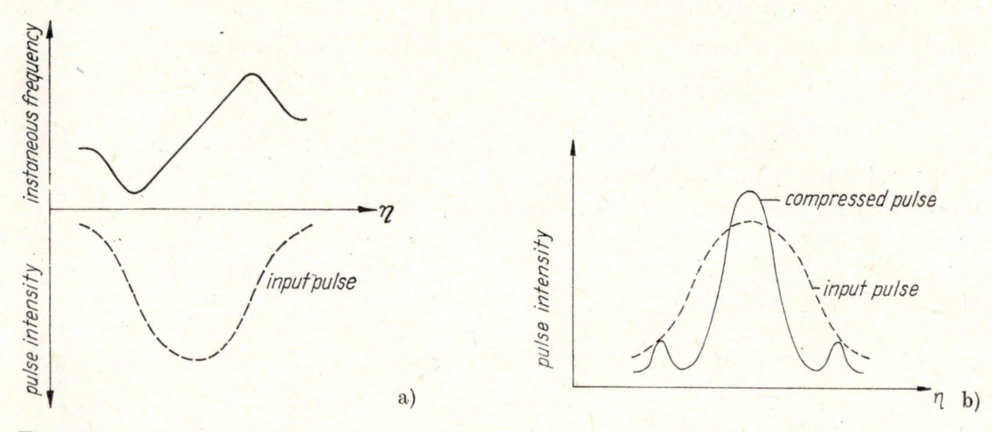

Fig. 8.12.
a) Chirp generation in a nonlinear optical medium
b) Compression of a chirped pulse after passing through a linear dispersive medium

8.3.4 *Dispersive Nonlinear Optical Media*

In many experimental situations such as pulse propagation through optical fibers, it is necessary to account simultaneously for the dispersion of the group velocity and the nonlinearity of the refractive index (see [8.36]). Although pulses do not propagate in fibers as plane waves, the pulse propagation in a single mode fiber can be described by the equation

$$\mathrm{i}\,\frac{\partial A}{\partial z}(z,\eta) = -\frac{k_{\mathrm{L}}''}{2}\,\frac{\partial^2 A}{\partial \eta^2}(z,\eta) + \varkappa\,|A(z,\eta)|^2\,A(z,\eta), \tag{8.57}$$

where k_{L}'' and $\varkappa$ must be replaced by the corresponding fiber parameters. For fibers with positive group velocity dispersion and positive nonlinear refractive index Grischkowsky and Balant [8.38] numerically solved equation (8.57), and Meinel [8.40] solved it analytically using the inverse scattering method. After the pulse has passed over a sufficiently long distance

$$z \gg \frac{0.6\tau_{\mathrm{L0}}}{\sqrt{\varkappa k_{\mathrm{L}}'' A_{0\max}}} \tag{8.58}$$

it obtains a nearly rectangular shape which is independent of its initial shape. The pulse duration is given by [8.40]

$$\tau_{\mathrm{Lr}} = 2.9\sqrt{k_{\mathrm{L}}''\varkappa}\,A_{0\max}z \tag{8.59}$$

and the electric field is

$$A_r \approx 0.6 \sqrt{\frac{A_{0\max}\tau_{L0}}{\sqrt{k_L''\varkappa\, z}}}. \tag{8.60}$$

The pulse exhibits a positive chirp that is constant over nearly the entire pulse which means that the frequency increases linearly with time where the total frequency sweep is

$$\Delta\omega_r \approx 1.4 A_{0\max} \sqrt{\varkappa}/\sqrt{k_L''}. \tag{8.61}$$

The instantaneous frequency of such chirped pulses has been measured directly in [8.62].

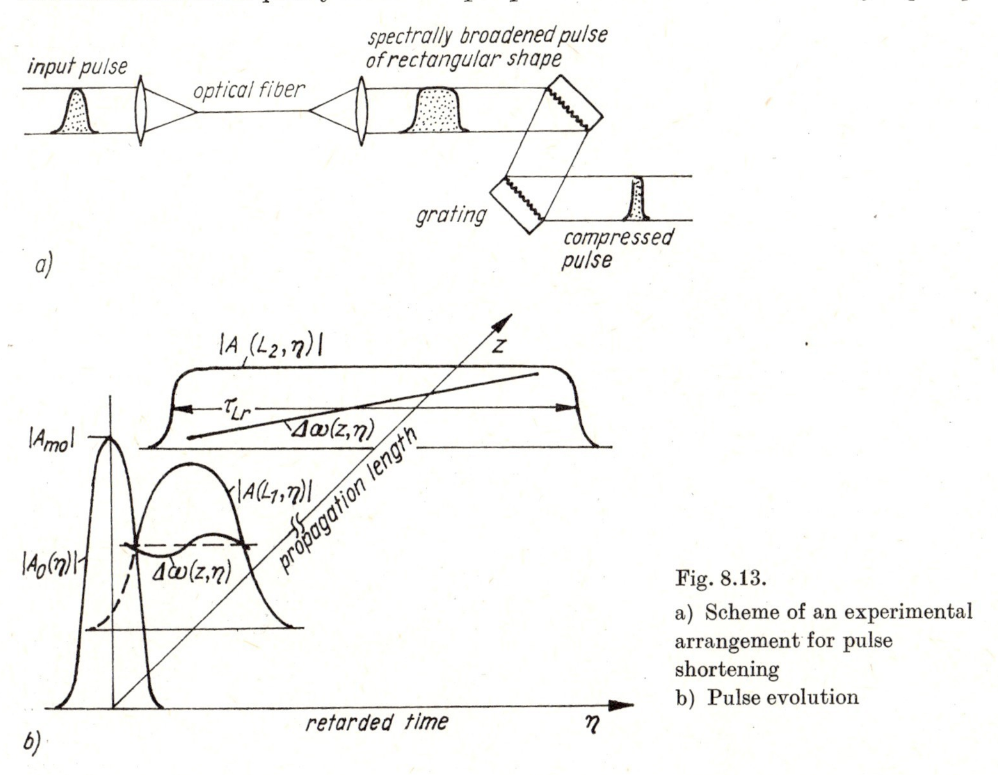

Fig. 8.13.
a) Scheme of an experimental arrangement for pulse shortening
b) Pulse evolution

Thus, the pulse is first broadened within the fiber (compare Fig. 8.13). However, then it can be very favorably compressed, in accordance with section 8.3.2, by passing through a linear medium with negative group velocity dispersion (e.g. in a grating compressor). The resulting pulse duration at the output of the compressor is given by

$$\tau_L \approx \frac{2.78}{\Delta\omega_r} \approx 2 \sqrt{\frac{k_L''}{\varkappa}} \frac{1}{A_{0\max}}. \tag{8.62}$$

The pulse length at the end of the shortening process thus becomes smaller with increasing amplitude of the input pulse and is independent of the input pulse duration. 95% of the maximum possible pulse shortening occurs after a fiber length

$$z = 5.6 \frac{\tau_{L0}}{\sqrt{\varkappa k_L''}\, A_{0\max}}.$$

This method of pulse shortening has been used successfully in a series of experiments ([6.31], [8.39], [8.41]). Using this method Nakatsuka et al. [8.39] shortened 5.5 ps pulses by the factor $s = 3.7$ by generating a self-phase modulation in a 70 m long optical fiber with subsequent compression in a near resonant atomic sodium vapor delay path. Given the typical values in single-mode fibers $\tilde{n}_2 \approx 1.5 \times 10^{-22}\ \mathrm{m^2/V^2}$, $\varkappa = 6.5 \times 10^{-16}\ \mathrm{m/V^2}$ and $k_L'' \approx 6.5 \times 10^{-26} s^2\ \mathrm{m^{-1}}$ we obtain a theoretical value of $s = 3$ according to (8.62) with a pulse of 10 W peak power and a fiber of 4 μm core diameter.

Johnson, Stolen and Simpson even successfully shortened 33 ps pulses by a factor of 80 ([8.57]). A similar technique was also successfully used for pulse shortening in the femtosecond range. Shank et al. ([6.31]) chirped a 90 fs pulse from a CPM laser (compare section 6.3.4) in a 15 cm fiber and subsequently shortend it to 30 fs in a grating

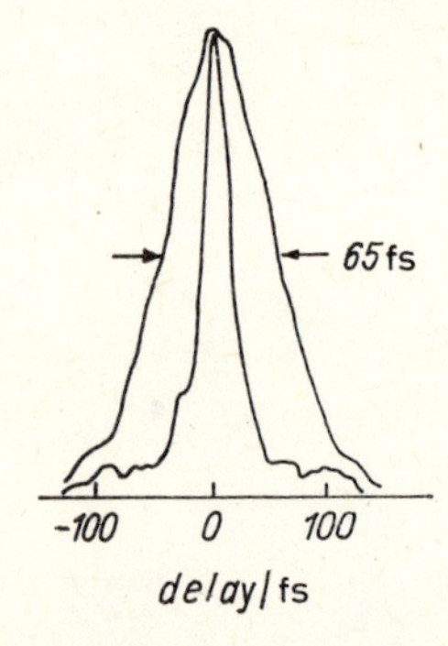

Fig. 8.14. SHG-autocorrelation trace of the input pulse as well as of the shortened pulse which has a pulse duration of 16 fs (from Fujimoto, Weiner, and Ippen [8.41a])

compressor. Ippen et al. succeeded in achieving 16 fs pulses (8.41a) using this method. Their recorded autocorrelation function of the initial and compressed pulse is shown, for example, in Fig. 8.14. Recently, pulses as short as 12 fs (8.41b) and 8 fs (8.41c), respectively were obtained in the described manner. Note that the shortest pulses produced up to now with durations of 8 fs contain only about 4 light oscillations. The possible limits of pulse compression have been discussed in [8.61].

Let us now discuss the solution of equation (8.57) for the case of negative group velocity dispersion ($k_L'' < 0$), which at long wavelengths occurs in many transparent media due to the influence of the IR absorption. For example single mode silicate glass fibers have a negative group velocity dispersion at low losses in the range $\lambda > 1.3$ μm. The pulse chirp produced in such media, as a result of the nonlinear refractive index, can now be simultaneously compensated by the linear optical processes, where under certain conditions bandwidth-limited and shortened pulses develop. The propagation of a pulse of the form

$$A(z = 0, \eta) = \frac{a}{\tau_L'} \sqrt{\frac{|k_L''|}{|\varkappa|}}\ \mathrm{sech}\left(\frac{\eta}{\tau_L'}\right) \tag{8.63}$$

was investigated analytically and numerically in [8.42, 8.43] where is a dimensionless amplitude factor and τ_L' is related to the pulse duration by $\tau_L = 1.76\tau_L'$. It is surprising that under such conditions soliton-like solutions result if a is an integer ($a = N$). For $N = 1$ the pulse shape remains unchanged as it propagates through the medium. For greater N, however, the pulse does not retain its shape but the solutions have a periodic behavior with the length z_0. After the period $z_0 = \dfrac{\pi}{2} \dfrac{(\tau_L')^2}{|k_L''|}$ the pulses reproduce their shape, though within this interval they split up, as shown in Fig. 8.15. Mollenauer, Stolen and Gordon confirmed this behavior experimentally [8.43]. They observed the shortening and splitting of 7 ps pulses from a synchronously pumped color center laser (F_2 centers in NaCl) using a 700 m single mode silicate glass fiber. At certain critical intensities of the input laser pulse the soliton-like behavior of the output pulse described above occurred. Mollenauer and Stolen [8.44] recently constructed a soliton laser, in which a single mode fiber is used as an additional feedback component (see

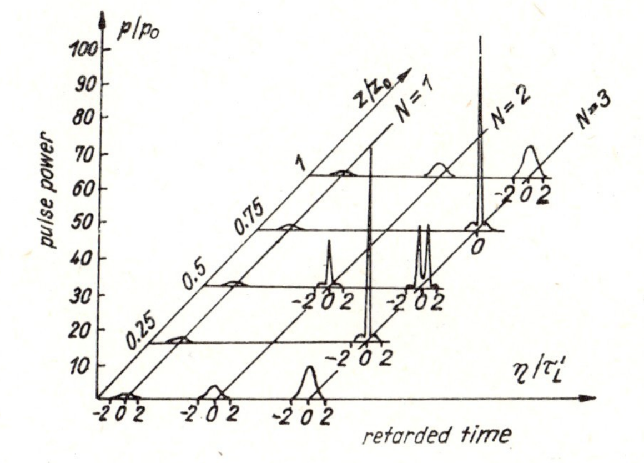

Fig. 8.15. Formation of soliton-like pulses in a dispersive nonlinear medium with negative group velocity dispersion $k_L'' < 0$ (from Mollenauer, Stolen and Gordon [8.43]). Note that with higher-order solitons substantial pulse shortening can be achieved at appropriate length of the sample [8.59].

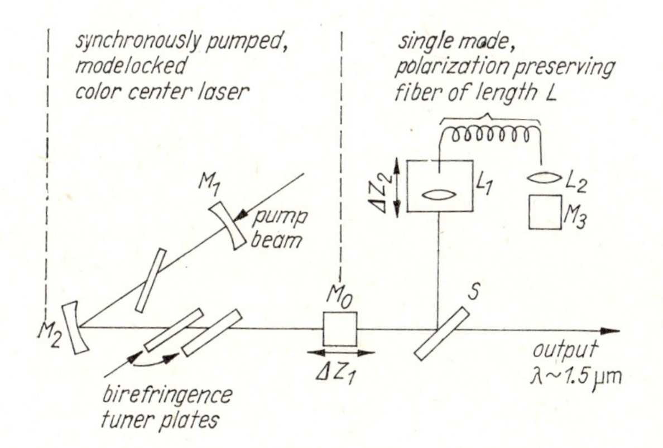

Fig. 8.16. Scheme of a soliton laser (from Mollenauer and Stolen [8.44])

Fig. 8.16). The soliton laser consists of a synchronously pumped color center laser, which is tunable from 1.4 to 1.6 μm. A portion of the laser output radiation is coupled in a single mode fiber, from which it is reinjected into the laser. If the length of the feedback loop is an integral multiple of the main cavity length, the back-coupled pulse superposes exactly with those already present in the main cavity. Without the feedback loop the laser pulse duration is about 8 ps. During the build up of the steady-state regime, the sub-pulse separated by the feedback loop is shortened as it passes through the fiber. After it is coupled back into the main cavity it causes the laser to generate shorter pulses. Eventually, after several round trips the pulse becomes soliton-like, and the profile of the output pulse corresponds to that of the one reinjected from the fiber, where ($N = 2$)-solitons were experimentally generated. The shortest pulses obtained from such lasers have a pulse duration of about 200 fs.

8.4 Resonant Nonstationary Processes

In the investigation of typical nonstationary processes with respect to the excitation of an atomic system with an ultrashort pulse, let us restrict ourselves to one photon processes. In previous discussions such processes have already played an important role at various points. In particular, saturable absorption in passive modelocking and saturation phenomena in the stimulated emission in all laser types proved to be fundamental for understanding these operational principles of the lasers. The typical time scales on which these processes occurred were given mostly by the steady-state conditions ($\tau_L \gg \tau_{21}, T_{21}$) or quasi-steady-state conditions ($\tau_L \gg \tau_{21}, \tau_L \lesssim T_{21}$). Thus, in both cases a description of these processes using rate equations was justified. In this section, however, let us deal with the investigation of such processes, which occur essentially in the non-steady-state range ($\tau_L < T_{21}, \tau_{21}$) and which cannot be described using rate equations. Consequently, in the equation for the off-diagonal elements of the density matrix (1.65) we may, in general, no longer neglect the time derivative (compare also 6.2.3.4 and 8.2.4).

8.4.1 *Optical Nutation and Free Optical Induction Decay*

In this and in the next sub-sections we want to investigate the emission of atomic systems in a thin sample, which results from an extremely short-time pulse excitation:

The pulse duration τ_L is assumed to be short compared to the transverse and longitudinal relaxation times τ_{21} and T_{21}, respectively, of the atomic transitions. The electric field is further assumed to be linearly polarized and nearly in resonance ($\omega_L \approx \omega_{21}$) with the atomic transition. Under these conditions we can neglect the relaxation processes during the light-matter interaction and obtain from (1.61), (1.62) and (1.65)

$$\frac{\partial}{\partial t} \tilde{\varrho}_{12}(t) = \frac{i}{2\hbar} \mu_{12} \Delta\varrho(t) A_L(t) \tag{8.64a}$$

and

$$\frac{\partial}{\partial t} \Delta\varrho(t) = -\frac{i}{\hbar} \mu_{12}[\tilde{\varrho}_{21}(t) A_L(t) - \tilde{\varrho}_{12} A_L^*(t)] \tag{8.64b}$$

where

$$\mu_{21} = (\vec{\mu}_{21}\vec{e}_L)$$

and

$$\Delta\varrho(t) = \varrho_{22}(t) - \varrho_{11}(t).$$

For the sake of simplification we assume that the amplitude of the incident radiation is real and changes only slightly as the pulse passes through the ensemble of atomic systems. With the initial condition $\tilde{\varrho}_{12} = 0$, which is always satisfied in thermal equilibrium, the real part of $\tilde{\varrho}_{12}$ disappears at all times. If we designate the corresponding imaginary part by Im $\{\tilde{\varrho}_{12}\} = \tilde{\varrho}_{12}''$, the equations (8.64a,b) can be reduced to

$$\frac{\partial}{\partial t} \tilde{\varrho}_{12}''(t) = \frac{1}{2\hbar} \mu_{12} \Delta\varrho(t) A_L(t), \tag{8.65a}$$

$$\frac{\partial}{\partial t} \Delta\varrho(t) = -\frac{2}{\hbar} \mu_{12} \tilde{\varrho}_{12}''(t) A_L(t). \tag{8.65b}$$

Thus, this system of equations relates the off-diagonal element of the density matrix $\tilde{\varrho}_{12}''$, which is phase-shifted against the electric field by 90°, to the probability of the occupation inversion between the upper and lower levels $\Delta\varrho = \varrho_{22} - \varrho_{11}$. The macroscopic meaning of these equations can easily be seen: In a homogeneously broadened transition $\Delta N = N_2 - N_1 = N\Delta\varrho$ represents the density of the occupation inversion, and $\overline{P}_2 = 2N\mu_{12}\tilde{\varrho}_{12}''$ the amplitude of a polarization wave that is phase-shifted by 90° against the exciting electric field. In terms of these macroscopic quantities (8.65) can be rewritten as

$$\frac{\partial}{\partial t} \overline{P}_2(t) = \frac{1}{\hbar} \mu_{12}^2 \Delta N(t) A_L(t), \tag{8.66a}$$

$$\frac{\partial}{\partial t} \Delta N(t) = -\frac{1}{\hbar} A_L(t) \overline{P}_2(t). \tag{8.66b}$$

With the intitial conditions $\Delta N(t_0) = \Delta N^e = -N$ and $\overline{P}_2(t_0) = 0$ this system has the solution

$$\overline{P}_2(t) = -N\mu_{12} \sin \sigma(t) \tag{8.67a}$$

and

$$\Delta N(t) = -N \cos \sigma(t), \tag{8.67b}$$

where

$$\sigma(t) = \frac{1}{\hbar} \mu_{12} \int_{t_0}^{t} \mathrm{d}t' A_L(t'),$$

which can easily be verified by substitution. $\sigma(t)$ is the pulse area. If the amplitude $A_L(t)$ is constant after the turn-on at the moment t_0, the occupation inversion and the polarization amplitude vary periodically with time, where the corresponding angular frequency

$$\Omega = \frac{\mathrm{d}}{\mathrm{d}t} \sigma(t) = \frac{1}{\hbar} \mu_{12} A_L \tag{8.68}$$

is called the Rabi frequency. With deviation from the exact resonance the general expression

$$\Omega = \sqrt{(\omega_L - \omega_{21})^2 + \frac{1}{\hbar^2} \mu_{12}^2 A_L^2} \tag{8.68'}$$

holds for the Rabi frequency. Under the conditions given here the atomic ensemble has a completely different behavior from that described by the rate equations. A pulse having the pulse area $\sigma = \pi$ converts the ensemble into a completely inverted state ($N_2 = N, N_1 = 0$), whereas a 2π-pulse restores the system to the initial state again. Accordingly the polarization amplitude reaches its maximum after the passage of a $\pi/2$-pulse. The energy is obviously exchanged periodically between the field and the atomic ensembles with the Rabi frequency $\Omega = \mu_{12}A_L/\hbar$. This process is called optical nutation or Rabi oscillation. No actual absorption takes place here, because it only occurs when the neglected relaxation processes become active. After the pulse excita-

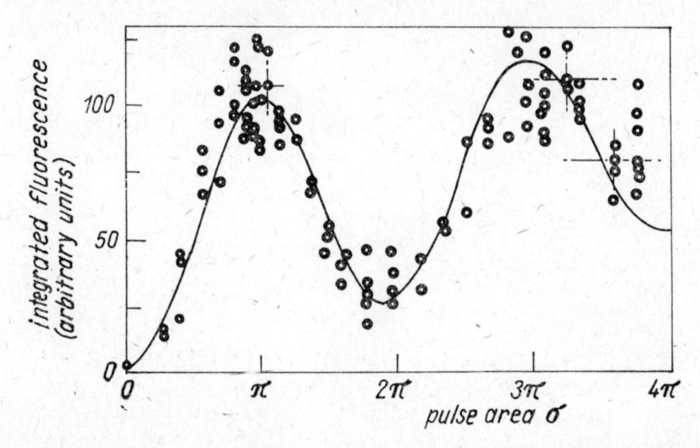

Fig. 8.17. Integrated fluorescence as a function of the pulse area of the input pulse (from [9.25]) Zeeman transition $5p^2P_{1/2}(F = 2, M_J = 1/2, M_I = 3/2) \to 5s^2S_{1/2}(F = 2, M_J = -1/2, M_I = 3/2$ of ^{87}Rb under the influence of a magnetic field of 7.45 Tesla; excited by a ^{202}Hg laser ($\lambda_L = 794.466$ nm)

tion we can use the corresponding equations of motion including the relaxation terms, but not the radiation influence, to describe the atomic systems and we obtain

$$\overline{P}_2(t) = \overline{P}_2(t_E)\, e^{-(t-t_E)/\tau_{21}} \tag{8.69a}$$

and

$$\Delta N(t) = -N + [\Delta N(t_E) + N]\, e^{-(t-t_E)/T_{21}}, \tag{8.69b}$$

where t_E correspond to the end of the excitation pulse — compare 1.2.3.

The periodic dependence of the occupation inversion on the pulse area of the excitation pulse can be favorably measured, in that after the coherent excitation the power of the incoherently emitted fluorescence and its temporal integral, respectively are measured. Fig. 8.17 shows an experimental result. The number of incoherently emitted fluorescence photons thus increases not monotonically with the exciting power, as is usual, but exhibits the typical Rabi oscillations. The fluorescence does not decrease to zero because the assumptions of $\tau_L \ll T_{21}$ and of exact resonance $\omega_L = \omega_{21}$ are not completely satisfied. In particular, the condition of exact resonance for all atomic transitions is violated, because the ensemble is not only homogeneously broadened, which was assumed here for the sake of simplification.

In the measurements described, care has to be taken to record only the incoherently emitted fluorescence. Besides the incoherent process coherent light is also emitted, which is caused by the polarization wave of the amplitude $\overline{P}_2$. Let us assume that the atomic systems are in a volume element, whose dimensions are small compared to the light wavelength. In this case all transition dipoles oscillate with the same phase. The oscillation of this super dipole, whose amplitude $\hat{\mu}_g$ is proportional to the particle number, leads to a collective radiation effect which is connected with a particularly strong emission, the so-called free optical induction decay. We can calculate approximately the emission of this super dipole within the framework of classical electrodynamics, according to which the emitted power is proportional to the temporal average of the square of the second derivative of the dipole moment, hence $\overline{(d^2\mu_g/dt^2)^2}$. Consequently, the emitted power is proportional to the square of the number of coherently excited atomic systems, whereas the incoherent fluorescence is proportional to the particle number itself. A more detailed quantum-theoretical treatment of this problem is found for instance in [24]. Under the conditions assumed here (homogeneously broadened transition, neglect of the radiative decay with respect to other relaxation processes), the collective polarization and emission decay after the excitation with the relaxation time τ_{21}. However, in inhomogeneously broadened media with the linewidth $\Delta\omega_{inh}$ which are assumed to be large as opposed to the homogeneous width $2/\tau_{21}$, the polarization as well as the super dipole already decay after times of the order $(\Delta\omega_{inh}^{-1})$. This occurs because the individual atoms oscillate at their (various) eigenfrequencies after the excitation, whereby the inital phase synchronism caused by the coherent excitation is destroyed. After the relaxation of the polarization the incoherent emission processes may continue since, like the occupation inversion, they decay after times of the order T_{21}. It is possible to distinguish between coherent and incoherent emission by their different temporal evolution and radiation characteristics (see, e.g., [31]–[35]).

8.4.2 Photon Echo

In contrast to the irreversible decay of the polarization due to relaxation processes, its destruction resulting from different resonance frequencies in inhomogeneously broadened media is a reversible process. Therefore, as long as the duration of the process is small compared to τ_{21}, it is possible to reproduce the original state. For this purpose we can proceed as follows: a maximum polarization amplitude is generated by irradiating a $(\pi/2)$-pulse. For an accurate calculation of the action of the pulse we must generally begin with the basic equations (1.61) and (1.65), which include a detuning between the pulse mid-frequency ω_L and the atomic transition frequencies ω_{21} and can therefore be used to describe the individual particle groups. If we assume for the sake of simplification that the pulse is extremely short ($\tau_L < 1/\Delta\omega_{inh}$) and that the laser frequency ω_L corresponds to the center of the inhomogeneously broadened line ($\omega_L = \omega_{21}^0$), we can neglect the deviations $(\omega_L - \omega_{21})$ during the pulse excitation. Under these assumptions it is sufficient to solve (8.66). At the moment $t_0 = 0$ at the end of the extremely short $(\pi/2)$-pulse, $\overline{P}_1(0) = 0$, $\overline{P}_2(0) = -N\mu_{12}$ and $\Delta N(0) = 0$ hold. Let us now trace the temporal evolution subsequent to the pulse over a greater time interval, in which a neglect of the resonance deviations is not justified. Therefore, for the further treatment of the particle ensemble with the resonance frequency ω_{21} we start with the relation (1.65), which in the zero-field case ($A_L = 0$) yields the solution

$$\tilde{\varrho}_{12}(t) = \tilde{\varrho}_{12}(t_0)\, e^{-\left[\frac{1}{\tau_{21}} + i(\omega_L - \omega_{21})\right](t - t_0)} \tag{8.70}$$

where

$$\tilde{\varrho}_{12}(t_0 = 0) = -\frac{i}{2}\mu_{12}.$$

For the difference of the diagonal elements of the density matrix, $\Delta\varrho(t) \equiv 0$ holds after the action of the $(\pi/2)$-pulse, which produced $\Delta\varrho(0) = 0$. We obtain the polarization amplitude of the total ensemble by adding up the contributions of all the partial ensembles, hence by integrating over the inhomogeneously broadened transition with the line shape factor $g_{\text{inh}}(\omega_{21} - \omega_{21}^0)$ according to $\bar{P}(t) = 2\mu_{12}N \int_{-\infty}^{\infty} d\omega_{21}\tilde{\varrho}_{12}(t, \omega_{21})\, g_{\text{inh}}(\omega_{21} - \omega_{21}^0)$.

Setting $\omega_{21} - \omega_{21}^0 = \omega'$ and $\omega_L = \omega_{21}^0$ and inserting $\tilde{\varrho}_{12}(t)$ we obtain from (8.70)

$$\bar{P}(t) = \bar{P}(0)\, e^{-t/\tau_{21}} \int_{-\infty}^{\infty} d\omega' g_{\text{inh}}(\omega')\, e^{i\omega' t}, \tag{8.71}$$

where $\bar{P}(0) = -iN\mu_{12}$.

This amplitude is destroyed in times of the order $2\pi/\Delta\omega_{\text{inh}}$ because the monotonically decreasing Fourier transform of $g_{\text{inh}}(\omega')$ for $t > 0$ has already taken on small values for $t > 2\pi/\Delta\omega_{\text{inh}}$.

Now, an extremely short π-pulse is irradiated with a delay t_D with respect to the first pulse $\left((1/\Delta\omega_{\text{inh}}) \ll t_D \lessgtr \tau_{21}\right)$. First, we calculate again its influence on particles with the resonance frequency ω_{21}, for which the temporal evolution according (8.70) is valid up to the interaction with the π-pulse. During the short pulse duration we neglect again the frequency detunings. It is obvious from (8.64) that for real A_L the π-pulse leaves $\tilde{\varrho}'_{12}$ unchanged, whereas it reverses the sign of $\tilde{\varrho}''_{12}$. Hence, immediately after the π-pulse we obtain

$$\tilde{\varrho}_{12}(t_D) = -\tilde{\varrho}_{12}(0)\, e^{-t_D/\tau_{21}}\, e^{i(\omega_L - \omega_{21})t_D}. \tag{8.72}$$

After the excitation by the π-pulse, hence for $t > t_D$, $\tilde{\varrho}_{12}$ varies according to (8.70), where the value $\tilde{\varrho}_{12}(t_D)$ from (8.72) has to be substituted for the initial value $\tilde{\varrho}(t_0)$. Thus, we obtain for $t > t_D$,

$$\tilde{\varrho}_{12}(t) = -\tilde{\varrho}_{12}(0)\, e^{-t/\tau_{21}}\, e^{-i(\omega_L - \omega_{21})(t - 2t_D)}. \tag{8.73}$$

As described in connection with (8.71), we calculate from (8.73) the macroscopic polarization, which for $t > t_D$, i.e. after the action of the π-pulse, is thus given by

$$\bar{P}(t) = -\bar{P}(0)\, e^{-t/\tau_{21}} \int_{-\infty}^{\infty} d\omega' g_{\text{inh}}(\omega')\, e^{i\omega'(t - 2t_D)}. \tag{8.74}$$

After the doubled delay time $t = 2t_D$ the polarization becomes

$$\bar{P}(2t_D) = -\bar{P}(0)\, e^{-t_D/\tau_{21}} \tag{8.75}$$

where the normalization condition $\int_{-\infty}^{\infty} d\omega' g_{\text{inh}}(\omega') = 1$ was used. Obviously the initial state is reproduced apart from the relaxation factor $\exp(-t_D/\tau_{21})$. Therefore, a strong emission can again occur after the time $2t_D$ due to a collective radiation effect. This phenomenon is called photon echo. If we increase the delay time t_D, we can determine the relaxation time τ_{21} directly from the decrease in the echo signal according to (8.72). This method provides one possible mean of distinguishing homogeneous line broadening processes from inhomogeneous ones. The photon echo occurs, because due to the π-pulse the phase values of the dipole oscillations of the individual groups of atoms which were achieved up to t_D are replaced by the same phase values but with the opposite sign. This means that the advance of the fastest oscillating dipole is inverted in an equally large delay. After the time $2t_D$ the fast atomic systems have again overtaken this delay. To imagine this let us consider the following model. After the passage of the $\pi/2$ pulse all atoms start simultaneously, like runners in a stadium. After several rounds this synchronization is completely destroyed, because the fastest runners have reached advantage on the order of one round. The π-pulse acts as a signal, upon which all runners reverse their direction. In this manner the synchronization is reproduced after the doubled transit time.

Kurnit, Abella and Hartmann detected photon echoes for the first time in 1964 by radiating ruby crystals with ruby laser pulses [8.46]. In this solid the conditions mentioned initially could already be achieved at low temperature for ns pulses. In the meantime numerous gaseous and solid substances were investigated using pulses from the μs to the ps range [8.47, 8.48, 8.60]. Figure 8.18 shows a basic arrangement and the dependence of the echo signal on the pulse delay t_D which enables us to determine the transverse relaxation time τ_{21}.

8.4.3 Self-Induced Transparency

In the last two subsections we have mainly considered the emission of a very thin sample, and therefore we could neglect in particular the influence on the pump pulse. At greater interaction lengths in the sample or high densities of the resonant interacting atoms this influence can, however, no longer be neglected, i.e. for the change in the field amplitude, equation (1.50) must be taken into account. Thus, under the same conditions as in (8.66) the following equations hold:

$$\frac{\partial}{\partial t} \bar{P}_2(z, t) = \frac{1}{\hbar} \mu_{12}^2 \Delta N(t, z) A_L(t, z), \tag{8.76a}$$

$$\frac{\partial}{\partial t} \Delta N(t, z) = -\frac{1}{\hbar} A_L(t, z) \bar{P}_2(t, z), \tag{8.76b}$$

$$\left(\frac{\partial}{\partial z} + \frac{1}{v}\frac{\partial}{\partial t}\right) A_L(z, t) = -\frac{\mu_0 \omega_L{}^2}{2k_L} \bar{P}_2(t, z), \tag{8.76c}$$

where $\bar{P}_2$ is the imaginary part of the nonlinear polarization. We have already given the solution of the first two equations according to (8.67), where now the z-dependence of the pulse area must be taken into consideration:

$$\sigma(t, z) = \frac{1}{\hbar} \mu_{12} \int_{t'}^{t} dt' A_L(t', z). \tag{8.77}$$

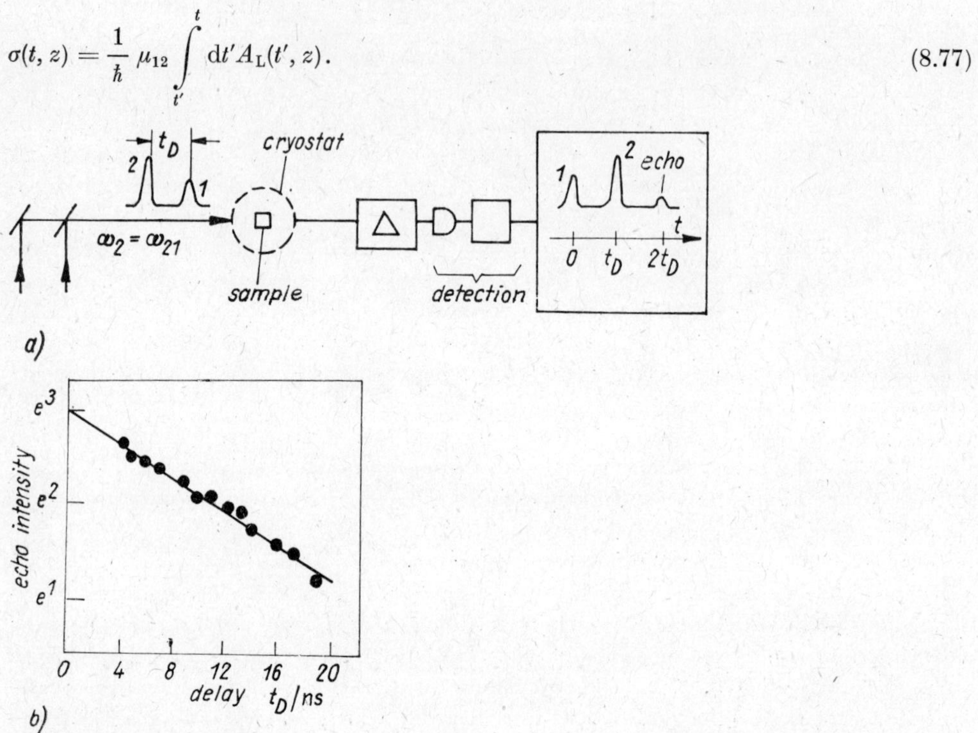

Fig. 8.18. Photon echo

a) Scheme

A π-pulse (2) passes through the sample after the delay t_D with respect to the passage of a $\pi/2$-pulse. Both pulses have the mid-frequency $\omega_L = \omega_{21}$. The sweep of the oscilloscope is triggered by the first pulse. Besides the signals at the instants $t = 0$ and $t = t_D$ a photon echo signal can be observed at $t = 2t_D$ at the oscilloscope.

b) Intensity of the echo signal (relative units) as a function of the delay t_D between the first and the second excitation pulses (from [8.45]).

The sample was a p-terphenyl crystal doped with pentacene ($\mathcal{T} = 4$ K).

If we substitute

$$A_L(t, z) = \frac{\hbar}{\mu_{12}} \frac{\partial}{\partial t} \sigma(t, z)$$

in (8.76c) and use $\overline{P}_2(t, z)$ from (8.76a) we obtain

$$\left(\frac{\partial^2}{\partial z\, \partial t} + \frac{1}{v} \frac{\partial_2}{\partial t^2}\right) \sigma(t, z) = -\beta \sin \sigma(t, z) \tag{8.78}$$

where $\beta = \frac{\mu_0 \omega_L{}^2 \mu_{12}^2}{2k_L \hbar} N$.

Let us now investigate the question of whether there are solutions for equation (8.78), which describe a form-stable and loss-free pulse propagation in a resonant medium. A light pulse propagates through the medium without changing its shape, if its amplitude only varies with the coordinate z and time t with respect to the "retarded time" $\eta = t - \frac{z}{v_S}$. v_S is here the velocity of the pulse envelope that deviates from the usual defined group velocity. These kinds of pulses, which propagate without changing their envelope, are designated as solitons.

For a soliton we can substitute $\frac{d}{d\eta}$ for $\frac{\partial}{\partial t}$ and $-\frac{1}{v_S}\frac{d}{d\eta}$ for $\frac{\partial}{\partial z}$ whereby (8.78) reduces to a common differential equation

$$\frac{d^2}{d\eta^2}\sigma(\eta) = \tilde{\omega}^2 \sin\sigma(\eta), \tag{8.79a}$$

where

$$\tilde{\omega}^2 = \beta v \left(\frac{v}{v_S} - 1\right)^{-1}. \tag{8.79b}$$

Thus, we have obtained an ordinary nonlinear differential equation similar to the pendulum equation, whose solution is well-known. Since we are interested in pulse-like solutions, $A(\eta)$ must vanish for $\eta \to \pm\infty$ as well as $\frac{d\sigma}{d\eta}$ and $\frac{d^2\sigma}{d\eta^2}$ due to (8.77) and (8.79a). From (8.79a) and our discussion of the Rabi-oscillation we know that the final state of the resonant medium corresponds to the initial state if $\sigma(\infty) = 2\pi m$ (m is an integer). Due to this requirement on the pulse area at $\eta = \infty$ and its derivatives, the solution of the pendulum equation is uniquely determined by

$$\sigma(\eta) = 4 \arctan(e^{\tilde{\omega}\eta}). \tag{8.80}$$

Hence the field amplitude is given by

$$A\left(t - \frac{z}{v_S}\right) = A_{max} \operatorname{sech}\left[\frac{\mu_{21}}{2\hbar} A_{max}\left(t - \frac{v}{v_S}\right)\right], \tag{8.81}$$

where $A_{max} = \frac{2\hbar\tilde{\omega}}{\mu_{12}}$.

From (8.79b) it is obvious that the propagation velocity of the soliton depends on its field amplitude. From (8.79b) and the definition of β and A_{max} we have

$$\frac{v}{v_S} - 1 = \frac{2\hbar\omega_L N}{\varepsilon_0 A_{max}^2}. \tag{8.82}$$

The important characteristics of such a pulse are as follows:

(1) After the pulse has passed through the medium the atomic system returns to its initial state, i.e. $\Delta N(t \to \infty) = -N, \overline{P}_2(t \to \infty) = 0$. Consequently, an energy exchange between the atomic system and the electromagnetic field occurs only intermittently; the medium is completely transparent regarding the total energy.

(2) The "total area" of the pulse $\sigma(\eta \to \infty) = \frac{\mu_{12}}{\hbar}\int_{-\infty}^{\infty} dt' A(t')$ is equal to 2π. We have already shown that usually such a pulse does not exchange energy with the medium. This phenomenon is called self-induced transparency.

(3) The passage of the soliton through the medium is loss-free and proceeds without the pulse changing its shape, i.e. its field amplitude depends only on the retarded time $\eta = t - \frac{z}{v_S}$. v_S is the propagation velocity of the pulse which deviates considerably from the phase velocity ω/k and the linear optical group velocity $v = \frac{d\omega}{dk}$ and depends not only on the characteristics of the medium but also on the peak amplitude A_{max}.

The phenomenon of self-induced transparency and the formation of solitons was first investigated by McCall and Hahn ([8.50]). The treatment of homogeneously broadened transitions made here can be extended without difficulty to inhomogeneously broadened atomic systems.

McCall and Hahn also conducted the first experiments on self-induced transparency ([8.50]). They irradiated a ruby crystal, which had been cooled by liquid helium, with the pulses of a ruby laser that had been cooled by liquid nitrogen. Through the temperature difference coincidence between the frequency of the laser transition $\bar{E}(2E) \leftrightarrow 4A_2\left(\pm\frac{3}{2}\right)$ and the absorption transition $4A_2\left(\pm\frac{1}{2}\right) \leftrightarrow \bar{E}(2E)$ could be achieved. The transverse relaxation time at the temperature of liquid helium was about 50 ns. Accordingly, it was considerably larger than the pulse length of the ruby laser (5 to 10 ns). Numerous investigations in gases and solids have been conducted since these first experiments, where the transmission and the propagation velocity have been measured with respect to the pulse area (see e.g. [11], [24], [3], [8.51], [8.52]). It was found that a pulse of any given arbitrary area changes into $(m \times 2\pi)$-pulses (m is an integer) as it propagates through the sample and eventually splits into a series of stable 2π-pulses. Up to now most experiments in optical nutation, free induction decay and photon echo were conducted in narrow transitions in solids at low temperatures and in gases, where it is possible to work with relatively long pulses of high quality and slow detection systems. With the ps-pulses and sub-ps-pulses of passively modelocked and continuously pumped dye lasers and nonlinearly optical detection methods it is now also possible to observe these effects in samples with strong line broadenings — for example, in liquids (see e.g. [28]—[35]).

9. Ultrafast Spectroscopy

One of the most important applications of ultrashort light pulses is in the area of spectroscopy where it has become possible for the first time to observe directly fast microphysical, chemical and biological processes.

The measuring methods of ultrafast spectroscopy are composed of two steps. The first step consists of exciting the sample to be investigated with an intense ultrashort

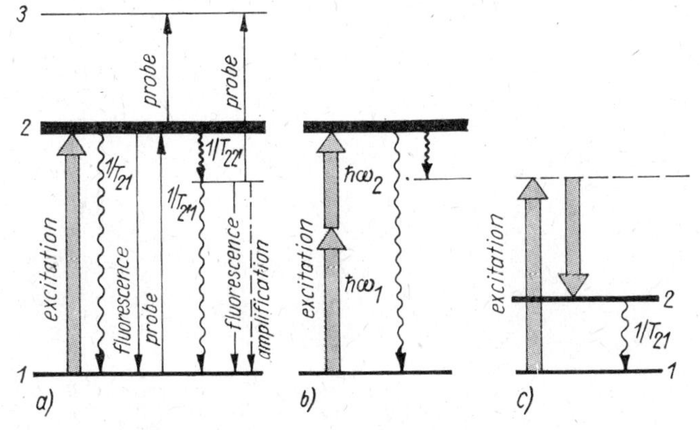

Fig. 9.1. Energy level scheme with radiative transitions (excitation, probe, fluorescence) and radiationless transitions

a) Excitation through one-photon absorption
b) Excitation through two-photon absorption
c) Excitation through stimulated Raman scattering

light pulse, whereby energy is transferred to the sample during the interaction. The excitation of the sample can occur by means of a one- or multi-photon absorption or an inelastic photon scattering, such as the Raman scattering (see Fig. 9.1). As a result of the transient interaction with the radiation field the sample passes from a state of thermodynamic equilibrium into a state of nonequilibrium. After the passage of the excitation pulse balancing processes occur which can be of local or non-local nature, and as a result the sample strives toward the initial or a changed equilibrium or quasi-equilibrium state. Various kinds of relaxation processes and reactions, energy and charge transfer as well as diffusion processes may be involved in these balancing processes. During the course of the balancing processes various measurable parameters of the sample generally undergo a change, through which an observation becomes possible. The second step of the ultrashort spectroscopy consists of measuring the temporal evolution of such parameters. In ultrafast spectroscopy the time dependence of an optical quantity is measured; for example, the intensity of the fluorescence, the spectral absorptivity or reflectivity or the refractive index. With a suitable choice of the wavelength of the exciting radiation as well as of the detection, these optical methods are very informative, because the various types of physical and chemical processes generally produce a time varying absorptivity in certain spectral ranges and are sometimes accompanied by an emission of light with a characteristic spectral distribution.

Two manufacturers (The Center for Scientific Instruments of the Academy of Sciences of the GDR [9.59] and Applied Photophysics [9.50]) already offer modular systems for ultrashort spectroscopy, which are based on industrially manufactured noble gas ion lasers (from Carl Zeiss Jena and Spectra Physics) and which make the construction of different fluorescence and absorption probe beam spectrometers possible.

9.1 Fluorescence Measurements

9.1.1 Nanosecond Techniques

The time-resolved measuring of fluorescence spectra belongs to the oldest short-time measuring techniques. Even with the use of classical light sources, the sub-ns range could be reached with techniques of this kind (see e.g. [15, 9.1, 9.2]). Once lasers had been developed these techniques could be applied to them as well. In Fig. 9.2 the sim-

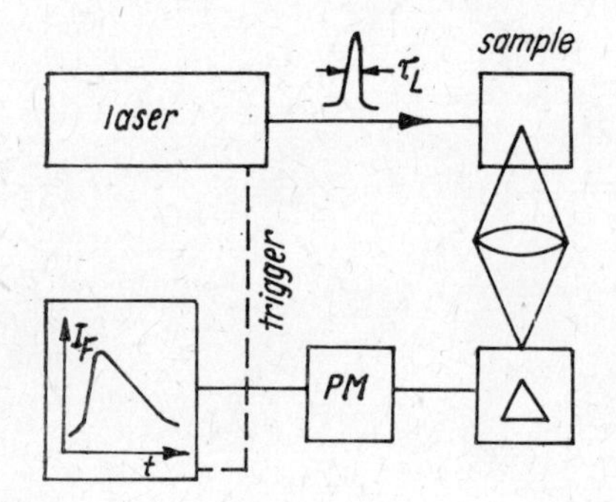

Fig. 9.2. Electronic recording of the fluorescence decay process

A laser pulse of duration τ_L excites the sample. The fluorescent light passes through a monochromator and strikes a fast photomultiplier PM whose output signal appears on the oscilloscope. The sweep of the oscilloscope can be triggered by the laser.

plest measuring principle is shown. A pulsed laser excites a sample to fluoresce. The emitted light is detected time-resolved by a detector with subsequent amplification and displayed on an oscilloscope. The time resolution is determined by the detector and the detection electronics and in favorable cases attains values of some 10^{-10} s. A monochromator can be introduced in the fluorescence beam path. The same fundamental arrangement can be modified by applying a sampling technique, in which a boxcar integrator is used in the electronic detecting system. As described in chapter 3 the sampling technique is specially suited for the measuring of periodically occurring signals. The short excitation pulse triggers the boxcar integrator, which then measures and records the signal from the detector after a certain delay time with a given gate width. The time delay can be changed from pulse to pulse, and consequently the fluorescence is recorded at different moments. By using a suitable electronic signal storage we can repeat the measuring process without difficulty, take an average and thus increase the accuracy. Let us again emphasize that in the sampling technique the time resolution is determined exclusively by the photoelectric detector and the boxcar integrator,

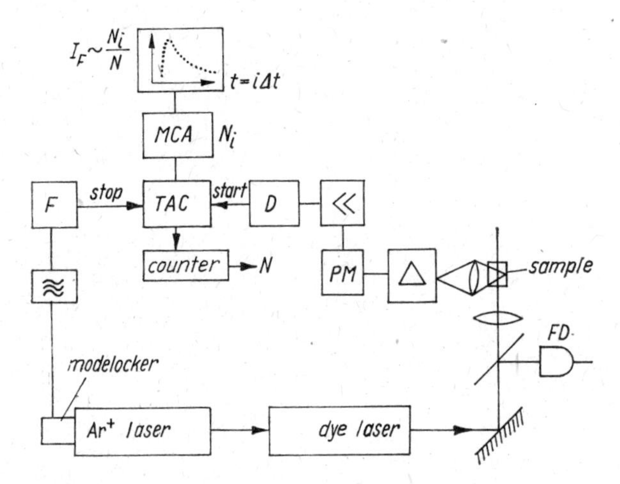

Fig. 9.3. Time correlated photon counting for measuring fluorescence decay processes (compare e.g. [9.5, 9.6])

FD is the photodiode for controlling the laser; PM is the photomultiplier; TAC is a time-to-amplitude converter; MCA is the multichannel analyzer; F is the pulse shaping stage, D the discriminator stage.

whereby very good boxcar integrators allow us to measure time intervals of about 0.1 ns [9.3]. An example of such a set-up based on the sampling technique is the fluorescence measuring set-up LIF 200 of the Center for Scientific Instruments of the Academy of Sciences in the GDR [9.4], in which a nitrogen laser excites the sample to fluoresce and the signal is recorded using a boxcar integrator. With this device a time resolution of $\lesssim 1$ ns is achieved.

Another very favorable detection technique for fluorescence spectroscopy is photon counting. Fig. 9.3 provides an example in which the sample is excited by a synchronously pumped dye laser (compare 5.). After spectral filtering in a monochromator the fluorescence enters a photomultiplier. The excitation is kept so weak that for each laser pulse with high probability only one or no fluorescence photon is detected by the photomultiplier. After amplification the photoelectronic pulses are fed to a discriminator that only allows the passage of pulses whose amplitudes are within a certain interval, by which means the number of noise pulses is reduced [9.1]. The photoelectronic pulse then starts the timer of a time-to-amplitude converter, which is stopped by the following excitation pulse or a pulse that is obtained from the driving generator of the laser by means of a pulse shaper F. After subtracting from the interval between the excitation pulses the stopped time gives the time interval between excitation and the elementary process of the fluorescence photon emission. The time-to-amplitude converter (TAC) converts the stopped times into proportional voltage signals, which are fed to a multichannel analyzer (MCA). When the signal whose amplitude lies within a definite interval i arrives, the multichannel analyzer increases the memory contents belonging to this amplitude or time interval by one. After a large number N of fluorescence pulses the memory contents N_i/N provide a good approximation for the probability of the fluorescence photon emission with respect to the time after the excitation.

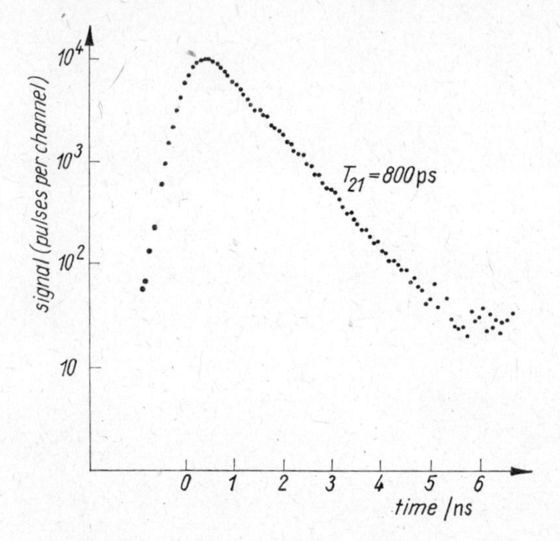

Fig. 9.4. Histogram of a fluorescence decay process

The numbers N_i of fluorescence photons measured in the time interval $it \cdots (i + 1)t$ are plotted (from [9.5]) (fluorescence from a defined rotational-vibrational level of tetrazine vapor, relaxation time $T_{21} = 800$ ps).

Fig. 9.4. [9.5] represents a recorded fluorescence decay curve of this kind. The time resolution of this technique is mainly given by the temporal fluctuation, the so-called jitter, of the photoelectronic pulses and is at present limited to about 50 ps using the best photomultipliers [9.6]. Compared with the measuring methods mentioned earlier, the photon counting technique has the special advantage of a high signal-to-noise ratio, high accuracy and a very large dynamic range, by which the decay process can be traced over several orders of magnitude. Thus, with carefully carried out measurements it is possible to recognize several superimposed decay processes — for example, several exponential decays with different time constants — and to evaluate them with regard to their parameters.

9.1.2 Picosecond Techniques

With the methods discussed in the previous section — using the excitation pulses of modelocked lasers — the time resolution was determined essentially by the photoelectric detection technique. We can increase considerably the time resolution, if instead of the photomultiplier and the detection electronics depicted in Fig. 9.2 we use a streak camera for the detection (compare 3.). At the excitation with a single laser pulse, this pulse triggers the time sweep of the camera. The fluorescence light is directed over an optical system to the entrance slit of the camera and produces the streak pic-

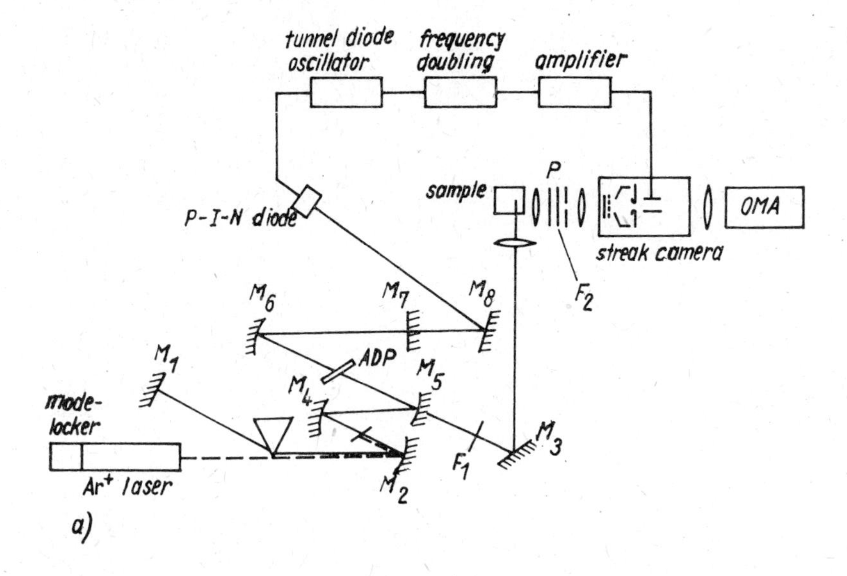

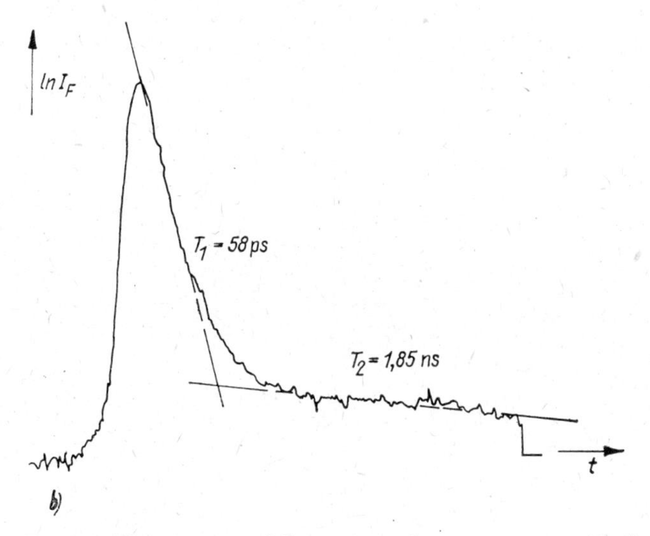

Fig. 9.5. Measurement of fluorescence decay processes with the synchroscan technique [9.8]

a) Arrangement: M_i are mirrors; P is the polarizer; F_1 is the filter; OMA is the optical multichannel analyzer. The modelocked argon ion laser pumps a dye laser ($\lambda_{Fl} = 565$ to 630 nm, $\tau_L = 2$ ps). The second harmonic of the dye laser radiation ($\lambda_2 = 282 \cdots 315$ nm) generated in an ADP crystal excites the fluorescence. (The mirrors M_5 and M_6, in whose common focal point the ADP crystal is located, possess the reflectivity 1 at 600 nm; at 300 nm M_5 possesses the reflectivity 0). The sinusoidal deflection voltage for the synchroscan operation of the streak camera is obtained by delivering about 10% of the dye laser power to the *p-i-n* photodiode, which generates the trigger signal for the tunnel diode oscillator. The 140 MHz output voltage of this oscillator is amplified and applied to the deflecting plates of the streak camera.

b) Semi-logarithmic depiction of a fluorescence decay profile (stilbene 5×10^{-4} mol/l in 85% ethanol/15% glycerol). Two decay processes could be resolved [9.8].

ture on the display screen, which is photographed or recorded with an optical multichannel analyzer. The streak picture can be evaluated with a time resolution of several picoseconds. As an example Fig. 9.5b shows the densitometer trace of a streak camera recording [9.8]. As described in chapter 3., it is possible to use the coordinate in the slit direction for the representation of a spectrum. For this purpose a spectroscope can be placed between the sample and the streak camera.

A time-resolved fluorescence spectrometer with a high level of automation and software, in which the streak camera SCS 185 (with a time resolution of 10 ps) attached to a vidicon is used in the detecting electronics, is produced by VEB Carl Zeiss Jena under the name of "Laser-Impuls Spektrometer" LIS 201 [9.56]. The excitation of the sample occurs by means of a distributed feedback dye laser (DFDL, compare 2.8) pumped with a nitrogen laser, which emits pulses of a duration $\lesssim 100$ ps in the wavelength range between 400 and 700 nm (with SHG module that enables the generation of radiation from 265 nm on). Adjustable, high excitation energy, high-performance optics and spectral dispersion of the fluorescent light with imaging, holographic gratings ensure a high detection sensitivity.

A particularly high resolving power and a relatively large dynamic range are achieved by excitation with continuously pumped modelocked lasers using the synchroscan technique. Here the streak camera beam is deflected periodically with the pulse repetition frequency and phase-locked to these excitation pulses [16, 9.7], whereby a permanent picture of the intensity distribution appears on the screen (with suitable excitation lasers and very stable electronics the time jitter can be reduced to the subpicosecond range.) Fig. 9.5a shows a typical arrangement [9.8].

In addition to the streak camera nonlinear optical gates are employed for the fluorescence detection with picosecond and subpicosecond resolution. As described in 3.3, the cross correlation function between the fluorescence signal to be measured and a short laser pulse is measured here by means of a sampling technique. If the duration of the laser pulse is small compared to the fluorescence lifetime the cross correlation function represents directly the fluorescence decay curve. Frequently, a Kerr cell is used as the nonlinear optical element in fluorescence investigations, where the experimental set-up corresponds to those in Fig. 3.14a and b. With the Kerr cell the signal is attenuated during its passage, which has an unfavorable influence particularly with weak fluorescence signals. This disadvantage can be avoided by using controllable amplifiers, in which the stimulated emission in pumped dyes [9.9, 9.10] or the parametric amplification (see e.g. [9.11]) can be utilized. The highest possible amplification is limited in both cases by the necessity to avoid self-oscillation and by the linearity requirement. In table 9.1 the characteristics of some nonlinear optical gates are compared. It is evident from the table that the Kerr cell is distinguished particularly by its large spectral bandwidth, and the parametric amplifier by the high time resolution.

9.1.3 *Applications*

9.1.3.1 *Dyes*

Using the methods of time-resolved fluorescence spectroscopy numerous investigations were conducted on organic dyes, whereby deactivation rates of up to several 10^{11} s^{-1} were found for the fluorescent S_1 level (see e.g. [16 to 20 and 28 to 35, 39]). A series of papers attempted to explain the variation of rates with respect to the molecular constitution and the solvent interaction. Extremely short lifetimes are found particularly in molecules that alter their steric configuration in the electronically excited state and

Table 9.1. Detection of fluorescence signals by means of optical gates

Arrangement	Switched transmission	Not switched transmission		Opening time	Bandwidth
	τ_g	τ_u	τ_g/τ_u	τ_A/ps	$\Delta\lambda$/nm
Optical Kerr effect	10^{-1}	10^{-4}	10^3	2	700
Saturable absorber	10^{-1}	10^{-4}	10^3	10	50
Dye amplifier	10^3	10^{-2}	10^5	10	50
Optical parametric amplifier	10^6	10^{-1}	10^7	10^{-2}	[1]

[1] 1 nm for fixed crystal orientation and 300 nm for variable crystal orientation

which can thus pass rapidly into a nonfluorescent state (see e.g. [9.12, 9.57] and references there). Due to this competing process the fluorescence quantum efficiency is reduced. Therefore, as active materials for dye lasersitis preferable to use substances in which isomerizations of this kind are prevented through suitable substitutions in the molecules.

In certain types of molecules the transition probability from the singlet to the triplet system (intersystem crossing) is so great that it also influences considerably the fluorescence lifetime (see e.g. [9.13]).

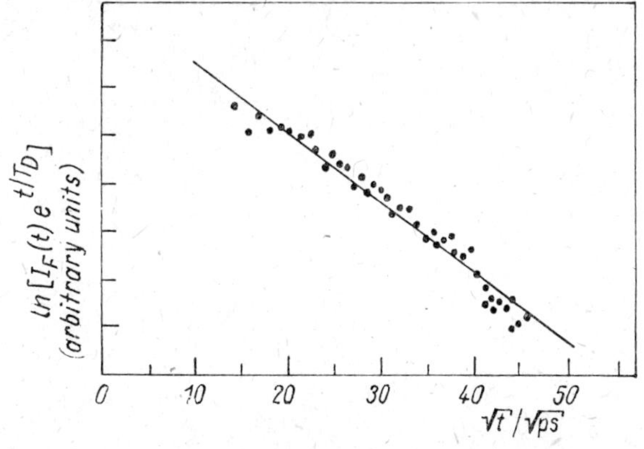

Fig. 9.6. Fluorescence kinetics under the influence of energy transfer from the fluorescent donor molecules (DODCI, 10^{-4} molar in ethanol) to the rapidly relaxing acceptor molecules (malachite green, 5×10^{-4} molar). T_D is the fluorescence decay of the donor molecules in the absence of acceptors. From the curve a decay law according to $\ln [I_F(t)\, e^{t/T_D}/I_F(0)] = -C\sqrt{t}$ can be evaluated. The ascent of the curve yields the parameter C from which using (1.35) the Förster radius can be determined ($R_0 = 7$ nm).

By adding suitable acceptors to the fluorescent dye solution the lifetime of molecules acting as donors in the excited level is shortened due to energy transfer, whereby — as explained in chapter 1 — a decay law of the type $\exp\left[-t/T_{\mathrm{D}} - C \cdot \sqrt{t}\right]$ emerges — see Fig. 9.6.

Furthermore molecular reactions, especially dimerizations and aggregate formations, have been observed (see e.g. [20, 16] and [33]).

9.1.3.2 *Influence of the Orientational Relaxation*

In the investigations on dissolved molecules, it should always be noted that the time dependence of the fluorescence is determined not only by the lifetime limiting processes, but also by the reorientation of the excited molecule (see e.g. [3, 16 to 20, 9.15, 9.16, 9.17]). At the excitation with linearly polarized light, primarily those molecules whose

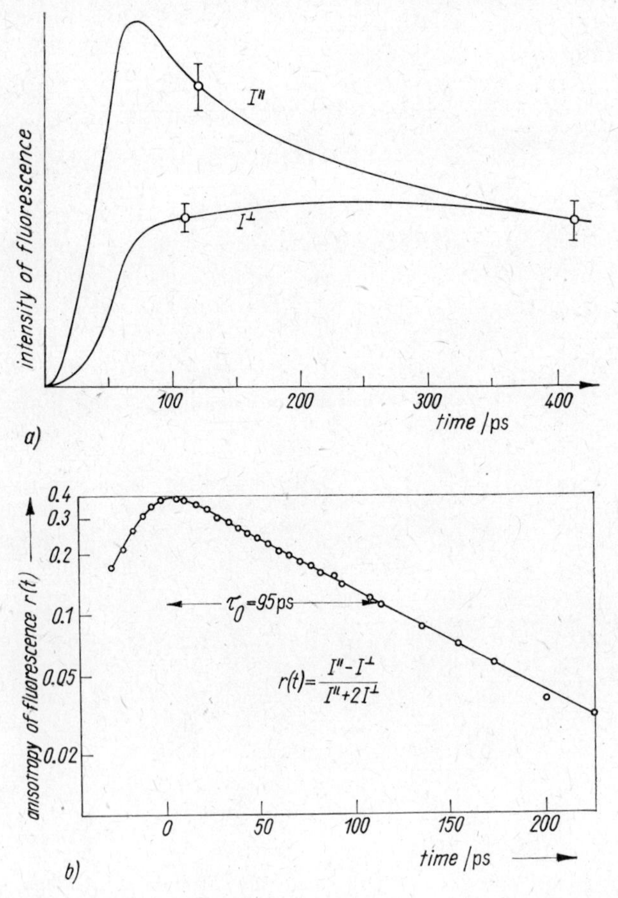

Fig. 9.7. Time-resolved fluorescence of a 10^{-3} molar solution of rhodamine in methanol (from [9.18])

a) $I^{\parallel}$, $I^{\perp}$ denote the fluorescence intensities polarized parallel and perpendicular to the pump radiation

b) $r = (I^{\parallel} - I^{\perp})/(I^{\parallel} + 2I^{\perp})$ is the fluorescence anisotropy

transition matrix element only forms a small angle with the vector of the electric field participate in the absorption — compare (1.68). Thus, after the excitation with a short light pulse, there is an anistropic distribution of the molecules in the excited state — and consequently in the ground state, too. If the state excited by the absorption is identical with the fluorescent state, this anisotropy initially favors the emission of fluorescent light that is polarized parallel to the excitation light. If the molecule rotates in the solution after the excitation, the anisotropy diminishes (see e.g. Fig. 9.7.). By observing the fluorescent light polarized parallel and perpendicular to the excitation

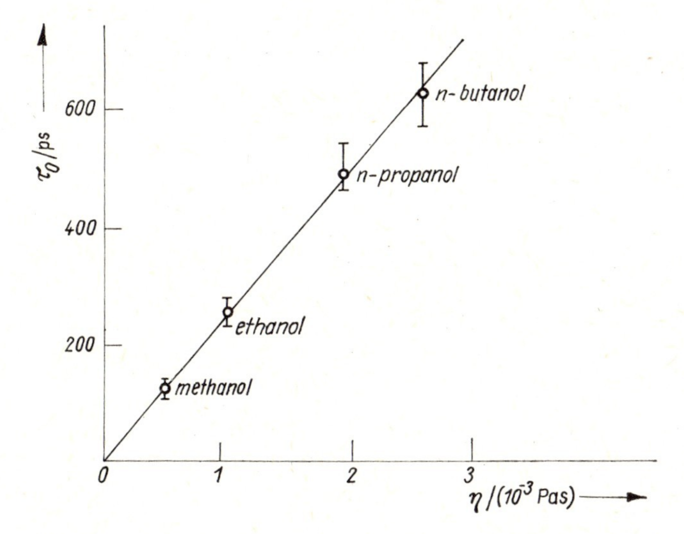

Fig. 9.8. Orientational relaxation time τ_0 of cresyl-violet with respect to the solvent viscosity η (from [9.19])

radiation $I^{||}$ and $I^{\perp}$, respectively, it is therefore possible to obtain information about the reorientation of the dye molecule, the so-called orientational relaxation. A closer inspection shows that the value $I^{\text{L}}(t) = I^{||}(t) + 2I^{\perp}(t)$ is determined in its time dependence only by lifetime limiting processes, and $r(t) = [I^{||}(t) - I^{\perp}(t)]/[I^{||}(t) + 2I^{\perp}(t)]$ only by the reorientation (see e.g. [9.18]). The quantity $I^{\text{L}}(t)$ can be directly measured in experiment, if the fluorescent light is observed in a polarizing direction at an angle of 54.7° to the excitation light. The measuring of the reorientational processes of dye molecules in solvents provides information about the specific interaction processes in the solution. In the simplest model we deal with the dye molecule within the scope of hydrodynamics as a body that rotates in a liquid of the viscosity η, where the liquid adheres to the surface of the body. Specifically, for the reorientational time τ_0 of a spherical molecule of volume V, we obtain

$$\tau_0 = \frac{\eta V}{k_{\text{B}}\mathcal{T}}.$$

Fig. 9.8 gives an example, in which a proportionality between τ_0 and η is found in good approximation [9.19]. Of course, conclusions about the applicability of the model mentioned are not justified by this proportionality, since various other models could also lead to this result. A closer treatment is found e.g. in [9.20, 9.21]. In addition to measuring the polarized fluorescent light the orientational relaxation can also be determined by means of the optical Kerr effect. The refractive index changes according to this effect due to the influence of an intense electromagnetic field, where the change in the refractive index Δn under radiation with a short laser pulse can be described by (3.14). With small molecules in low viscous solvents, orientational relaxation times were measured by means of this effect from some picoseonds to some tens of picoseconds, e.g. $\tau_0 \approx 2$ ps was found for CS_2. This molecule was investigated again recently with a femtosecond time resolution [9.51]. Fig. 9.9 shows the phase shift of a weak probe pulse, measured in a Mach-Zehnder interferometer. The probe pulse was delayed with respect to the intense excitation pulse of 70 fs duration. The phase shift measured is proportional to the change in the refractive index Δn and varies with the delay time t_D. Obviously, the decay consists of a slow and a fast reorientational process, whose relaxation times could be determined at 2.1 ps and 360 fs.

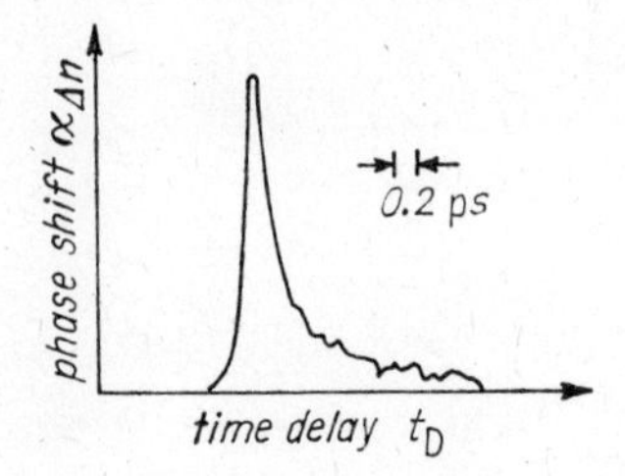

Fig. 9.9. Measurement of the orientational relaxation in CS_2 by means of pump-probe spectroscopy (from Tang and Halbout [9.51])

9.1.3.3 *Biological Substances*

Fluorescence investigations with ps resolution are successfully applied not only to dye molecules in solution but also to biological substances. In biological application the high detection sensitivity of fluorescence spectroscopy plays an important role, because it enables the excitation energy to be held relatively low. This conclusion refers to single pulse investigations, such as those conducted with solid state lasers, as well as investigations with high repetition frequencies, as is typical with modelocked cw dye lasers. Up to now experiments have concentrated particularly on the primary energy transfer and charge separation processes in photosynthesis, the primary reactions in lumirhodopsin and hemoglobin in the visual process and in the oxygen uptake, respectively, and the conversion between different forms of phytochrome and biliverdin dyes (see e.g. [16, 18, 20, 28, 29, 36, 39]).

9.1.3.4 *Solids*

In solids ps-fluorescence investigations are used to investigate the deactivation processes in molecular crystals (see e.g. [9.22]), in impurities and to investigate color centers in dielectrics (see e.g. [9.23, 9.24]) as well-as in semiconductors [9.23], [28] to [36], [39], [9.59].

9.2 Probe Pulse Spectroscopy

9.2.1 *Probe Pulse Spectrometer*

The principle of probe pulse spectroscopy (pump-and-probe spectroscopy) consists in exciting a sample with a strong pump pulse and subsequently after a defined adjustable delay time measuring its capability for absorption, amplification, reflection or polarization rotation of a probe pulse. The probe pulse can have the same wavelength as the excitation pulse or any other wavelength, by which we can probe different transitions of the sample. In this respect probe pulse spectroscopy is more widely applicable and, with a suitable choice of the probe wavelengths, more informative than time-resolved fluorescence spectroscopy. In particular, the occupation number of non-fluorescent levels can also be measured with respect to time.

The forerunner of probe pulse spectroscopy was flash spectroscopy. Here, a photoelectric detector is used to measure the transmission of a sample with respect to time for a continuously radiating light source at a certain wavelength, whereby the sample is exposed to an intense short-time flash at time t_0. The time resolution of this method is

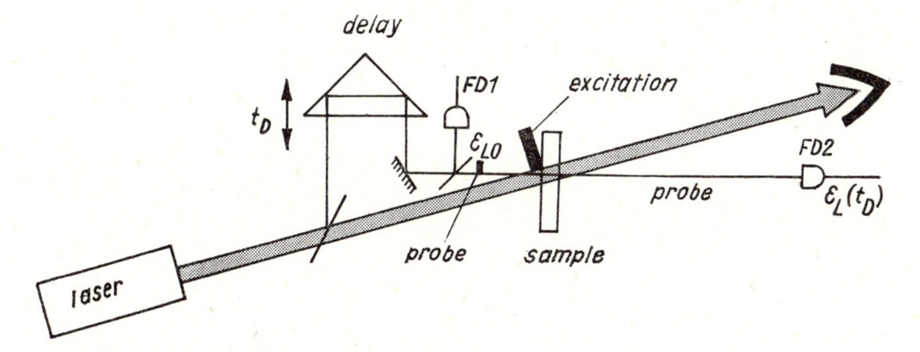

Fig. 9.10. Arrangement for measuring the probe pulse absorption
FD1, FD2 are photodiodes: $\mathcal{E}_{L0}$ and $\mathcal{E}_L(t_D)$ are the energies in front of and behind the sample, respectively.

given by the duration of the excitation flash and the response time of the photoelectric detection system. With suitable flashlamps efficient pulses of several nanoseconds are achieved. Instead of such flashlamps Q-switched or modelocked lasers can also be used (laser flash spectroscopy). With sufficiently short pulses the smallest resolvable time interval is determined solely by the detector and the electronics and is at least about 100 ps [9.25]. In the detection with streak cameras it is several picoseconds [9.26].

The first probe pulse experiments were conducted by Shelton and Armstrong in 1967 [9.27]. They focussed a train of intense ultrashort light pulses from a modelocked solid state laser into a saturable absorber (see Fig. 9.10). These pulses saturate the absorption, which — provided that a simple model is applicable — decays with the relaxation time T_{21}, where T_{21} is assumed to be small compared to the pulse spacing in the train. The weak probe pulses are separated from the strong excitation pulses at a beam splitter and strike the sample after travelling through a delay line of variable length. The probe pulse energy is measured before and after the sample at various delays t_D, which gives the transmission with respect to t_D. In this experiment the re-occupation of the absorbing ground state is measured (for $T_{21} \gg \tau_L$) directly after the saturation (ground-state-recovery spectroscopy). One can imagine the probe pulse spectrometer as a modification of this simple basic device. In the device described the method of exciting and testing with a pulse train of finite duration, which contains pulses of various energy and quality (compare 7.), is unfavorable, because these variations in the pulse characteristics complicate the evaluation and interpretation of the results and reduce the accuracy. Therefore, two types of spectrometers have been developed, which avoid this disadvantage: in the one the sample is excited and probed with single pulses selected from the train; in the other the arbitrarily long, reproducible pulse sequence of modelocked, continuously pumped lasers is employed for this purpose. In the following, both types of spectrometers will be described briefly.

9.2.1.1 *Probe Pulse Spectrometer with Single Pulse Excitation*

In Fig. 9.11 a spectrometer is shown, in which the initial pulses are generated by a passively modelocked solid state laser (ruby, Nd:glass or Nd:YAG laser). A single pulse is selected from the pulse train and amplified (compare 7.3.3). The amplified single pulse is divided into an excitation and a probe pulse. In the excitation channel the wavelength

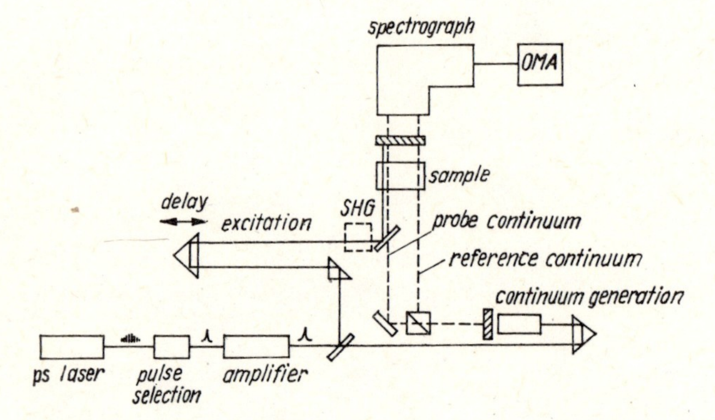

Fig. 9.11. Pump and probe beam spectrometer with single pulse excitation and probe continuum

of the pulse can be transformed into other spectral ranges through a nonlinear optical process — e.g. the generation of the second harmonic (SHG), parametric generation or stimulated Raman scattering (compare 8.) — and in parametric generation specifically it can also be continuously tuned. In the probe beam path the laser pulse is focussed on a nonlinear material in order to generate a ps-continuum (compare 8.2.4). The white light pulse that emerges is divided into a probe pulse and a reference pulse. The probe pulse passes through the sample in a region that is pumped by the excitation pulse, whereas the reference pulse passes through an unaffected part. At the output of the sample the excitation light is absorbed by a filter, while the probe continuum and reference continuum are dispersed spectrally in a spectrograph and recorded with temporal integration in different traces by means of a photoplate or an optical multichannel analyzer. The computer of the optical multichannel analyzer calculates the energy transmission T as the ratio of the pulse energies per wavelength interval of the probe pulse and the reference pulse $\mathsf{T}(\lambda, t_D) = \mathscr{E}_\lambda^T(\lambda, t_D)/\mathscr{E}^R(\lambda)$. This quantity is displayed on the plotter, whereby the delay time t_D between the excitation and the test can again be varied from pulse to pulse by means of an optical delay line (arranged here in the excitation channel) (see e.g. [16 to 20, 9.28]). As an example, Fig. 9.12 shows probe pulse spectra that are recorded at three different delay times. It can be clearly seen that the structure of the spectrum varies with time. (For example, at $\lambda = 533$ nm the absorption decreases rapidly, whereas at $\lambda = 480$ nm a new absorption band builds up, which decays very slowly — not recognizably here — on a nanosecond time scale.) One advantage of single pulse excitation with low repetition frequency is that — if irreversible processes occur during the interaction — the radiated sample material can be renewed from pulse to pulse (for example, by means of a recirculating system in liquid samples or by shifting solid samples). The excitation power can be varied over a broad range to suit the problem. Furthermore, the parameters of each single pulse can be measured. Compared with continuously pumped excitation systems on the basis of dye lasers a certain disadvantage consists in the greater fluctuation of the pulse parameters and in the greater pulse duration. The effect of para-

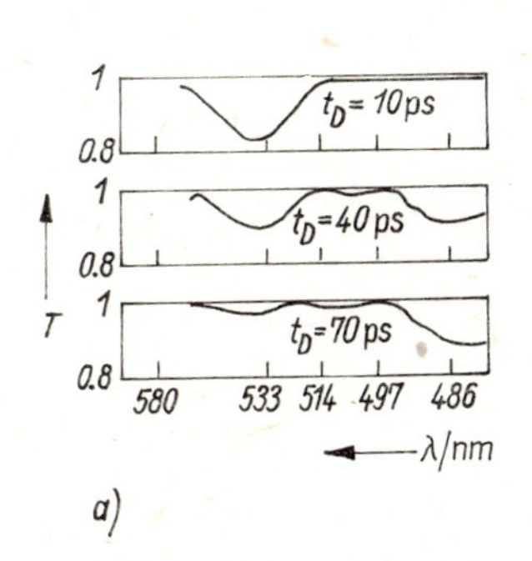

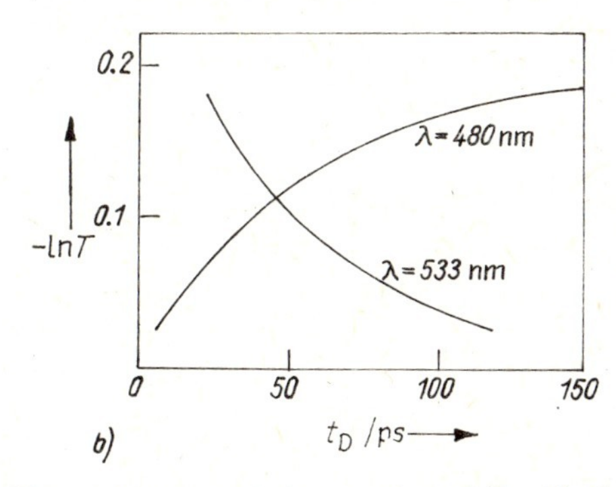

Fig. 9.12. Probe pulse transmission T as a function of the wavelength λ with the delay time t_D as parameter (a) and as a function of t_D with λ as parameter (b) (sample Cu-phthalocyanin in a water/acetonitrile mixture (0.5:0.5)) (from [9.29])

meter fluctuations can be compensated by measuring parameters such as the excitation energy for each pulse and accounting for this actual value in the evaluation [9.29]. Moreover, techniques were developed in order to reduce the influence of the finite pulse duration through deconvolution procedures [9.30].

Finally, let us point out that Laubereau and Kaiser (see e.g. [9.31] and references there) as well as Piskarskas (see e.g. [9.11]) succeeded in generating excitation and probe pulses having a duration up to the subpicosecond range by means of suitable parametric generators, even when starting with the single pulses from solid state lasers (compare 8.) As opposed to pulses from dye lasers the pulses generated here are distinguished particularly by their steep edges, where the power more than exponentially decays by several orders of magnitude with regard to the maximum.

Thus, even small probe signals can be very accurately measured and it is possible to determine relaxation times that are up to five times shorter than the pulse duration [9.31]. Such parametric generators can be employed in the excitation as well as in the probe beam path, through which it is possible to choose the excitation and probe transition over a wide range (see Fig. 9.13). In particular, it is also possible to choose one or both wavelengths in the near infrared range, and thereby to excite directly and to probe vibrational transitions. It should be pointed out that the photometric accuracy for measurements of probe beam absorption using narrow band parametric signals is greater than the accuracy obtained using the strongly fluctuating ps-continuum whereas the latter has the advantage that it immediately provides an overall view of the spectrum at a certain delay time. For this reason, devices were constructed in which alternatively a narrow band parametric signal or the ps-continuum is available as probe signal [9.11, 9.37].

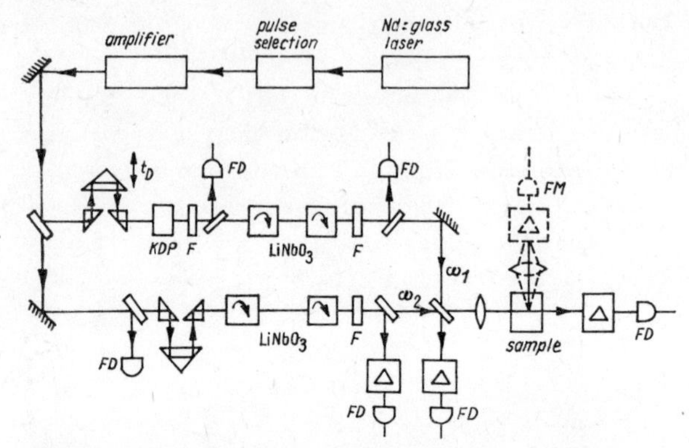

Fig. 9.13. Ultrashort-time spectrometer with optical parametric generators on the basis of $LiNbO_3$ crystals for the generation of the pump and probe pulses (from [9.31])
One generator is pumped by the pulses at the fundamental wavelength $\lambda = 1.06$ μm; the other is pumped by the second harmonic (generated in KDP) at $\lambda = 0.53$ μm (compare 8.). The wavelength is tuned by rotating the crystals. By means of several photodiodes the pulse parameters can be monitored. The generators can also be used for a step excitation of the sample. In this case the excitation of the upper energy level can be determined by measuring the temporally integrated fluorescence with respect to the delay between the two excitation pulses (dotted part of the figure).

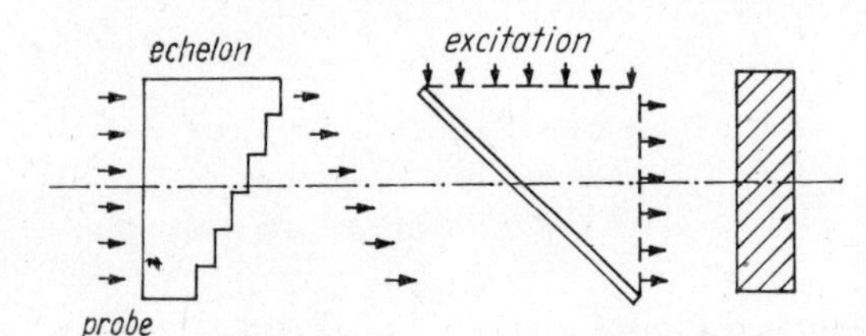

Fig. 9.14. Pump-probe pulse configuration with several delay channels

If we want to obtain a general view of the probe beam absorption with respect to the delay time by means of a single excitation, techniques similar to the optical gate with transverse excitation can be used (Fig. 3.14b). We must only replace the active material of the gate, i.e. the Kerr cell, with the sample to be investigated [9.32]. A similar effect, i.e. variation of the delay time along a space coordinate, can be achieved by sending the probe radiation through delay paths of different lengths — for instance, through an echelon (see Fig. 9.14) [9.33, 9.34].

9.2.1.2 Probe Pulse Spectrometer with High Pulse Repetition Frequency

As described in the chapters on pulse generation, subpicosecond pulses of high repetition frequency (up to 10^8 Hz) with high reproducibility can be generated using continuously or synchronously pumped dye lasers. At these high repetition frequencies we generally do not measure the single pulses before or after the sample, but rather we measure a signal averaged over many pulses. The excitation beam path is interrupted periodically by means of a modulator (e.g. a chopping disk) with a suitable, relatively low frequency, and thus a sequence of probe pulses, which has passed through the excited sample enters the detector (see Fig. 9.15). The electronic detection system is selectively matched to the chopping frequency and thus detects a signal that is proportional to the mean probe beam power with and without excitation. If necessary, it is possible to reduce the extremely

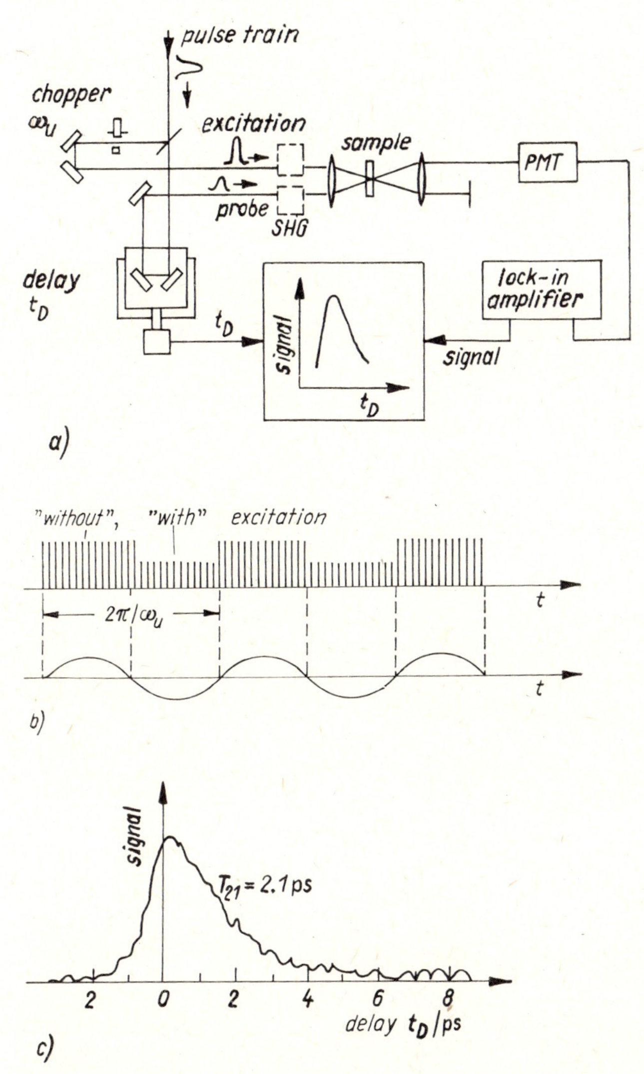

Fig. 9.15. Pump-probe measurements with pulses of high repetition frequency (from [9.35])

a) Experimental arrangement

b) Probe pulses entering the detector with and without excitation by the pump pulses (switching period $2\pi/\omega_u$) and the corresponding electronic signal after a narrow-band amplifier

c) Probe pulse signal with respect to the delay time t_D (malachite green in ethanol)

high repetition frequency using a cavity dumper in the laser resonator (compare 5.) which also results in an increase of the pump power. This is necessary especially if the recovery time of the sample to the ground state is large. The time interval between the excitation pulse and the next probe pulse can be adjusted by means of an optical delay line with a stepper motor. Crystals for the generation of the second harmonic may be arranged alternatively in the excitation and probe beam paths. For other nonlinear optical conversion processes the intensity is generally too low. (If using pulsed amplifiers — compare 5. — through which pulses in the Gigawatt range are obtained, the pulse repetition frequency must be so greatly reduced that the methods described in single pulse excitation are again applicable). An advantage of the method depicted is the high resolution of some 10 fs [9.61] and detection sensitivity for absorption changes ($|\Delta \ln T/\ln T| \gtrsim 10^{-6}$). Thus, it is possible to record probe pulse spectra even with relatively small mean powers (in the mW-range). Very high time resolution and sensitivity has also been achieved with correlations of equally strong pump and probe pulses [9.60].

9.2.1.3 Probe Pulse Spectrometer for Measuring the Raman Gain

With sophisticated detection techniques, particularly that of raising the modulation frequency in the excitation beam path to 10 MHz, it is possible to increase the detection sensitivity by about two orders of magnitude [9.36]. Thus, a device similar to that in Fig. 9.15 could be employed with very good results for the measuring of the Raman gain in extremely thin layers. In this case, two synchronously pumped dye lasers whose frequency spacing is tuned to a Raman transition generate the pump and probe pulses, which pass through the sample at optimal temporal overlapping, since in these measurements only the gain is to be sensitively measured and time resolution is not necessary. Due to the stimulated Raman effect the higher frequency pump pulses are weakened in the thin (monomolecular!) sample while the low-frequency probe pulses, i.e. the Stokes pulses, are amplified. Interfering fluorescent light can be suppressed by an additional slow wavelength modulation of one of the two dye lasers. This example should illustrate that those picosecond techniques with high repetition frequency can also be employed successfully for the purposes of stationary spectroscopy.

9.2.1.4 Probe Pulse Spectrometer Using Self-Induced Gratings

In single as well as multi-pulse excitation the build up of grating structures, which develop due to the spatially coherent superposition of two light pulses propagating in different directions in the sample, can be employed for measuring energy as well as phase relaxation times by probe beam techniques. Stepanov et al. [9.38] conducted the first investigations on gratings of this kind and their diffraction efficiency for probe pulses. Phillion, Kuizenga and Siegman [9.39] were the first to apply these methods in the picosecond range.

Fig. 9.16 shows the scheme of such an arrangement. Two excitation pulses superimpose in the sample (delay $\vartheta \geqq 0$), whereby a standing wave develops. This standing wave can lead to the spatial modulation of material properties such as the absorption coefficient or the refractive index. At this grating a probe pulse is diffracted, whereby the diffraction signal decreases with increasing delay time, from which certain lifetimes can be determined.

Of interest is that a grating effect can also build up, if both excitation pulses no longer superimpose because their relative delay is greater than the pulse duration. The preceding excitation pulse generates a polarization wave that decays with the phase relaxation time τ_{21}. As long as $\vartheta \lesssim \tau_{21}$ holds for the delay time, the second excitation pulse still finds a polarization in the material. Due to the superposition of the coherent polarization and the electric field waves a grating can again develop, at which the much later arriving probe pulse is diffracted. The phase relaxation time τ_{21} can accordingly be determined from the dependence of the diffraction signal on the delay time ϑ. More accurate investigations and literature dealing with these problems is found in [9.40], [11]. By use of this experimental device one can excite and detect low-frequency coherent optical phonons [9.64]. The two strong light pulses excite coherent optical phonons with wave vectors $\pm(\vec{k}_1 - \vec{k}_2)$, which form a vibrational standing wave at the phonon frequency. If the pump pulses are short compared to the vibrational period, the mode is driven "impulsively". This "impulsively" excited stimulated Raman scattering (ISRS) of the probe pulse as a function of the delay t_D shows phonon oscillations and decay, and if several modes are excited, phonon beating. This is the first direct observation of phonon oscillations.

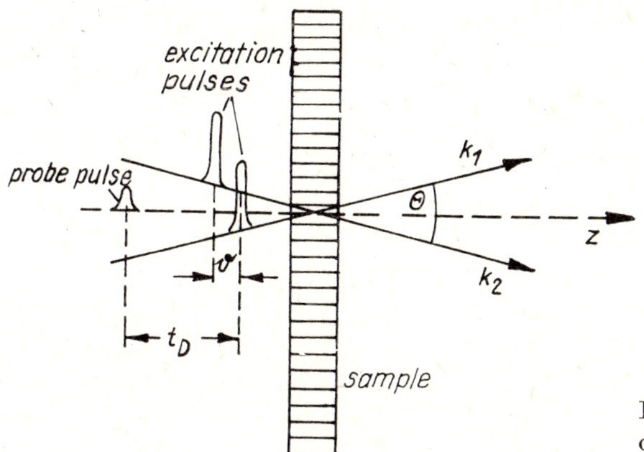

Fig. 9.16. Pump-probe technique based on self-induced gratings

9.2.2 *Application*

9.2.2.1 *Electronic Relaxation Processes*

Picosecond probe beam spectroscopy has already found a wide range of applications, which we can only briefly mention here. For more detailed discussion of these applications we would like to refer to the extensive literature in this field (see especially [16 to 20, 28 to 36] and other references there). Almost all processes that we have discussed in regard to fluorescence spectroscopy can also be pursued with probe pulse methods, whereby a greater degree of freedom exists in probe pulse spectroscopy because the occupation of various levels (see Fig. 9.1) can be probed. In simple cases, for example, when the molecules only return directly from the excited level S_1 to the ground level S_0, the measuring of the reoccupation of the ground state leads to the same simple exponential time law as the measuring of the decay of the fluorescence or the probe beam absorption from the excited level into higher states as well as probe beam gain due to the fluorescence transition from S_1 to S_0. On the other hand, the differences that these various methods provide can indicate additional relaxation channels — e.g. transitions into the triplet system or rapid conformation changes. An example of an extremely fast direct transition from the excited state S_1 to the ground state S_0 is shown in Fig. 9.15c); the recovery time comes to about 2.1 ps. In contrast the probe pulse spectrum from Fig. 9.12 shows more complex behavior. Here, the rapidly decaying absorption band at 533 nm is to be attributed to a transition from the S_1 to the S_x level, and the band building up at 480 nm to a $T_1 \rightarrow T_x$ transition within the triplet system.

Internal conversion, intersystem crossing, isomerizations due to internal rotations as well as proton transfer, and energy and electron transfer could be measured in many molecular systems. The corresponding relaxation times range to within a few picoseconds. Shank, Fork and Yen [6.31] measured a very short relaxation time up to now in polyacetylene; it came to 160 fs. Among higher electronic levels of molecules even shorter relaxation times up to 10^{-14} s occur. Until now, however, these could only be determined indirectly.

Specific details of fast electronic relaxation processes could also be investigated in semiconductors by means of a probe pulse spectrometer with femtosecond time resolution or more indirect methods. Thus, for example, the lifetime of electrons in GaAs, which were excited into the conduction band 0.5 eV above the band gap, was determined at 60 fs [9.52]. This relaxation is mainly determined by electron-phonon interactions. In [9.53] relaxation processes in ε-GaSe were resolved using a 3 pulse diffraction method with pulses of two different frequencies ω_1 and ω_2 and adjustable polarizing directions, where an interband relaxation time of $T_{CV} = 380$ ps, an intraband energy relaxation time of $T_\varepsilon = 36$ fs and an orientational relaxation time of the electronic wave vector of $T_\Theta = 69$ fs were measured. With $\frac{1}{\tau} = \frac{1}{T_\varepsilon} + \frac{1}{T_\Theta}$ we obtain a phase decay time of $\tau = 23$ fs. In [9.60] a double exponential decay ($T_1 = 60$ fs, $T_2 = 200$ fs) has been observed with GaAlAs, and in [9.61] relaxational processes have been found with GaAs, which decay in less than 30 fs.

9.2.2.2 *Vibrational Relaxation Processes*

Especially fast relaxation processes also occur with vibrational transitions in the condensed phase. Kaiser, Laubereau et al. [9.41, 9.31] and Alfano and Shapiro [9.42] were the first to develop suitable methods for measuring the longitudinal and transverse relaxation times T and τ of vibrational transitions in fluids and solids. Various Raman processes proved to be suitable here. Thus, for measuring the energy relaxation time T the sample was excited with an intense ultrashort single pulse at the frequency ω_L, which due to the stimulated Raman effect generates a Stokes pulse at the frequency $\omega_S = \omega_L - \omega_M$ and excites molecules from the vibrational ground state to the excited vibrational state corresponding to the energy $\hbar\omega_M$. A delayed, weak light pulse of frequency $2\omega_L$ serves for probing the excited molecules. It generates spontaneous, incoherent Raman scattering and the anti-Stokes intensity produced by the excited molecules is detected at the frequency $\omega_A = 2\omega_L + \omega_M$. The intensity of this anti-Stokes radiation is proportional to the occupation number in the excited vibrational level. Therefore T can be determined from the decrease in the anti-Stokes intensity with respect to the delay time between both pulses (see Fig. 9.17). The τ-measurement can be carried out similarly. Here we make use of the fact that in the stimulated Raman effect not only the occupation numbers change, but simultaneously an intense polarization wave having the frequency ω_M and the wave vector $\vec{k}_M = \vec{k}_L - \vec{k}_S$ is generated. The formation of this coherent wave occurs in a similar way to that described in section 9.1.2 in the discussion of one photon phenomena. The polarization wave decays with the phase relaxation time τ after the light pulses pass. This decay can be probed by means of a coherent anti-Stokes scattering process, in which the incident light of a probe pulse having the wave number vector $\vec{k}_L'$ and the frequency ω_L' is scattered at the already present polarization wave. The coherent anti-Stokes scattering occurs preferably in that direction, for which the matching condition for the wave number vectors $\vec{k}_A = \vec{k}_L' + \vec{k}_M = \vec{k}_L' + (\vec{k}_L - \vec{k}_S)$ is satisfied (compare Figs. 9.17a and 8.2). Through this pronounced direction dependence a clear distinction from the incoherent radiation is possible.

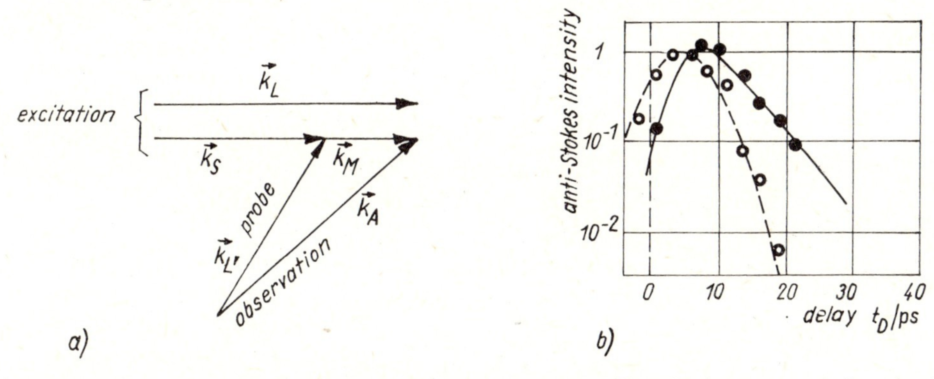

Fig. 9.17. Measurement of the vibrational relaxation times T and τ by means of the stimulated Raman effect (from [9.41])

a) Relation between the wave number vectors for the τ-measurement

b) Normalized anti-Stokes signal from the valence vibration $\left(\omega_m/(2\pi c) = 2939\ \mathrm{cm}^{-1}\right)$ of 1,1,1-trichlorethane as a function of the delay time t_D for incoherent scattering (solid curve) and coherent scattering (dotted curve)

For the relaxation times we obtain $T = (5 \pm 1)$ ps and $\tau = 2.6$ ps. The value for τ is in satisfactory agreement with the value calculated from the line width $\tau = 2/\Delta\omega$, from which we can conclude that this vibrational transition is predominantly homogeneously broadened.

In the meantime the energy and phase relaxation times for numerous vibrational transitions were measured with these and similar methods. In addition, the combined action of various relaxation channels was verified. For example, the relaxation via other, low-frequency normal vibrations, the energy transfer to neighboring molecules and the influence of Fermi-resonances were observed [9.31, 28]. The Raman methods have the disadvantage of a relatively low detection sensitivity, and therefore it is difficult to use them for measuring fast relaxation processes in low concentrations in gases or solutions. In such cases, the successive excitation methods with subsequent fluorescence detection described in connection with figure 9.13 have been employed successfully [9.43, 28].

In the excited electronic states the vibrational relaxation occurs more rapidly than in the electronic ground state due to the stronger coupling between the vibrational modes. In large molecules the intramolecular vibrational relaxation can be described by the transition from a selectively excited vibrational state to a quasi-continuum which arises from the great number of vibrational modes. From this diversity alone an energy relaxation in the environment occurs due to phonon coupling (see e.g. [9.54]). Erskine et al. [9.55] investigated single large molecules, such as nile blue, rhodamine 640, DODCI, cresyl violet, oxacine 725 etc. in solutions and found that the first relaxation step took less than 30 fs, which was equal to the time resolution of their set-up. It is obvious from such experiments that the redistribution of the vibrational energy proceeds very rapidly in the excited electronic states.

9.2.2.3 Selective Excitation

Double pulse and multiple pulse methods are employed successfully not only for spectroscopic investigations, but also for the selective excitation of substances with short relaxation times. Fig. 9.18 shows that the first laser pulse excites level 1, from where another absorption into 2 occurs, whereas no excitation occurs at the intermediate levels 2′ and 2″. Thus it is possible to excite certain groups in a molecule with specific aims and, subsequently, to observe specific reactions from the excited state. In addition,

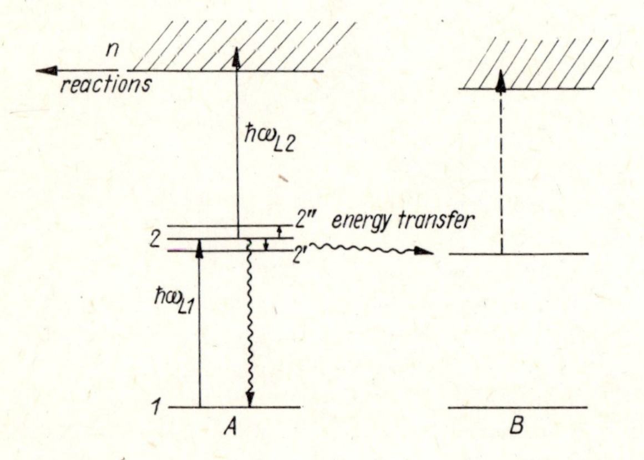

Fig. 9.18. Selective excitation

in a molecular mixture specific types of molecules can be excited, whereby only these molecules participate in further processes. In both cases, it is necessary that the selectivity of the excitation into the intermediate levels is not destroyed due to rapid relaxation processes within the molecule or due to energy transfer to neighboring molecules. For this reason pulses that are short compared with the corresponding time constants are necessary [9.44, 16 to 20]. For example, in [9.45] it was shown that with typical dye molecules in diluted gases the energy stored selectively in a normal vibration is itself distributed within a few ps to other normal vibrations inside the molecule even under collisionless conditions. Therefore it is only possible to concentrate the excitation energy on a certain part of the molecule for the duration of picosecond and subpicosecond pulses.

Kriukov, Letokhov et al. [9.46] succeeded in selectively exciting nucleic acid bases in this manner. The bases were decomposed through a two stage excitation, where the lifetime of the S_1 level serving as intermediate stage was also in the picosecond range:

	$T_{S_1S_0}$/ps
Thymin	5 ± 2
Uracil	5 ± 3
Cytosin	5 ± 4
Adenin	1 ± 1
Guanin	2 ± 2

Starting from the S_n-level, which is excited in the second step, at first reversible intermediate products and eventually stable irreversible final products occur.

Letokhov [9.47, 28] developed ideas about how one can utilize such multi-stage excitation using picosecond pulses that lead to an ionization in order to pursue selective excitation with high spatial resolution in a field electron or field ion microscope. In this way it seems possible to build devices that combine high temporal, spectral and spatial resolution.

Investigations of phase transitions with femtosecond time resolution have been performed for silicon crystals [9.62]. The electron distribution in a thin surface layer was measured to completely loose its crystal symmetry within 150 fs after irradiation by an intense ultrashort light pulse. The transition from the crystalline state into the new state was measured by observing the second harmonic radiation generated at the surface by a weak probe pulse. In this way it is possible to produce a new phase very fast, whereby the temperature is rather low immediately after irradiation. Furthermore the ultrafast melt and material ejection from the silicon surface has been observed by excite- and probe photography (compare 3.) using an amplified femtosecond laser pulse for excitation and variably delayed femtosecond spectral continuum (compare 8.2.2) for observation [9.63].

References

General Literature on Lasers and the Generation, Measurement and Application of Ultrashort Light Pulses

[1] W. Kleen, R. Müller: Laser. Springer, Berlin, Heidelberg, New York 1969.
[2] O. Svelto: Principles of Lasers. Heyden, London, New York 1976.
[3] F. T. Arecchi, E. O. Schulz-Dubois (Editors of Vol. 1 and 2), M. L. Stitch (Editor of Vol. 3): Laser Handbook. North Holland, Amsterdam 1972 (Vol. 1 and 2), 1979 (Vol. 3).
[4] W. Köchner: Solid State Laser Engineering. Springer, New York, Heidelberg, Berlin 1976.
[5] H. Paul: Laser Theory (in German). Akademie-Verlag, Berlin 1969.
[6] W. Brunner, W. Radloff, K. Junge: Quantum Electronics. An Introduction to the Physics of Lasers (in German). VEB Deutscher Verlag der Wissenschaften, Berlin 1975.
[7a] K. Guers: Laser Light Amplifier and Oscillators (in German). Akademische Verlagsgesellschaft, Frankfurt/M. 1966.
[7b] K. Guers: Laser. Fundamentals, Properties and Applications (in German). Frankfurt/M. 1970.
[8] D. Röss: Laser, Light Amplifiers and Oscillators (in German). Akademische Verlagsgesellschaft, Frankfurt/M. 1966.
[9] F. P. Schäfer (Editor): Dye Lasers. Springer, Berlin, Heidelberg, New York 1978.
[10] N. Bloembergen: Nonlinear Optics. Benjamin, New York, Amsterdam 1965.
[11a] M. Schubert, B. Wilhelmi: Introduction to Nonlinear Optics (in German). Teubner, Leipzig 1971 (Vol. 1), 1978 (Vol. 2).
[11b] M. Schubert, B. Wilhelmi: Nonlinear Optics and Quantum Electronics. Wiley, New York 1986.
[12] V. S. Letokhov: Laser Spectroscopy (in German). Akademie-Verlag, Berlin 1977.
[13] W. A. Subow: Methods for Measuring the Properties of Laser Radiation (in Russian). Nauka, Moscow 1973.
[14] A. A. Vainstein: Introduction to Waveguides and Resonators (in Russian). Sovetskoe Radio, Moscow 1966.
[15] K. Vollrath, G. Thomer: High-Speed Physics. Springer, Wien, New York 1967.
[16] S. L. Shapiro (Editor): Ultrashort Light Pulses. Springer, Berlin, Heidelberg, New York 1977.
[17] C. V. Shank, E. P. Ippen, S. L. Shapiro (Editors): Picosecond Phenomena I. Springer, Berlin, Heidelberg, New York 1978.
[18] R. M. Hochstrasser, W. Kaiser, C. V. Shank (Editors): Picosecond Phenomena II. Springer, Berlin, Heidelberg, New York 1980.
[19] Ultrafast Phenomena in Spectroscopy I (Vol. 1 and 2). Tallin 1978.
[20] Ultrafast Phenomena in Spectroscopy II (Vol. 1 and 2). Reinhardsbrunn 1980.
[21] S. A. Akhmanov, A. S. Chirkin: Statistical Phenomena in Nonlinear Optics (in Russian). Moskow State University, Moscow 1971.

[22] F. Zernike, J. E. Midwinter: Applied Nonlinear Optics. Wiley, New York, Sydney, Toronto, London 1973.
[23] A. M. Prokhorov (Editor): Encyclopedia of Lasers (Vol. 1, 2) (in Russian). Sovetskoe Radio, Moscow 1978).
[24] L. Allen, J. H. Eberly: Optical Resonance and Two-Level Atoms. Wiley, New York, London, Sydney, Toronto 1975.
[25] W. Demtröder: Laser Spectroscopy. Springer, Berlin, Heidelberg, New York 1981.
[26] Author group: Handbook of Laser Technology (in German). Verlag Technik, Berlin 1982.
[27] B. H. Pantell, H. E. Puthoff: Fundamentals of Quantum Electronics. Wiley, New York, London, Sydney, Toronto 1969.
[28] K. B. Eisenthal, R. M. Hochstrasser, W. Kaiser, A. Laubereau (Editors): Picosecond Phenomena III. Springer, Berlin, Heidelberg, New York 1982.
[29] R. R. Alfano (Editor): Biological Events Probed by Ultrafast Laser Spectroscopy. Academic Press, New York, London, Toronto, Sydney, San Francisco 1982.
[30] Proceedings of "Ultrafast Phenomena in Spectroscopy" (UPS), Minsk 1983.
[31] D. H. Auston, K. B. Eisenthal (Editors): Ultrafast Phenomena IV. Springer, Berlin, Heidelberg, New York, Tokyo 1984.
[32] Technical Digest of CLEO, Anaheim 1984.
[33] V. B. Eisenthal (Editor): Applications of Picosecond Spectroscopy to Chemistry. Reidel, Dordrecht, Boston, Lancaster 1983.
[34] E. Klose, B. Wilhelmi (Editors): Ultrafast Phenomena in Spectroscopy. Proceedings of UPS 85, Teubner, Leipzig 1986.
[35] Technical Digest of CLEO, San Francisco 1986.
[36] Technical Digest of IQEC, San Francisco 1986.
[37] C. H. Lee (Editor): Picosecond Optoelectronic Devices, Academic Press, Orlando 1984.
[38] G. A. Mourou (Editor): Picosecond Electronics and Optoelectronics. Springer, Berlin, Heidelberg, New York, Tokyo 1986.
[39] G. R. Fleming, A. E. Siegman (Editors): Ultrafast Phenomena V. Springer, Berlin, Heidelberg, New York, Paris, Tokyo 1986.

References to Chapter 1

[1.1] V. S. Letokhov, V. P. Chbotaev: Nonlinear Laser Spectroscopy. Springer, Berlin, Heidelberg, New York 1977.
[1.2] J. Wanter, in: Quantum Optics and Electronics (eds. C. De Witt, A. Blandin, C. Cohen-Tannoudji). Gordon and Breach, New York, London, Paris 1965.
[1.3] J. B. Birks: Photophysics of Aromatic Molecules. Wiley-Interscience, New York 1970.
[1.4] K. F. Herzfeld, T. A. Litovitz: Absorption and Dispersion of Ultrasonic Waves. Academic Press, New York 1959.
[1.5] Author group: Introduction to Photochemistry (in German). Berlin 1976.
[1.6] N. J. Turro: Molecular Photochemistry. W. A. Benjamin, New York, Amsterdam 1965.
[1.7] A. Weller, Z. Physik. Chem. **NF 13** (1957) 335.
[1.8] Th. Förster, Ann. Physik **2** (1948), 55, Z. Naturforsch. **4a** (1949) 321.
Th. Förster: Fluorescence of Organic Substances (in German). Göttingen 1951.
V. M. Agranovich, M. D. Galanin: Energy Transfer in Condensed Matter (in Russian). Nauka, Moscow 1978.
[1.9] A. Lösche: Nuclear Induction (in German). Teubner, Leipzig 1957.
[1.10] A. Abragam: The Principles of Nuclear Magnetism. Clarendon Press, Oxford 1961.
[1.11] L. D. Landau, E. M. Lifschitz: Electrodynamics of the Continua. Akademie-Verlag, Berlin 1974 (in German), Addison-Wesky, New York, 1961 (in English).
[1.12] M. Born, E. Wolf: Principles of Optics. Pergamon, Oxford 1970.

[1.13] H. Haken, W. Waidlich, in: Quantum Optics (ed. R. J. Glauber). Academic Press, New York 1969.
[1.14] V. M. Fain, J. I. Chanin: Quantum Electronics (in German). Teubner, Leipzig 1969.

References to Chapter 2

[2.1] A. Sommerfeld: Lectures on Theoretical Physics (in German). Vol. IV, Teubner, Leipzig 1964.
[2.2] H. Kogelnik, T. Li, Appl. Opt. **5** (1966), 1550 or
H. Kogelnik, T. Li, in: Handbook of Lasers with Selected Data on Optical Technology (ed. R. J. Pressley). Chemical Rubber Co, Cleveland 1971.
[2.3] H. Kogelnik, E. P. Ippen, C. V. Shank, A. Dienes, IEEE J. Quant. Electr. **QE 8** (1972) 373.
[2.4] Documents for Noble Gas Ion Lasers from Spectra Physics (USA), Coherent (USA) and VEB Carl Zeiss Jena (GDR).
[2.5] W. Dietel, E. Döpel, D. Kühlke; in: Ultrafast Phenomena in Spectroscopy (Proceedings of UPS 80). Reinhardsbrunn 1980, p. 42.
[2.6] Documents for Nd: YAG Laser pumped Dye Lasers from Quantel (USA).
[2.7] Documents for continuously pumped Dye Lasers from Spectra Physics (USA), Coherent (USA) and Center for Scientific Instruments (GDR).
[2.8] N. Kempe, H. Orzegowski, C. Peschel, G. Thiede, in: Ultrafast Phenomena in Spectroscopy (Proceedings of UPS 80). Reinhardsbrunn 1980, p. 89.
[2.9] E. Mollvo, W. Kaule: Maser and Laser (in German). Bibliografisches Institut, Mannheim 1966.
[2.10] Documents for Laser Diodes from Laser-Optronic (FRG).
[2.11] C. H. Kittel: Quantum Theory of solids (in German). Springer, Wien 1970.
[2.12] L. Valenta, E. Jäger: Lectures on the Theory of Solids, (Vol. 1, 2) (in German). VEB Deutscher Verlag der Wissenschaften, Berlin 1980.
[2.13] B. A. Lengyel: Lasers. Wiley, New York 1971.
[2.14] G. Litfin, H. Welling, Laser und Optoelektronik **14** (1982) 17.
[2.15] L. Esak, R. Tsu, IBM Journal of Research **14** (1970) 61.
[2.16] R. L. Fork, C. V. Shank, B. I. Greene, F. K. Reinhalt, R. A. Logan, in [19] p. 280.
[2.17] J. Katz, E. Kapon, C. Lindsey, S. Margalit, U. Shreter, A. Yariv, Appl. Phys. Lett. **43** (1983) 521.
[2.18] H. Kogelnik, C. V. Shank, Appl. Phys. Lett. **18** (1971) 152, J. Appl. Phys. **43** (1972) 2327.
[2.19] Zs. Bor, IEEE J. Quant. Electr. QE-**16** (1980) 512.
[2.20] Zs. Bor, A. Müller, B. Racz, F. P. Schäfer, Appl. Phys. **27B** (1982) 9, Appl. Phys. **27B** (1982) 77.
[2.21] G. Szabo, Zs. Bor, A. Müller, Appl. Phys. **31B** (1983) 1.
[2.22] G. Szabo, B. Racz, A. Müller, B. Nikolaus, Zs. Bor, Appl. Phys. **34B** (1984) 145.
[2.23] P. L. Liu, C. Lin, T. C. Damen, D. J. Eilenberger, in [18] p. 30.
[2.24] S. Szatmari, B. Racz: Appl. Phys. B (to be published).
[2.25] S. Szatmari, B. Racz: Opt. Quant. Electr. (to be published).

References to Chapter 3

[3.1] M. Schubert, B. Wilhelmi, in: Progress in Optics, Vol. XVII (ed. E. Wolf). North-Holland Amsterdam, New York, Oxford, 1980, p. 165.
[3.2] H. P. Weber, R. Dändliker, Phys. Rev. Lett. **28A** (1968) 77.
[3.3] Z. Baumann, IEEE J. Quant. Electr. **QE 13** (1977) 875.

[3.4] P. R. Bird, D. J. Bradley, W. Sibbett, Proc. 11. Congr. High Speed Photography (ed. P. J. Rolls). Chapman and Hall, London 1974, p. 112; and D. J. Bradley, W. Sibbett, Appl Phys. Lett. **27** (1975) 382.
T. Hiruma, E. Inuzuka, V. E. Postovalov, A. M. Prokhorov, M. Ya. Schelev, Yu. N. Serdynchenko, Y. Tsuchiya, in [20] p. 105.
[3.5] Documents for the Streak Camera "Temporal disperser C 979". Hamamatsu, Japan 1982.
[3.6] Documents for the Streak Camera "Agat", USSR 1981.
[3.7] J. A. Armstrong, Appl. Phys. Lett. **10** (1967) 16.
H. P. Weber, J. Appl. Phys. **38** (1967) 2231.
M. Maier, W. Kaiser, J. A. Giordmaine, Phys. Rev. Lett. **17** (1966) 1275.
[3.8] J. A. Giordmaine, P. M. Rentzepis, S. L. Shapiro, K. W. Wecht, Appl. Phys. Lett. **11** (1967) 216.
J. R. Klauder, M. A. Duguay, J. A. Giordmaine, S. L. Shapiro, Appl. Phys. Lett. **13** (1968) 174.
[3.9] E. P. Ippen, C. V. Shank, Appl. Phys. Lett. **27** (1975) 488.
[3.10] R. Kühmstedt, D. Schubert, W. Triebel, in [20] p. 114.
[3.11] W. Dietel, D. Kühlke, W. Rudolph, B. Wilhelmi, Kvant. Electr. **10** (1983) 79.
[3.12] J. C. Diels, in [20] p. 527.
[3.13] R. C. Eckardt, C. H. Lee, Appl. Phys. Lett. **15** (1969) 425.
[3.14] J. Herrmann, M. Palme, K. E. Süsse, Opt. Quant. Electr. **10** (1978) 195.
[3.15] T. Damm, B. Grünberg, W. Triebel: unpublished results, Jena 1981.
[3.16] N. J. Frigo, T. Daly, H. Mahr, IEEE J. Quant. Electr. **QE 13** (1977) 101.
[3.17] H. Mahr, M. D. Hirsch, Opt. Commun. **13** (1975) 96.
[3.18] A. Yariv, in: Current Trends in Optics (eds. F. T. Arecchi, F. R. Aussenegg), Taylor and Francis, London 1981.
[3.19] M. A. Duguay, A. T. Mattik, Appl. Optics **10** (1971) 2162.
[3.20] D. H. Auston, Appl. Phys. Lett. **26** (1975) 101.
[3.21] D. H. Auston, P. R. Smith, A. M. Johnson, W. M. Augustiniak, J. C. Bean, D. B. Fraser, in [18] p. 71.
[3.22] D. H. Auston, P. Lavallard, N. Sol, D. Kaplan, Appl. Phys. Lett. **36** (1980) 66, **37** (1980) 371, **38** (1981) 47.
[3.23] A. M. Johnson, D. H. Auston, P. R. Smith, J. C. Bean, J. P. Harbison, D. Kaplan, in [18] p. 285.
[3.24] G. Veith, Elektronikschau **6** (1981) 38.
[3.25] P. LeFur, D. H. Auston, Appl. Phys. Lett. **28** (1976) 21.
[3.26] G. Mourou, W. Knox, in [18] p. 75.
[3.27] A. M. Johnson, D. H. Auston, IEEE J. Quant. Electr. **QE 11** (1975) 283.
[3.28] C. H. Lee, P. S. Mak, A. P. DeFonzo, IEEE J. Quant. Electr. **QE 16** (1980) 277.
C. H. Lee, P. S. Mak, in [18] p. 88.
[3.29] J. J. Wiczer, H. Merkelo, Appl. Phys. Lett. **27** (1975) 397.
[3.30] J. A. Valdmanis, G. A. Mourou, C. W. Gabel, IEEE J. Quant. Electr. **QE 17** (1983) 664.
[3.31] K. E. Meyer, G. A. Mourou, in [31] p. 406 and in [37], p. 249.

References to Chapter 4

[4.1] S. E. Harris, R. Targ, Appl. Phys. Lett. **5** (1964) 202.
[4.2] S. E. Harris, O. P. McDuff, IEEE J. Quant. Electr. **QE 1** (1965) 245.
[4.3] Di Domenico, J. E. Geusic, H. M. Marcuse, R. G. Smith, Appl. Phys. Lett. **8** (1966) 180.
[4.4] L. M. Osterink, J. D. Forster, J. Appl. Phys. **39** (1968) 4163.

[4.5] D. J. Kuizenga, A. E. Siegman, IEEE J. Quant. Electr. **QE 6** (1970) 694.
[4.6] D. J. Kuizenga, A. E. Siegman, IEEE J. Quant. Electr. **QE 6** (1970) 709.
[4.7] M. H. Crowell, IEEE J. Quant. Electr. **QE 1** (1965) 12.
[4.8] A. J. De Maria, D. A. Stetzer, Appl. Phys. Lett. **7** (1965) 71.
[4.9] L. M. Ostering, R. Targ, Appl. Phys. Lett. **10** (1967) 115.
[4.10] G. Horn, D. Schubert, J. Schwarz, in [5.23] p. 4.
[4.11] M. G. Cohen, SPIE **322** (1982) 44.
[4.12] A. Seilmeier, B. Kopainsky, W. Kranitzky, W. Kaiser, in [28] p. 23.
[4.13] L. F. Mollenauer, D. M. Bloom, Opt. Lett. **4** (1979) 247.
[4.14] D. J. Kuizenga, D. W. Phillion, T. Lund, A. E. Siegmann, Opt. Commun. **9** (1973) 221.

References to Chapter 5

[5.1] D. J. Bradley, A. J. F. Durrant, Phys. Lett. **27 A** (1968) 73.
[5.2] W. H. Glenn, M. J. Brienza, A. J. De Maria, Appl. Phys. Lett. **12** (1968) 54.
[5.3] B. H. Soffer, J. W. Linn, J. Appl. Phys. **39** (1968) 5859.
[5.4] C. K. Chan, S. O. Sari, Appl. Phys. Lett. **25** (1974) 403.
[5.5] C. K. Chan, S. O. Sari, R. E. Foster, Appl. Phys. Lett. **47** (1976) 1139.
[5.6] H. Mahr, M. D. Hirsch, Opt. Commun. **13** (1975) 96.
[5.7] J. M. Harris, R. M. Chrisman, F. E. Lytle, Appl. Phys. Lett. **26** (1975) 16.
[5.8] J. Heritage, R. Jain, Appl. Phys. Lett. **32** (1978) 101.
[5.9] J. Kuhl, H. Klingenberg, D. von der Linde, Appl. Phys. Lett. **18** (1979) 279.
[5.10] N. Frigo, C. Hemenwey, H. Mahr, Appl. Phys. Lett. **37** (1980) 981.
[5.11] J. Herrmann, U. Motschmann, Appl. Phys. **B 27** (1982) 27.
[5.12] J. Herrmann, U. Motschmann, Opt. Commun. **40** (1982) 379.
[5.13] Z. A. Yasa, O. Teschke, Opt. Commun. **15** (1975) 169.
[5.14] A. Scavannec, Opt. Comm. **17** (1976) 14.
[5.15] N. J. Frigo, T. Daly, H. Mahr, IEEE J. Quant. Electr. **QE 13** (1977) 100.
[5.16] D. M. Kim, J. Kuhl, R. Lambrich, D. von der Linde, Opt. Comm. **27** (1978) 123.
[5.17] C. P. Ausschnitt, R. K. Jain, Appl. Phys. Lett. **32** (1978) 727.
[5.18] C. P. Ausschnitt, R. K. Jain, J. D. Heritage, IEEE J. Quant. Electr. **QE 15** (1979) 912.
[5.19] Documents for the synchronously pumped Dye Laser from Spectra Physics, USA 1978.
[5.20] Documents for the synchronously pumped Dye Laser from Coherent, USA 1981.
[5.21] Documents for the synchronously pumped Ringlaser from Coherent, USA 1981.
[5.22] Documents for the Argon Ion/Laser ILA 120 from VEB Carl Zeiss Jena, GDR 1982.
[5.23] Special issue on "Ion Laser Application 1" of VEB Carl Zeiss Jena, Jena 1982.
[5.24] D. Schubert, J. Schwarz, in [5.23] p. 9.
[5.25] N. Kempe, C. Peschel, in [5.23] p. 14.
[5.26] G. Horn, D. Schubert, J. Schwarz, in: Proceedings of the 4th International Conference on "Laser and their Applications", Leipzig 1981, p. 93.
[5.27] G. Horn: Diploma Thesis, Jena 1981.
[5.28] J. Kuhl, R. Lambrich, D. von der Linde, Appl. Phys. Lett. **31** (1977) 657.
[5.29] I. M. Clemens, J. Nojbar, I. Bronstein-Bonte, R. M. Hochstrasser, Opt. Commun. **47** (1983) 271.
[5.30] S. R. Rotman, C. B. Roxlo, O. Bebelaar, T. K. Yee, M. M. Salour, in [18] p. 50.
[5.31] D. von der Linde, D. Wiechert, J. Kluge, M. Kemmler in [30].
[5.32] U. Stamm: Diploma Thesis, Jena 1983.
D. Schubert, U. Stamm, B. Wilhelmi, Opt. Quant. Electr. **17** (1985) 337.
[5.33] J. P. Ryan, L. S. Goldberg, D. J. Bradley, Opt. Comm. **27** (1978) 127.
[5.34] G. A. Mourou, T. Sizer, Opt. Comm. **41** (1982) 47.

References to Chapter 6

[6.1] W. Schmidt, F. P. Schäfer, Phys. Lett. **26 A** (1968) 558.
[6.2] D. J. Bradley, F. O'Neill, Opto-Electronics **1** (1969) 69.
[6.3] E. P. Ippen, C. V. Shank, D. Dienes, Appl. Phys. Lett. **21** (1972) 348.
[6.4] F. O'Neill, Opt. Commun. **6** (1972) 360.
[6.5] J. L. Diels, J. Menders, H. Salaba, in [18] p. 41.
[6.6] R. L. Fork, B. I. Greene, C. V. Shank, Appl. Phys. Lett. **38** (1981) 197.
[6.7] W. Dietel, Opt. Commun. **43** (1982) 64.
[6.8] G. H. C. New, Opt. Commun. **6** (1972) 188.
[6.9] G. H. C. New, IEEE J. Quant. Electr. **QE 10** (1974) 115.
[6.10] H. A. Haus, IEEE J. Quant. Electr. **QE 13** (1975) 736.
[6.11] J. Herrmann, F. Weidner, Appl. Phys. **B 27** (1982) 105.
[6.12] D. J. Bradley, in [16] p. 17.
[6.13] J. Herrmann, F. Weidner, B. Wilhelmi, Appl. Phys. **B 26** (1981) 197.
[6.14] J. S. Ruddock, D. J. Bradley, Appl. Phys. Lett. **29** (1976) 296.
[6.15] C. V. Shank, E. P. Ippen, Appl. Phys. Lett. **24** (1974) 373.
[6.16] E. G. Arthurs, D. J. Bradley, A. G. Roddie, Appl. Phys. Lett. **19** (1971) 480.
[6.17] E. G. Arthurs, D. J. Bradley, P. N. Puntambekar, J. S. Ruddock, Opt. Commun. **12** (1974) 360.
[6.18] R. S. Adrain: Ph. D. Thesis, The Queens University of Belfast 1974.
[6.19] E. P. Ippen, C. V. Shank; Appl. Phys. Lett. **27** (1975) 488.
[6.20] J. C. Diels, E. van Stryland, D. Gold, in [17] p. 117.
[6.21] W. Dietel, E. Döpel, D. Kühlke, W. Rudolph, B. Wilhelmi, in [28] p. 45, and Opt. Commun. **43** (1982) 433.
[6.22] B. K. Garside, T. K. Lim, J. Appl. Phys. **44** (1973) 2335.
[6.23] G. H. C. New, D. H. Rea, J. Appl. Phys. **47** (1976) 3107.
[6.24] G. H. C. New, K. E. Orkney, M. J. Nock, Opt. Quant. Electr. **8** (1976) 425.
[6.25] R. Müller, Opt. Commun. **28** (1979) 259.
[6.26] J. P. Ryan, L. S. Goldberg, D. J. Bradley, Opt. Commun. **27** (1978) 127.
[6.27] G. W. Fehrenbach, K. J. Gruntz, R. G. Ulbricht, Appl. Phys. Lett. **33** (1978) 159.
[6.28] Y. Ishida, T. Yajima, K. Naganuma, Jap. J. of Appl. Phys. **19** (1980) L 717.
[6.29] D. Kühlke, W. Rudolph, B. Wilhelmi, Appl. Phys. Lett. **42** (1983) 325, and IEEE J. Quant. Electr. **QE 19** (1983) 526.
[6.30] G. A. Mourou, T. Sizer, Opt. Commun. **41** (1982) 47; and in [28] p. 107.
[6.31] C. V. Shank, R. L. Fork, R. T. Yen, in: Proceedings of CLEO, 1982, p. 32, and in [28] p. 2. C. V. Shank, R. L. Fork, R. Yen, R. H. Stolen, W. J. Tomlinson, Appl. Phys. Lett. **40** (1982) 761.
[6.32] W. Dietel, W. Rudolph, B. Wilhelmi, J. C. Diels, J. J. Fontaine, in: Ultrafast Phenomena in Spectroscopy III, Minsk 1983.
[6.33] D. Kühlke, W. Rudolph, Opt. Quant. Electr. **16** (1984) 57.
[6.34] J. C. Diels, J. J. Fontaine, I. C. McMichael, B. Wilhelmi, W. Dietel, D. Kühlke, Kvant. Elektr. **10** (1983) 2398.
[6.35] W. Rudolph, B. Wilhelmi, Appl. Phys. **B 35** (1984) 37.
[6.36] W. Dietel, W. Rudolph, B. Wilhelmi, J. C. Diels, J. J. Fontaine, Isvestija Akademii Nauk SSSR, **48** (1984) 480, and J. Opt. Soc. Am. B **2** (1985) 681.
[6.37] W. Dietel, J. Fontaine, J.-C. Diels, Opt. Lett. **8** (1983) 4.
[6.38] J. M. Halbout, C. L. Tang, IEEE J. Quant. Electr. **QE 19** (1983) 487.
[6.39] M. S. Stix, E. P. Ippen, IEEE J. Quant. Electr. **QE 19** (1983) 520.
[6.40] M. Yoshizawa, T. Kobayashi, IEEE J. Quant. Electr. **QE 20** (1984) 797.
[6.41] A. E. Siegman, Opt. Lett. **6** (1981) 334.

[6.42] H. Vanherzeele, J. L. van Eck, A. E. Siegman, Appl. Opt. **20** (1981) 3484.
[6.43] J. C. Diels, H. Vanherzeele, R. Torti, in [32] paper WC 5 and Opt. Lett. **9** (1984) 549.
[6.44] J. A. Valdmanis, R. L. Fork, J. Opt. Soc. Am. **A 1** (1984) 1337 (A).
[6.45] R. L. Fork, C. V. Shank, R. T. Yen, Appl. Phys. Lett. **41** (1982) 223.
[6.46] C. Rolland, P. B. Corkum, Opt. Commun., in press.
[6.47] W. H. Knox, J. Opt. Soc. Am. **2** (1985) 54.

References to Chapter 7

[7.1] H. W. Mocker, R. J. Collins, Appl. Phys. Lett. **7** (1965) 270.
[7.2] A. J. De Maria, D. A. Stetser, H. Heynau, Appl. Phys. Lett. **8** (1966) 174.
[7.3] V. S. Letokhov, Zh. exp. teor. fis. **55** (1968) 1077.
[7.4] P. G. Kriukov, V. S. Letokhov, IEEE J. Quant. Electr. **QE 8** (1972) 766.
[7.5] J. A. Fleck, Phys. Rev. **B 1** (1970) 84.
[7.6] B. J. Seldovich, T. J. Kusnetzova, Usp. Fiz. Nauk **106** (1972) 47.
[7.7] J. G. Lariontzev, V. N. Serkin, Kvant. Electr. **1** (1974) 2166.
[7.8] W. H. Glenn, IEEE J. Quant. Electr. **QE 11** (1975) 8.
[7.9] R. Wilbrandt, H. Weber, IEEE J. Quant. Electr. **QE 11** (1975) 186.
[7.10] G. H. C. New, Proc. IEEE **67** (1979) 380.
[7.11] M. S. Demokan, D. A. Lindsey, Int. J. Electron. **41** (1976) 421.
[7.12] J. Herrmann, F. Weidner; Opt. Quant. Electron. **11** (1979) 119.
[7.13] J. Herrmann, F. Weidner, B. Wilhelmi, Appl. Phys. **20** (1979) 237.
[7.14] D. J. Bradley, B. Liddy, W. E. Sleat, Opt. Commun. **2** (1971) 391.
[7.15] M. A. Duguay, J. W. Hansen, S. L. Shapiro, IEEE J. Quant. Electr. **QE 6** (1970) 725.
[7.16] D. J. Bradley, W. Sibbett, Opt. Commun. **9** (1973) 17.
[7.17] D. J. Bradley, G. H. C. New, S. J. Caughey, Opt. Commun. **2** (1970) 41.
[7.18] D. von der Linde, IEEE J. Quant. Electr. **QE 8** (1972) 328.
[7.19] K. H. Drexhage, G. A. Reynolds, IEEE J. Quant. Electr. **QE 9** (1973) 960.
[7.20] M. E. Mack, IEEE J. Quant. Electr. **QE 4** (1968) 1015.
[7.21] R. Cubeddu, R. Polloni, C. A. Sacchi, O. Svelto, IEEE J. Quant. Electr. **QE 5** (1969) 470
[7.22] A. R. Clobes, M. J. Brienza, IEEE J. Quant. Electr. **QE 6** (1970) 651.
[7.23] D. von der Linde, O. Bernecker, A. Lauberau, Opt. Commun. **9** (1973) 1173.
[7.24] D. J. Bradley, T. Morrow, M. S. Petty, Opt. Comm. **2** (1970) 1.
[7.25] E. B. Treacy, Phys. Lett. **28 A** (1968) 34.
[7.26] H. F. Rowe, T. Li, IEEE J. Quant. Electr. **QE 6** (1970) 49.
[7.27] R. C. Eckardt, IEEE J. Quant. Electr. **QE 10** (1974) 48.
[7.28] F. De Martini, C. H. Townes, T. K. Gustafson, P. L. Kelley, Phys. Rev. **164** (1967) 312.
[7.29] A. N. Sherichin, P. G. Kriukov, Yu. A. Matveetz, S. V. Chekalin, Kvant. Electr. **4** (1974) 956.
[7.30] S. V. Chekalin, P. G. Kriukov, Yu. A. Matveetz, O. B. Shatberashvili, Opto-electronics **6** (1974) 249.
[7.31] N. G. Basov, Yu. A. Drozhbin, P. G. Kriukov, V. B. Lebedev, V. S. Letokhov, Yu. A. Matveetz, Pisma Zh. teor. eksp. fiz. **9** (1969) 428.
[7.32] S. D. Zakharov, P. G. Kriukov, Yu. A. Matveetz, S. V. Chekalin, S. A. Churilova, O. B. Shatberashvili, Kvant. Electr. **5** (1973) 17, 52.
[7.33] A. N. Sherichin, V. A. Kovalenko, P. G. Kriukov, Yu. A. Matveetz, S. V. Chekalin, O. B. Shatberashvili, Kvant. Electr. **2** (1974) 377.
[7.34] P. G. Kriukov, Yu. A. Matveetz, S. A. Churilova, O. B. Shatberashvili, Zh. teor. eksp. fiz. **62** (1972) 2036.
[7.35] E. G. Arthurs, D. J. Bradley, Opt. Commun. **12** (1974) 136.
[7.36] G. J. Kachen, J. O. Kusilka, IEEE J. Quant. Electr. **QE 6** (1970) 84.
[7.37] W. Zinth, A. Laubereau, W. Kaiser, Opt. Commun. **22** (1977) 161.

[7.38] A. Penzkofer, Opto-Electronics **6** (1974) 87.
[7.39] J. Herrmann, J. Wienecke, B. Wilhelmi, Opt. Quant. Electron. **7** (1975) 337.
[7.40] M. S. Demokan, P. A. Lindsey, Int. J. Electron. **11** (1977) 417.
[7.41] G. H. C. New, IEEE J. Quant. Electr. **QE 14** (1978) 642.
[7.42] D. Kolmeder, W. Zinth, Appl. Phys. **24** (1981) 341.
[7.43] A. Leitner, M. E. Lippitsch, E. Roschger, S. R. Aussenegg, in: Proceedings of the 4th Conference on "Lasers and their Applications", Leipzig 1981, p. 96.
[7.44] W. Rudolph, H. Weber: Preprint No. 236105, University of Kaiserslautern.
[7.45] R. W. Dixon, W. D. Joyce, IEEE J. Quant. Electr. **QE 15** (1979) 470.
[7.46] T. C. Damen, M. A. Duguay, Electron. Lett. **16** (1980) 166.
P. L. Liu, C. Lin, T. C. Damen, D. J. Eilenberger, in [18] p. 30.
[7.47] J. A. Copeland, S. M. Abbott, W. S. Holden, IEEE J. Quant. Electr. **QE 16** (1980) 388.
[7.48] H. Ito, H. Yokoyama, S. Murata, H. Inaba, Electron. Lett. **15** (1979) 763.
[7.49] T. C. Damen, M. A. Duguay, J. M. Wiesenfeld, in [18] p. 38.
[7.50] P. T. Ho, L. A. Glasser, E. P. Ippen, H. A. Haus, Appl. Phys. Lett. **33** (1978) 241.
[7.51] M. B. Holbrock, W. E. Sleat, D. J. Bradley, in [18] p. 26.
[7.52] E. P. Ippen, D. J. Eilenberger, R. W. Dixon, in [18] p. 21.
[7.53] I. P. van der Ziel, W. T. Tsang, R. A. Logan, R. M. Mikulyak, W. M. Augustyniak, Appl. Phys. Lett. **39** (1981) 525.
[7.54] H. C. Casey, M. B. Panish: Heterostructure Lasers. Academic Press, New York 1978.
R. L. Hartman, R. A. Logan, L. A. Koszi, W. T. Tsang, J. Appl. Phys. **51** (1980) 1909.
[7.55] D. Middleton, J. Appl. Phys. **19** (1948) 817.
[7.56] D. Middleton: Introduction to Statistical Communication Theory. McGrac-Hill, New York 1960.
[7.57] V. Kempe: Analysis of Stochastical Systems (Vol. 1) (in German). Akademie-Verlag, Berlin 1976.
[7.58] V. J. Tichonov, Usp. Fis. Nauk **77** (1962) 449.
[7.59] B. Wilhelmi, E. Heumann, W. Triebel, Kvant. Electr. **3** (1976) 732.
[7.60] H. Vanherzeele, J. L. van Eck, A. E. Siegman, Appl. Optics **20** (1981) 3484.
[7.61] Y. Silberberg, P. W. Smith, D. J. Eilenberger, D. A. B. Miller, A. C. Gossard, W. Wiegmann, Opt. Lett. **9** (1984) 507.

References to Chapter 8

[8.1] P. A. Franken, A. E. Hill, C. W. Peters, G. Weinreich, Phys. Rev. Lett. **7** (1961) 118.
[8.2] E. Börner, R. Kühmstedt, W. Triebel, B. Wilhelmi, Exp. Techn. Phys. **23** (1975) 159.
[8.3] W. Glenn, IEEE J. Quant. Electr. **QE 5** (1969) 284.
[8.4] S. L. Shapiro, Appl. Phys. Lett. **13** (1968) 19.
[8.5] R. J. Orlov, T. Usmanov, A. S. Chirkin, Zh. teor. eksp. fiz. **57** (1969) 1079.
[8.6] J. Comly, E. Garmire, Appl. Phys. Lett. **12** (1965) 7.
[8.7] A. H. Kung, J. F. Young, G. C. Bjorklund, S. E. Harris, Phys. Rev. Lett. **29** (1972) 985.
[8.8] J. Reintjes, C. Y. She, R. C. Eckardt, N. E. Karangelen, R. C. Elton, R. A. Andrews, J. Opt. Soc. Am. **67** (1977) 251.
[8.9] S. L. Shapiro, J. A. Giordmaine, K. W. Wecht, Phys. Rev. Lett. **19** (1967) 1093.
[8.10] C. G. Bret, H. P. Weber, IEEE J. Quant. Electr. **QE 4** (1968), 807.
[8.11] M. J. Colles, Opt. Commun. **1** (1969) 169.
[8.12] R. L. Carman, M. E. Mack, F. Shimizu, N. Bloembergen, Phys. Rev. Lett. **23** (1969) 1327.
[8.13] R. L. Carman, F. Shimizu, C. S. Wang, N. Bloembergen, Phys. Rev. **A 2** (1970) 60.
[8.14] R. L. Carman, M. E. Mack, Phys. Rev. **A 5** (1972) 341.
[8.15] S. A. Akhmanov, K. N. Drabovich, A. P. Sukhorukov, A. S. Chirkin, Zh. teor. eksp. fiz. **59** (1970) 485.
[8.16] A. Piskarskas, in: Application of Lasers (in Russian). Nauka, Moscow 1979, p. 294.

[8.17] A. Seilmeier, K. Spanner, A. Laubereau, W. Kaiser, Opt. Commun. **24** (1978) 237.
[8.18] R. R. Alfano, S. L. Shapiro, Phys. Rev. Lett. **24** (1970) 584.
[8.19] A. Penzkofer, W. Kaiser, Opt. Quant. Electron. **9** (1977) 315.
[8.20] S. Belke, R. Gase, K. Vogler, Opt. Quant. Electron. **12** (1980) 9.
[8.21] J. Herrmann, Kvant. Electr. **2** (1975) 364, and **4** (1977) 1779.
[8.22] E. B. Treacy, Phys. Lett. **28 A** (1968) 34, IEEE J. Quant. Electr. **QE 5** (1969) 454.
[8.23] S. A. Akhmanov, IEEE J. Quant. Electr. **QE 4** (1968) 598.
S. A. Akhmanov, A. P. Sukhorukov, A. S. Chirkin, Zh. teor. eksp. fiz. **55** (1968) 1430.
[8.24] T. K. Gustafson, J. P. Taran, H. A. Haus, J. R. Lifschitz, P. L. Kelley, Phys. Rev. **177** (1969) 1196.
[8.25] M. M. T. Loy, Y. R. Shen, IEEE J. Quant. Electr. **QE 9** (1973) 409.
[8.26] A. Laubereau, Phys. Lett. **29 A** (1969) 539.
[8.27] R. H. Lehmberg, J. M. McMahon, Appl. Phys. Lett. **28** (1976) 204.
[8.28] L. F. Mollenauer, R. H. Stolen, J. P. Gordon, Phys. Rev. Lett. **45** (1981) 1095.
[8.29] K. Boyer, H. Egger, T. S. Luk, D. F. Müller, H. Pummer, C. K. Rhodes, in [28] p. 19.
[8.30] S. A. Akhmanov, K. N. Drabovich, A. P. Sukhorukov, A. K. Schednova, Zh. teor. eksp. fiz. **62** (1972) 525.
[8.31] S. A. Akhmanov, M. A. Bolshov, K. N. Drabovich, A. P. Sukhorokov, Pisma V Zh. teor. eksp. fiz. **12** (1970) 547.
[8.32] M. J. Colles, Opt. Commun. **1** (1969) 169.
[8.33] R. R. Alfano, S. L. Shapiro; Phys. Rev. **A 2** (1970), 2376.
[8.34] R. L. Carman, M. E. Mack, Phys. Rev. **A 5** (1972) 341.
[8.35] R. S. Adrain, E. G. Arthurs, W. Sibbett, Opt. Commun. **15** (1975) 290.
[8.36] B. Wilhelmi, in: "Laser Application in Physics", Proceedings of ISLA III, Vilnjus 1984.
[8.37] T. Damm, K. Kaschke, F. Noack, B. Wilhelmi, Opt. Lett. **10** (1985) 176.
[8.38] D. Grischkowsky, A. C. Balant, Appl. Phys. Lett. **41** (1982) 1.
[8.39] Nakatsuka, D. Grischkowsky, A. C. Balant, Phys. Rev. Lett. **47** (1981) 1910.
[8.40] R. Meinel, Opt. Comm. **47** (1983) 343.
[8.41a] I. G. Fujimoto, A. M. Weiner, E. P. Ippen, Appl. Phys. Lett. **44** (1984) 832.
[8.41b] J. M. Halbout, D. Grischkowsky, Appl. Phys. Lett. **45** (1984) 1281.
[8.41c] W. H. Knox, R. L. Fork, M. C. Downer, R. H. Stolen, C. H. Shank, J. A. Valdmanis, Appl. Phys. Lett. **46** (1985) 1120.
[8.42] J. Satsoma, N. Yajima, Progr. Theor. Phys. Suppl. **55** (1974) 106.
[8.43] L. F. Mollenauer, R. H. Stolen, J. P. Gordon, Phys. Rev. Lett. **45** (1980) 1095.
[8.44] L. F. Mollenauer, R. H. Stolen, in [30].
[8.45] H. M. Gibbs, Phys. Rev. **A 8** (1973) 446.
[8.46] N. A. Kurnitt, I. D. Abella, S. R. Hartmann, Phys. Rev. Lett. **12** (1964) 567.
[8.47] R. G. Brewer, R. G. Devoe, S. C. Rand, A. Schenzle, N. C. Wong, S. S. Kano, A. Wokaun, in: Laser Spectroscopy V (eds. A. R. W. McKellar, I. Oka, B. P. Stoicheff). Springer, Berlin, Heidelberg, New York 1981.
[8.48] A. N. Orajewski, Usp. Fis. Nauk **91** (1973) 181.
[8.49] J. B. W. Morsink, T. J. Aartsma, O. A. Wiersma, Chem. Phys. Lett. **49** (1977) 34.
[8.50] S. L. McCall, E. L. Hahn, Phys. Rev. Lett. **18** (1967) 908, Phys. Rev. **183** (1969) 457.
[8.51] I. A. Polujektow, J. M. Popow, W. S. Rojtberg, Usp. Fis. Nauk **114** (1974) 97.
[8.52] R. E. Slusher; in: Progress in Optics (ed. E. Wolf), Vol. XII. Amsterdam 1974.
[8.53] M. Bass, P. A. Franken, J. F. Ward, G. Weinreich, Phys. Rev. Lett. **9** (1962) 446.
[8.54] D. H. Auston, Appl. Phys. Lett. **43** (1983) 713.
[8.55] D. A. Kleinman, D. H. Auston, IEEE J. Quant. Electr. **QE 20** (1984) 964.
[8.56] D. H. Auston, K. P. Cheung, J. A. Valdmanis, D. A. Kleinman, in [31] p. 409.
[8.57] A. M. Johnson, R. H. Stolen, W. M. Simpson, in [31] p. 16.
[8.58] P. B. Corkum, P. P. Ho, R. R. Alfano, J. T. Manassah, Opt. Lett. **10** (1985) 624.

[8.59] A. Tomita, K. Tai, A. Hasegawa, J. R. Simpson, H. T. Shang in [35], p. 244.
[8.60] K. Duppen, D. P. Weitekamp, D. A. Wiersma, in [28], p. 179.
[8.61] B. Wilhelmi, Ann. Phys. **43** (1986) 355.
B. Wilhelmi, W. Rudolph, E. Döpel, W. Dietel, Optica Acta **32** (1985) 1175.
[8.62] J. E. Rothenberg, D. Grischkowsky, J. Opt. Soc. Am. B2 (1985) 626.

References to Chapter 9

[9.1] K. R. Naqvi, A. R. Holzwarth, U. P. Wild, Appl. Spectr. Rev. **12** (1976) 131.
[9.2a] Documents for the Nanosecond Fluorescence Measuring Set-up, from ORTEC, USA.
[9.2b] Documents for the Nanosecond Fluorescence Measuring Set-up, from Edinburgh Instruments, Scotland.
[9.3] W. Becker: Boxcar Integrator BCI 280. Preprint of the Center for Scientific Instruments of the Academy of Science in the GDR, 1981.
[9.4] Documents for the "Laser-Impulse Fluorimeter" LIF 200. Center for Scientific Instruments of the Academy of Sciences in the GDR, 1982.
[9.5] D. Bebelaar, J. J. F. Ramackers, M. W. Leeuw, R. P. H. Rettschnick, H. Langelaar, in [20] p. 212.
[9.6] U. P. Wild, A. R. Holzwarth, H. P. Good, Rev. Sci. Instr. **48** (1977) 1621.
[9.7] M. C. Adams, W. Sibbett, D. J. Bradley, Opt. Commun. **26** (1978) 273.
[9.8] J. R. Taylor, M. C. Adams, W. Sibbett, J. Photochemistry **12** (1980) 127.
[9.9] G. L. Olson, G. E. Busch, Appl. Phys. Lett. **27** (1975) 684.
[9.10] S. Belke, M. Fritsche, J. Herrmann, B. Wilhelmi, in: Nonlinear Optics. SOAN, Novosibirsk 1979, p. 99.
[9.11] A. S. Piskarskas, A. J. Stabinis: Parametric Light generation and Picosecond Spectroscopy (in Russian). Mokslas, Vilnius 1984.
[9.12] B. Wilhelmi, Chem. Physics **66** (1982) 351, and in: Mitteilungsbl. Chem. Ges. GDR, Beiheft 67, 1982, p. 63.
[9.13] K. J. Kaufmann: CRC Critical Reviews, in: Solid State Sciences (1979), p. 265.
[9.14] M. C. Adams, D. J. Bradley, W. Sibbett, J. R. Taylor, in: Spectroscopy IV (eds. H. Walther, K. W. Rothe). Springer Series in Optical Sciences, Vol. 21, 1979, p. 639.
[9.15] K. B. Eisenthal, Accounts Chem. Res. **8** (1975) 118.
[9.16] G. R. Fleming, J. H. Morris, G. W. Robinson, J. Chem. Phys. **17** (1970) 91.
[9.17] A. v. Jena, H. N. Lessing, Chem. Phys. **78** (1981) 187.
[9.18] H. J. Eichler, U. Klein, D. Langhans, Chem. Phys. Lett. **67** (1979) 21.
[9.19] D. Schubert, H. Wabnitz, B. Wilhelmi, Exp. Techn. Phys. **28** (1980) 435.
[9.20] D. Schubert, H. Wabnitz, B. Wilhelmi, Exp. Techn. Phys. **30** (1982) 153.
[9.21] B. Wilhelmi, in: Application of Laser in Atomic, Molecular and Nuclear Physics (in Russian) (ed. V. S. Letokhov). Nauka, Moscow 1983, p. 79.
[9.22] J. Aaviksoo, P. Saari, T. Tamm, in [20] p. 479.
K. K. Rebane, in [20] p. 449.
[9.23] D. von der Linde, in [16] p. 202.
[9.24] M. Schubert, K. Vogler, in [20] p. 413.
[9.25] G. H. Mc Call, Rev. Sci. Instr. **43** (1972) 865.
[9.26] A. Müller, J. phys. Chem. **101** (1976) 361.
[9.27] J. W. Armstrong, J. W. Shelton, IEEE J. Quant. Electr. **QE 3** (1967) 302.
[9.28] B. Wilhelmi, in: Quantum Electronics and Nonlinear Optics. University of Poznan, Poznan 1980, p. 103.
[9.29] S. Belke, I. Kapp, W. Triebel, B. Wilhelmi, in [18] p. 367.
[9.30] T. Damm, W. Triebel, B. Wilhelmi, Exp. Techn. Phys. **32** (1984) 155.
[9.31] A. Laubereau, W. Kaiser, Ann. Rev. Phys. Chem. **26** (1975) 83.
A. Laubereau, W. Kaiser, Rev. Mod. Phys. **50** (1978) 607.

A. Laubereau, W. Kaiser, in: Application of Lasers in Atomic, Molecular and Nuclear Physics. Nauka, Moscow 1979, p. 257.
[9.32] M. M. Malley, P. M. Rentzepis, Chem. Phys. Lett. **3** (1969) 534.
[9.33] T. L. Netzel, P. M. Rentzepis, Chem. Phys. Lett. **29** (1974) 337.
[9.34] M. R. Topp, P. M. Rentzepis, R. P. Jones, Chem. Phys. Lett. **9** (1971) 1.
[9.35] E. P. Ippen, C. V. Shank, A. Bergmann, Chem. Phys. Lett. **29** (1974) 337.
[9.36] J. P. Heritage, Appl. Phys. Lett. **34** (1979) 470.
B. F. Levine, C. V. Shank, J. P. Heritage, IEEE J. Quant. Electr. **QE 15** (1979) 1418.
J. P. Heritage, in [18] p. 343.
[9.37] H. Bergner, V. Brückner, R. Gase, A. Schlisio, B. Schröder; Exp. Techn. Phys. **30** (1982) 407.
[9.38] B. I. Stepanov, E. W. Ivakin, A. S. Rubanov, Dokl. AN SSSR Fis. **16** (1971) 46.
[9.39] D. W. Phillion, D. J. Kuizenga, A. E. Siegman, Appl. Phys. Lett. **27** (1975) 85.
[9.40] J. Herrmann, B. Wilhelmi, J. Opt. Soc. **50** (1980) 529, in [20] p. 179.
H. Paerschke, K. E. Süsse, B. Wilhelmi, Opt. Quant. Electron. **15** (1983) 325.
[9.41] D. von der Linde, A. Laubereau, W. Kaiser, Phys. Rev. Lett. **26** (1971) 954.
A. Laubereau, D. von der Linde, W. Kaiser, Phys. Rev. Lett. **28** (1972) 1162.
[9.42] R. R. Alfano, S. L. Shapiro, Phys. Rev. Lett. **26** (1971) 1247.
[9.43] A. Fendt, J. P. Maier, A. Seilmeier, W. Kaiser, in [18] p. 145.
[9.44] U. Köpf: Laser in Chemistry (in German). Salle, Sauerländer, Frankfurt/M. 1979.
[9.45] A. Laubereau, A. Seilmeier, W. Kaiser, Chem. Phys. Lett. **36** (1975) 232.
[9.46] P. G. Kriukov, V. S. Letokhov, D. N. Nikogosjan, A. V. Borodovkin, E. I. Budowsky, N. A. Simukova, Chem. Phys. Lett. **61** (1979) 375, in [20] p. 338.
[9.47] V. S. Letokhov, Kvant. Electr. **2** (1975) 930.
V. S. Antonov, V. S. Letokhov, A. N. Shibanov, Pisma Zh. teor. eksp. fiz. **31** (1980) 471.
V. S. Letokhov, in [20] p. 504, and in [28] p. 310.
[9.48] S. Dähne, F. Fink, E. Klose, K. Teuchner, S. Bach, J. V. Grossmann, J. Signal **AM 6** (1978) 105.
[9.49] Modular System for the Picosecond Laser Spectroscopy, Documents of the Center of Scientific Instruments of the GDR, Berlin 1983.
[9.50] Picosecond Spectrometers, Documents from Applied Photophysics Ltd., London and Spectra Physics (USA), 1983.
[9.51] C. L. Tang, J. M. Halbout, in [28] p. 212.
[9.52] C. L. Tang, P. J. Erskine; Phys. Rev. Lett. **51** (1983) 840.
[9.53] V. M. Petnikova, S. A. Pleshanov, V. V. Shuvalov: Preprint 18/1984, Physics Department of Moscow State University, Moskow 1984, and in [34], p. 98.
[9.54] K. Süsse, D. Welsch: Relaxation Phenomena in Atomic Systems (in German). Teubner, Leipzig 1984.
[9.55] D. J. Erskine, A. J. Taylor, C. L. Tang, Chem. Phys. Lett. **103** (1984) 430.
[9.56] Laser Pulse Spectrometer LIS 201, Technical Data from VEB Carl Zeiss Jena, 1985.
[9.57] M. Kaschke, S. Rentsch, B. Wilhelmi, Comments on Atomic and Molecular Spectroscopy **17** (1986) 309.
[9.58] B. Wilhelmi, in: SOS, Proceedings of the Third Symposium Optical Spectroscopy, (eds. D. Faßler, K. H. Feller, B. Wilhelmi). Teubner, Leipzig 1985, p. 9.
[9.59] E. Gornik, G. Bauer, E. Vass (editors), Proceedings of the Fourth International Conference on Hot Electrons, North Holland, Amsterdam 1985.
[9.60] A. J. Taylor, D. I. Erskine, C. L. Tang, J. Opt. Soc. Am. B**2** (1985) 663.
[9.61] W. Z. Lin, J. G. Fujimoto, E. P. Ippen, in [36], p. 50.
[9.62] C. V. Shank, B. Yen, C. Hirlimann, Phys. Rev. Lett. **50** (1983) 454.
[9.63] M. C. Downer, R. L. Fork, C. V. Shank, J. Opt. Soc. Am. B**2** (1985) 595.
[9.64] S. De Silvestri, J. G. Fujimoto, E. P. Ippen, E. B. Gamble, L. B. Williams, K. Nelson, Chem. Phys. Lett. **116** (1985) 146.

Subject Index

R

S